Selected Titles in This Series

66 **Yu. Ilyashenko and Weigu Li,** Nonlocal bifurcations, 1999

65 **Carl Faith,** Rings and things and a fine array of twentieth century associatiative algebra, 1999

64 **Rene A. Carmona and Boris Rozovskii,** Stochastic partial differential equations: Six perspectives, 1999

63 **Mark Hovey,** Model categories, 1999

62 **Vladimir I. Bogachev,** Gaussian measures, 1998

61 **W. Norrie Everitt and Lawrence Markus,** Boundary value problems and symplectic algebra for ordinary differential and quasi-differential operators, 1999

60 **Iain Raeburn and Dana P. Williams,** Morita equivalence and continuous-trace C^*-algebras, 1998

59 **Paul Howard and Jean E. Rubin,** Consequences of the axiom of choice, 1998

58 **Pavel I. Etingof, Igor B. Frenkel, and Alexander A. Kirillov, Jr.,** Lectures on representation theory and Knizhnik-Zamolodchikov equations, 1998

57 **Marc Levine,** Mixed motives, 1998

56 **Leonid I. Korogodski and Yan S. Soibelman,** Algebras of functions on quantum groups: Part I, 1998

55 **J. Scott Carter and Masahico Saito,** Knotted surfaces and their diagrams, 1998

54 **Casper Goffman, Togo Nishiura, and Daniel Waterman,** Homeomorphisms in analysis, 1997

53 **Andreas Kriegl and Peter W. Michor,** The convenient setting of global analysis, 1997

52 **V. A. Kozlov, V. G. Maz′ya, and J. Rossmann,** Elliptic boundary value problems in domains with point singularities, 1997

51 **Jan Malý and William P. Ziemer,** Fine regularity of solutions of elliptic partial differential equations, 1997

50 **Jon Aaronson,** An introduction to infinite ergodic theory, 1997

49 **R. E. Showalter,** Monotone operators in Banach space and nonlinear partial differential equations, 1997

48 **Paul-Jean Cahen and Jean-Luc Chabert,** Integer-valued polynomials, 1997

47 **A. D. Elmendorf, I. Kriz, M. A. Mandell, and J. P. May (with an appendix by M. Cole),** Rings, modules, and algebras in stable homotopy theory, 1997

46 **Stephen Lipscomb,** Symmetric inverse semigroups, 1996

45 **George M. Bergman and Adam O. Hausknecht,** Cogroups and co-rings in categories of associative rings, 1996

44 **J. Amorós, M. Burger, K. Corlette, D. Kotschick, and D. Toledo,** Fundamental groups of compact Kähler manifolds, 1996

43 **James E. Humphreys,** Conjugacy classes in semisimple algebraic groups, 1995

42 **Ralph Freese, Jaroslav Ježek, and J. B. Nation,** Free lattices, 1995

41 **Hal L. Smith,** Monotone dynamical systems: an introduction to the theory of competitive and cooperative systems, 1995

40.3 **Daniel Gorenstein, Richard Lyons, and Ronald Solomon,** The classification of the finite simple groups, number 3, 1998

40.2 **Daniel Gorenstein, Richard Lyons, and Ronald Solomon,** The classification of the finite simple groups, number 2, 1995

40.1 **Daniel Gorenstein, Richard Lyons, and Ronald Solomon,** The classification of the finite simple groups, number 1, 1994

39 **Sigurdur Helgason,** Geometric analysis on symmetric spaces, 1994

38 **Guy David and Stephen Semmes,** Analysis of and on uniformly rectifiable sets, 1993

37 **Leonard Lewin, Editor,** Structural properties of polylogarithms, 1991

(Continued in the back of this publication)

Nonlocal Bifurcations

Mathematical
Surveys
and
Monographs
Volume 66

Nonlocal Bifurcations

Yu. Ilyashenko
Weigu Li

American Mathematical Society

1991 *Mathematics Subject Classification*. Primary 34A47, 58F14.

ABSTRACT. The book studies nonlocal bifurcations that occur on the boundary of the domain of Morse–Smale systems in the space of all dynamical systems. These bifurcations provide a series of new fascinating scenarios for the transition from simple dynamical systems to complicated ones. The main effects are: generation of hyperbolic periodic orbits, nontrivial hyperbolic invariant sets and the elements of hyperbolic theory. All the results are rigorously proved and exposed in a uniform way. The foundations of normal forms and hyperbolic theories are presented from the very first steps. The proofs are preceded by heuristic descriptions of ideas. Most of the results have never been exposed in monographs; some of them are new.

The book is addressed to graduate students in mathematics, as well as specialists in pure and applied mathematics, physics and engineering.

Library of Congress Cataloging-in-Publication Data

Il'îashenko, ÎU. S.
Nonlocal bifurcations / Yu. Ilyashenko, Weigu Li.
p. cm. — (Mathematical surveys and monographs, ISSN 0076-5376 ; v. 66)
Includes bibliographical references.
ISBN 0-8218-0497-9 (acid-free)
1. Bifurcation theory. I. Li, Weigu. II. Title. III. Series: Mathematical surveys and monographs ; no. 66.
QA380.I38 1998
515′.35—dc21 — 98-40047
CIP

∞ The paper used in this book is acid-free and falls within the guidelines
established to ensure permanence and durability.
Visit the AMS home page at URL: http://www.ams.org/

10 9 8 7 6 5 4 3 2 1 — 04 03 02 01 00 99

Contents

Preface

The complete title of this book should be "Nonlocal bifurcations, normal forms, and elements of hyperbolic theory".

We present the modern theory of normal forms for local families of vector fields and diffeomorphisms. This presentation contains a complete list of integrable normal forms in the finitely smooth classification. This classification is the most suitable one for applications to nonlocal theory.

Hyperbolic and partial hyperbolic theory is the tool for the description of the invariant sets that occur under nonlocal bifurcations.

We restrict ourselves to the study of the bifurcations that occur on the boundary of the set of Morse–Smale systems and are generated by the loss of hyperbolicity of singular points and periodic orbits. Even this moderate goal leads to a variety of effects, which are only partly investigated up to now. A very rich domain of study is related to the homoclinic tangency of stable and unstable manifolds of singular points and periodic orbits. This subject, discussed in [**PT**], is beyond the scope of this book.

The most celebrated nonlocal bifurcations are those of the homoclinic orbit of a saddle. Bifurcations of this kind in space, both those that generate one periodic orbit and an arbitrary number of Smale horseshoes, are described in many books. The first occur at the boundary of the Morse–Smale set; the others are distant from this boundary. The general case of bifurcations of homoclinic orbits of saddles in spaces of arbitrary dimension was not described in detail in previous monographs on the same subject. All these results are presented below.

Less celebrated are bifurcations of homoclinic orbits of saddlenode singular points. One homoclinic curve generates one periodic orbit. But without increasing the rate of degeneracy, several homoclinic curves of the same point may occur. Their bifurcation gives rise to a nontrivial hyperbolic set: an Ω-explosion takes place.

The most complicated are the bifurcations of homoclinic surfaces of a saddle-node periodic orbit. It is not difficult to imagine a semistable cycle in a 2-torus filled by homoclinic orbits of this cycle. A homoclinic surface of this type may occur in a space of arbitrary dimension. But much more complicated homoclinic surfaces may occur as well. There may be a Klein bottle instead of a torus. Several homoclinic surfaces of the same periodic orbit may occur simultaneously. Their bifurcations lead to a new class of dynamical systems, whose investigation is now at the very beginning. The homoclinic surfaces may be topologically complicated. The so-called twisted homoclinic surfaces may occur in a higher-dimensional space (see Figure 1.15 below). The corresponding bifurcation gives rise to a hyperbolic attractor of solenoidal type.

Some new results related to bifurcations of homoclinic surfaces described above are presented in this book. Namely,

• smooth homoclinic tori and Klein bottles preserved under small perturbations;

• an invariant set with a random dynamical system on it occurring under the bifurcation of several homoclinic surfaces;

• a strange attractor generated under the bifurcation of the twisted homoclinic surface (Shilnikov and Turaev, 1995; the first complete proof was given by the first author during the work on this book).

The theory of normal forms drastically simplifies the study of nonlocal bifurcations. It provides integrable normal forms not only for the unperturbed equation near the equilibrium point, but for the perturbation as well. Therefore the map of the cross-sections transversal to invariant manifolds of the singular point along the orbits of the vector field may be explicitly calculated. Thus simple formulas replace the delicate estimates of the previous investigations.

There are two types of applications of these normal forms to nonlocal bifurcations.

The first is related to planar bifurcations. Here there are two directions of study as well: families with few and with many parameters.

Families with a small number of parameters may be investigated in full detail. The classical results of Andronov and his school are related to the nonlocal bifurcations in planar one-parameter families. The theory of normal forms transforms the proofs of these results into simple exercises. The general study of two- and three-parameter families in the plane was suggested by Arnold in 1985. At the beginning of the nineties Kotova collected the "zoo" of all polycycles that may occur in generic two- and three-parameter families. The cyclicity of these polycycles was investigated in a number of papers by Dumortier, Grozovskiĭ, Kotova–Stanzo, Morsalami, Mourtada, Roussarie, Rousseau, Sotomayor and others. The concluding paper was recently published by Trifonov.

The study of many-parameter families in the plane is mainly related to the so-called Hilbert–Arnold problem: to prove that the polycycle that occurs in a generic finite-parameter family generates no more than a finite number of limit cycles, and this number depends on the number of parameters. This problem is solved for polycycles with elementary singular points as vertexes, the so-called elementary polycycles. These results are included in two collections [**I2, IYa3**], where the other sources are quoted. Recently, Kaloshin obtained an explicit estimate of the number $E(k)$ of limit cycles generated by elementary polycycles in typical k-parameter families.

The second application of normal forms to nonlocal bifurcations is the study of spatial bifurcations. A systematic study is carried out for one-parameter families. This is the subject of this book. It seems that two-parameter spatial families may be studied in detail as well. The complete study of nonlocal bifurcations in three-parameter spatial and four-parameter planar families seems to be too complicated to be ever obtained.

The theory of normal forms for local families is presented in this book from the very beginning. We describe the homotopy method, which is the most convenient tool for the local smooth theory. The results we present are in a sense complete: we give the list of smooth integrable normal forms for the simplest families; smooth classification of more complicated families has functional moduli.

Elements of the hyperbolic theory are also presented from the very beginning and lead up to the newest results. The presentation of the nonlinear Smale horseshoe map deals with classical material. On the other hand, we study parallel results for partially hyperbolic maps. Roughly speaking, these maps have stable, unstable, and central subbundles of the tangent bundle. In general, stable and unstable foliations for such maps do exist, while the center-stable and center-unstable do not. The maps we study are subject to special geometric assumptions that guarantee the existence of all the invariant foliations mentioned above. This gives rise to a new kind of dynamics: random dynamical systems appear to be subsystems of smooth ones. The investigation of these systems is beyond the scope of this book. Here we only prove their birth under the bifurcations of several homoclinic surfaces of a saddlenode periodic orbit.

We describe all the bifurcations from a uniform point of view. Namely, all the proofs are obtained by studying the interaction of the theory of normal forms and hyperbolic theory. The main subject is the global Poincaré map represented as a product of singular and regular maps. The singular map is a map of cross-sections along the orbits passing near a singular point. This map is not everywhere defined and produces an unbounded distortion. It contracts in some directions and expands in others. The regular map is once more a map of cross-sections, this time along the orbits distant from a singular point. It is well defined, and bounded together with its derivatives.

The theory of normal forms gives explicit formulas for singular maps. The genericity assumptions guarantee that the regular map does not mix the contracting and expanding directions of the singular one. For the product of these maps hyperbolicity wins: the unbounded distortion of the singular maps overcomes the influence of the smooth regular map. Thus the global Poincaré map becomes the subject of hyperbolic theory, which provides the description of the invariant sets of this map.

Our uniform approach is illustrated in Figures 4.9, 4.12, 5.12, 7.7, 7.12, 7.18.

The idea of this book arose when the survey "Bifurcation theory" by Afraimovich, Arnold, Ilyashenko, and Shilnikov, 1986, was written. In the process, Arnold said that the survey should reflect the development of bifurcation theory for at least the twenty five years to come. The present book is a partial response to this challenge. Bifurcations, like torches, shed light on the transition from simple dynamical systems to complicated ones. Complicated systems occurring under bifurcations of homoclinic surfaces of nonhyperbolic periodic orbits partially described in Chapters 5 and 6 are the subject for promising future study.

The present book develops the third chapter of the survey [**AAIS**] (description of the bifurcations themselves) and the end of the second chapter of the same survey, where normal forms for local families were listed for the first time. In 1988 Professor Zhang Zhifen from Beijin organized a seminar on nonlocal bifurcations, where the proofs missing in the survey were reconstructed. The second author was an active participant of this seminar. In 1991–93 he had a scholarship in Moscow. During this period the first draft of the book was written. The present version is the result of several rewritings produced by both authors.

The first chapter contains all the necessary definitions beginning with very elementary ones. Its goal is to present the main results about nonlocal bifurcations. They are summarized in the main table at the end of the chapter.

Chapter 2 contains two sections that are in fact outside the main content of the book. The first discusses "prevalence": the concept of genericity different from the traditional "category" notion. The revision of the genericity assumptions throughout this book from the "prevalent" point of view is a challenging problem. The second large section is devoted to the Hausdorff dimension of attractors. The results of this section are applied only once, in Chapter 6. Yet the subject seems to be of independent interest, and was therefore included. The third section contains an elementary description of the Smale horseshoe, understandable for high school students. The rest of the chapter contains a summary of the hyperbolic and normal form theories used throughout the book.

The next five chapters present nonlocal bifurcations. Chapter 3 is elementary and deals with one-parameter planar families of vector fields. The only exception is the end of the chapter, where simultaneous occurrence of several saddle connections in two-parameter families is studied.

Chapter 4 contains the main ideas of our subsequent study. It demonstrates the mechanism of the occurrence of hyperbolicity via the singular and transversality properties of the Poincaré map.

The next three chapters are the main ones. Chapter 5 studies bifurcations of smooth homoclinic surfaces of saddlenode cycles. It contains the new results mentioned above. Chapter 6 deals with nonsmooth homoclinic surfaces. It presents the first detailed exposition of the study of the strange attractor begun by Afraimovich and Shilnikov in the seventies and followed by Newhouse, Palis, Takens in the eighties. Chapter 7 describes bifurcations of the homoclinic orbit of a saddle in the space of arbitrary dimension.

The hyperbolic and partial hyperbolic theory is presented in Chapter 8. The last two chapters are devoted to normal forms for local families. These three chapters contain proofs of the results stated in the last two sections of Chapter 2.

The dependence of chapters is shown below.

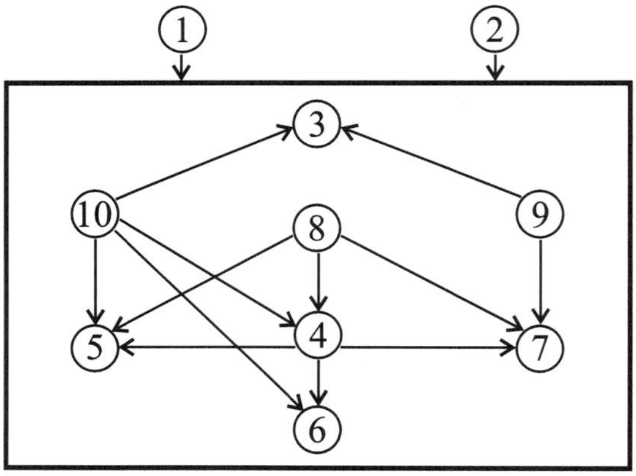

The system of references is as follows. Theorems, lemmas, propositions, and formulas have double numbers $a.b$, where a is the number of the section, and b is the number of the statement. The references inside the same chapter use this numeration. The reference to item $a.b$ in Chapter A looks like $A.a.b$.

The authors are grateful to Professor Zhang Zhifen who organized their cooperation; to Alexander Ilyashenko for the preparation of the computer version of the figures; to Helen Ilyashenko and Sergei Yakovenko, who keyboarded part of the manuscript; to Alexander Bufetov, Anton Gorodetskii, Maria Saprykina and other participants of the Moscow seminar on differential equations for helpful comments and technical assistance. We are grateful as well to Dr. Sergei Gelfand and Dr. Alexei Sossinskii who helped a great deal at the final stage of the preparation of the manuscript. The first author is grateful to Cornell University, where the large part of the final draft was written, to UNAM, Mexico, and CIMAT, Guanajuato, where the first author presented a course on nonlocal bifurcations. The second author had a fruitful stay at IMPA, Rio de Janiero, and at the University of São Paulo. Both authors are grateful to all the friends and colleagues for the cordial and stimulating atmosphere during these stays.

Further, the first author is grateful to Valya Afraimovich, who taught him nonlocal bifurcations while they were working together over the survey [**AAIS**].

CHAPTER 1

Introduction

This book is devoted to nonlocal bifurcations that may occur in typical one-parameter families crossing the boundary of the set of Morse–Smale systems in the space of vector fields. This approach allows us to build up a broad picture covering most of the results on nonlocal bifurcations obtained from the sixties to the present day. At the same time, some yet unsolved problems naturally arise.

Multidimensional structurally stable systems generally exhibit such complex behavior that it is hopeless to try to give a complete description of bifurcations occurring on the boundary of the set of such systems. The systematic study of these problems was never carried out. On the other hand, Morse–Smale systems are rather simple and constitute a sufficiently rich class of objects: most of the nonlocal bifurcations studied up to now occur on the boundary of that class. We shall consider only generic points of this boundary. This means that we study *generic one-parameter families* of vector fields and bifurcations occurring in such families.

§1. Structural stability and Morse–Smale systems

In this section we introduce the concept of structural stability and some basic properties of Morse–Smale systems.

1.1. Structural stability. The idea behind the definition of the structural stability of a vector field is the following: the field is structurally stable if a sufficiently small perturbation does not change the dynamical properties of the field. To give a precise definition, we need to specify what is meant by a small perturbation of a vector field f.

DEFINITION 1.1. The *C^r-topology* on the space of C^r smooth functions defined in a bounded domain $\Omega \subset \mathbb{R}^n$ is given by the following convergence rule: a sequence of functions converges to a certain limit in this topology if all derivatives of the functions from the sequence up to order r converge to the corresponding derivatives of the limit function uniformly in Ω.

The C^r-topology on the space of C^r smooth vector fields $\chi^r(\Omega)$ or maps from Ω to $\mathbb{R}^m$ (denoted by $C^r(\Omega, \mathbb{R}^m)$) is defined in terms of C^r-proximity of each component.

The convergence rule defines neighborhoods in the natural way: we say that the field $\tilde{f}$ is a C^r-small perturbation of the field f if the two fields are close in the C^r-topology.

Now we introduce an equivalence relation expressing the fact that two dynamical systems generated by equivalent vector fields have the same geometric structure. Roughly speaking, we say that two dynamical systems are equivalent if there exists a map taking the phase space of one system into that of the other and this map sends trajectories into trajectories. This definition can be made precise in several different ways.

DEFINITION 1.2. Two vector fields are *smoothly equivalent* if there exists a diffeomorphism taking one field into the other. If we denote the vector fields by v and w, and the diffeomorphism by H, then the latter condition means that

$$\frac{\partial H}{\partial x} \cdot v = w \circ H. \tag{1.1}$$

The smooth classification has numerical moduli, the eigenvalues of linearizations of vector fields at singular points: for smoothly equivalent vector fields the eigenvalues coincide. Recall that the set of eigenvalues of a vector field at a singular point is the spectrum of the linear operator linearizing the field at that point. Indeed, if both vector fields have a singularity at the origin $0 \in \mathbb{R}^n$ and $H(0) = 0$, then

$$CAC^{-1} = B, \tag{1.2}$$

where

$$C = \frac{\partial H}{\partial X}(0), \qquad v(x) = Ax + \cdots, \qquad w(x) = Bx + \cdots, \tag{1.3}$$

and the dots denote terms of order $\geqslant 2$. Thus it is clear that no vector field can be smoothly equivalent to all close fields if the given field has at least one singular point.

DEFINITION 1.3. Two vector fields are *topologically equivalent* if there exists a homeomorphism between the phase spaces of these fields that conjugates their flows.

Denote by g_v^t (resp., by g_w^t) the flows of the vector fields v and w respectively. If H is the homeomorphism from Definition 3, then

$$g_v^t \circ H(x) = H \circ g_w^t(x) \tag{1.4}$$

for all values x and t such that both parts of the equality make sense. The topological classification also has numerical moduli, one of them being the period of a closed periodic orbit. Thus one can see that a vector field with an isolated periodic orbit cannot be topologically equivalent to all perturbations.

DEFINITION 1.4. Two vector fields are *orbitally topologically equivalent* if there exists a homeomorphism between the phase spaces that takes phase trajectories (curves) of the first field into those of the second, preserving the natural orientation on the curves.

Finally we come to the principal definition.

DEFINITION 1.5. Let $\Omega \subset \mathbb{R}^n$ be a bounded domain. A vector field $f \in \chi^r(\Omega)$ is *structurally stable* if there exists an $\varepsilon > 0$ such that all ε-small perturbations in the sense of the C^1-topology are orbitally topologically equivalent to f.

1.2. Nonwandering points. What is meant when we say that a certain point in the phase space of a vector field has a nontrivial dynamic behavior? Naturally, points which return in time to an arbitrarily small neighborhood of their original position exhibit nontrivial dynamic behavior. Examples of such points are

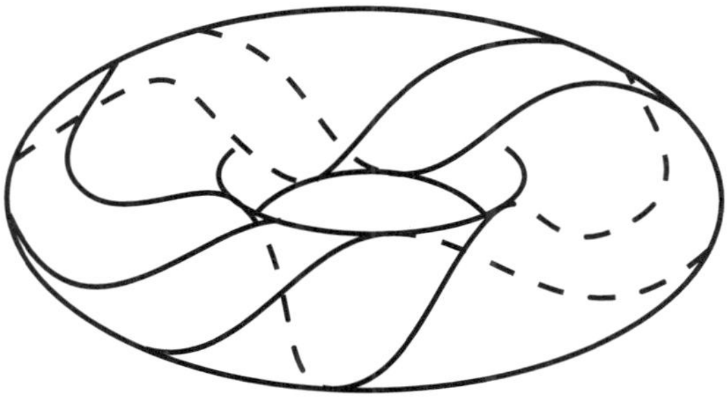

FIGURE 1.1. Irrational flow on a 2-torus.

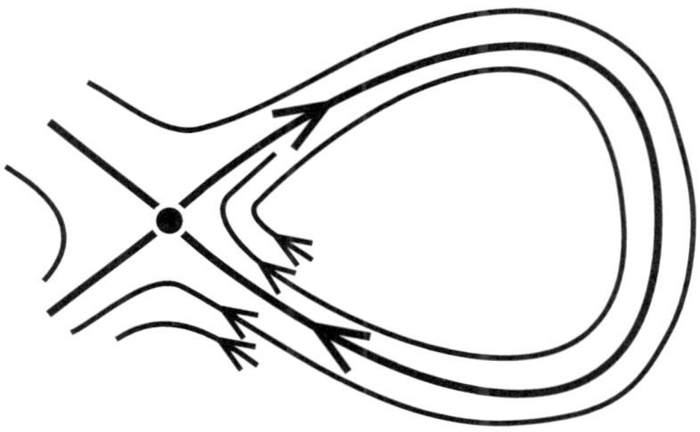

FIGURE 1.2. The homoclinic orbit of a saddle.

singular points or periodic orbits. There also may be more complex recurrence (see Figure 1.1).

On the other hand, a point may never return to its original position, though arbitrarily close points exhibit recurrent motion, as shown in Figure 1.2. This argument motivates the following definition.

DEFINITION 1.6. A point of the phase space is called *nonwandering* for the flow if any small neighborhood of that point, when moved by the flow, intersects its original position in any distant future.

In other words, the point p is nonwandering if for any neighborhood $U \ni p$ and for any $T < +\infty$ there exists a moment $t > T$ such that

$$g^t(U) \cap U \neq \varnothing. \tag{1.5}$$

An orbit of a nonwandering point is called a nonwandering orbit.

1.3. Hyperbolic singular points and cycles. In this section we describe singular points and periodic orbits which are *locally* structurally stable, i.e., there exists a small neighborhood of the point (resp., the cycle) such that the restriction of the field to that neighborhood is structurally stable.

DEFINITION 1.7. A singular point of a vector field is *hyperbolic* if it has no eigenvalues on the imaginary axis.

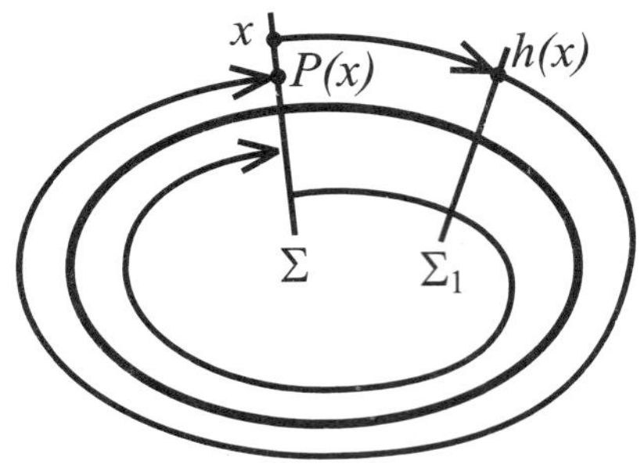

FIGURE 1.3. The Poincaré map of a cycle.

To give a similar definition for a periodic orbit, we recall the basic construction of the Poincaré return map. Take a hypersurface Σ transversal to the vector field at a certain point of the periodic orbit (cycle). For all points sufficiently close to the intersection of this surface with the cycle, the first return map is defined as taking a point x on Σ to the first intersection point of the positive semiorbit emanating from x with the surface Σ, as shown in Figure 1.3. We denote this map by P. It has numerous names: *Poincaré map, first return map, monodromy map*, etc. Note that the point of intersection of Σ with the cycle itself is a fixed point of the map P.

DEFINITION 1.8. Two C^r smooth maps F, G are C^k*-equivalent* if there exists a C^k smooth diffeomorphism h conjugating them: $h \circ F = G \circ h$. If $k = 0$, then such equivalence is called *topological equivalence.*

DEFINITION 1.9. The *multipliers* of a C^1 smooth map at a fixed point are the eigenvalues of the linearization of this map at that point. The multipliers of a periodic orbit (cycle) are the multipliers of its monodromy map at the fixed point corresponding to the cycle itself.

REMARK. Let Σ and Σ_1 be two transversal sections to the same cycle, and P and P_1 the two corresponding Poincaré maps. Let $h\colon \Sigma \to \Sigma_1$ be the map projecting one surface onto the other along flow curves, as shown in Figure 1.3. Then $P_1 \circ h = h \circ P$ (which means that the maps P and P_1 are C^r-equivalent if the field is of class C^r). From this simple observation we immediately conclude that the matrices of the linearization of P and P_1 are similar, hence have the same spectrum. In other words, the multipliers of a cycle are defined independently of the choice of the transversal section.

DEFINITION 1.10. A fixed point of a diffeomorphism is *hyperbolic* if it has no multipliers on the unit circle in the complex plane. A cycle is hyperbolic if its monodromy has a hyperbolic fixed point.

1.4. Topological classification of flows near hyperbolic singular points and cycles. For any finite-dimensional phase space, there is only a finite number of topological orbital normal forms for a vector field near a hyperbolic singular point. The same is true for hyperbolic fixed points of diffeomorphisms.

THEOREM 1.1 (Grobman–Hartman). *A C^1 smooth vector field (C^1-diffeomorphism) near a hyperbolic singular (resp., fixed) point is topologically equivalent to its linearization at that point.*

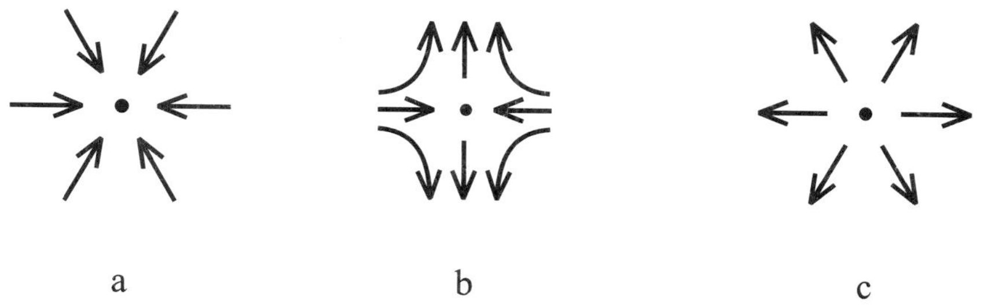

FIGURE 1.4. Topological classification of hyperbolic singularities in the plane.

FIGURE 1.5. Hyperbolic linear maps that preserve or reverse the orientation.

This result immediately implies that the problem of topological classification of hyperbolic singular points and cycles is reduced to the same problem for linear fields (maps). The last step is elementary. Let $A\colon \mathbb{R}^n \to \mathbb{R}^n$ be a linear operator, and denote by n_- (resp., n_+) the number of its eigenvalues in the left (resp., right) open half-plane so that $n = n_- + n_+$ due to hyperbolicity. Then the differential equation

$$\dot{x} = Ax, \qquad x \in \mathbb{R}^n, \tag{1.6}$$

is topologically equivalent to the *standard saddle*

$$\begin{aligned} \dot{y} &= y, \qquad & y \in \mathbb{R}^{n_+}, \\ \dot{z} &= -z, \qquad & z \in \mathbb{R}^{n_-}. \end{aligned} \tag{1.7}$$

This concludes the topological classification of vector fields near a hyperbolic singularity. One may see from Figure 1.4 that the topological equivalence is a very robust relation. However, this equivalence clearly distinguishes between sinks, saddles and sources.

The topological classification of hyperbolic linear maps is almost the same, the only difference being caused by the preservation or reversal of orientation (see Figure 1.5).

Denote by S the standard mirror symmetry of the space $\mathbb{R}^m$ in the hyperplane $x_1 = 0$: $S(x_1, x_2, \dots, x_m) = (-x_1, x_2, \dots, x_m)$. Suppose that a nonsingular linear operator $B\colon \mathbb{R}^n \to \mathbb{R}^n$ has n_- eigenvalues strictly inside the unit circle and n_+ eigenvalues strictly outside, again $n_- + n_+ = n$. Then the topological type of B depends on its restrictions to its contracting (stable) and expanding (unstable)

FIGURE 1.6. Poincaré map of a cycle on the Möbius band.

planes. Namely, different types occur for orientation-preserving and orientation-reversing restrictions. Let

$$B(y,z) = (y', z'), \qquad y \in \mathbb{R}^{n+},\ z \in \mathbb{R}^{n}. \tag{1.8}$$

The map B is topologically equivalent to one of the following four normal forms (the formulas give expressions for (y', z')):

$$(2y, z/2), \quad (2Sy, z/2), \quad (2y, Sz/2), \quad (2Sy, Sz/2). \tag{1.9}$$

Note that the monodromy map of a cycle in an orientable phase space is orientation-preserving. Therefore, two of the cases from the above list are ruled out for Poincaré maps on orientable manifolds. On nonorientable manifolds all four cases may occur, as the simplest example of the Möbius band shows (see Figure 1.6).

We conclude the topological classification of flows near a hyperbolic cycle by the following remark: two vector fields on a neighborhood of a cycle are topologically orbitally equivalent if and only if their monodromies are topologically equivalent.

1.5. Hadamard–Perron theorem. The Grobman–Hartman theorem implies that a neighborhood of a hyperbolic singular point (a hyperbolic cycle) has continuous "stable and unstable invariant manifolds".

DEFINITION 1.11. A manifold is said to be *invariant* for a vector field if, together with any point of the manifold, it contains the entire phase curve of the field passing through this point.

EXAMPLE. Let A be the same linear operator as in 1.4. Then the equation

$$\dot{x} = Ax, \qquad x \in \mathbb{R}^n, \tag{1.10}$$

has two invariant planes, $\mathbb{R}^{n-}$ and $\mathbb{R}^{n+}$; the phase curves lying on these planes exponentially tend to the origin as the time t tends to $+\infty$ or $-\infty$, respectively.

The Grobman–Hartman theorem implies that the differential equation

$$\dot{x} = Ax + \cdots, \qquad x \in \mathbb{R}^n, \tag{1.11}$$

with hyperbolic singular point 0 has two continuous invariant manifolds of dimensions $n_\pm$ exhibiting the above stability property for trajectories in direct or reverse time. The Hadamard–Perron theorem below asserts that these invariant manifolds,

FIGURE 1.7. Trajectories on stable and unstable manifolds.

whose existence is guaranteed by the Grobman–Hartman theorem, are in fact as smooth as the vector field is.

THEOREM 1.2 (Hadamard–Perron). *Assume that the right-hand side of equation* (1.11) *is of class* C^k *with* $k \leqslant \infty$ *or* $k = \omega$ (C^ω *stands for the class of analytic functions*), *and the singularity at the origin is hyperbolic. Suppose that the operator* A *has* n_- *eigenvalues to the left of the imaginary axis and* n_+ *of them to the right, and denote by* $T^\pm$ *the corresponding linear invariant subspaces for* A.

Then equation (1.11) *has two* C^k *smooth invariant manifolds* W^u (*unstable*) *and* W^s (*stable*) *tangent at the origin to* T^+ *and* T^-, *respectively. All orbits starting on* W^s *exponentially approach the origin as time increases to* $+\infty$; *trajectories on* W^u *exponentially converge to the singular point in the reverse time.*

Examples illustrating the behaviors of phase curves are given in Figure 1.7. In the planar case, trajectories different from the singular point and belonging to the invariant manifolds are called *separatrices*.

A similar assertion holds for diffeomorphisms near hyperbolic fixed points and for vector fields near hyperbolic cycles.

If the phase space is a compact manifold, then the flow of any vector field is defined globally, for all values of time. In particular, all orbits constituting the stable (unstable) manifold of a hyperbolic singular point (a cycle) may be infinitely continued. Saturation of locally defined stable and unstable manifolds by phase trajectories is a pair of immersed submanifolds of the phase space; their location may be rather complicated. In order to describe their mutual position, we need the concept of transversality.

1.6. Transversality.

DEFINITION 1.12. Two submanifolds of a smooth manifold intersect each other transversally if the following alternative holds:

(i) their intersection is empty, or
(ii) at each intersection point the tangent spaces to those manifolds together span the tangent space to the ambient manifold.

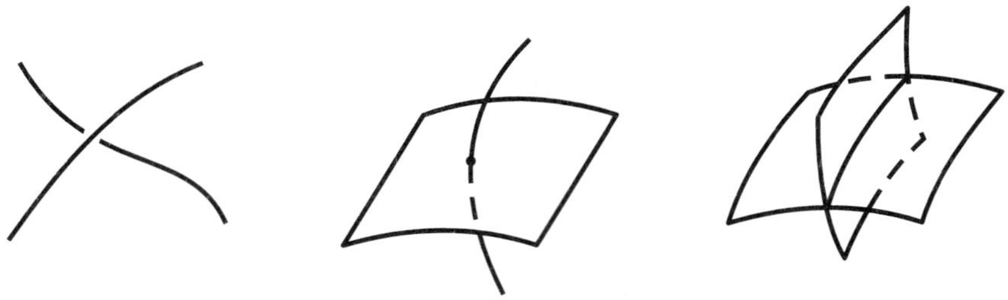

FIGURE 1.8. Transversally intersecting submanifolds in $\mathbb{R}^3$.

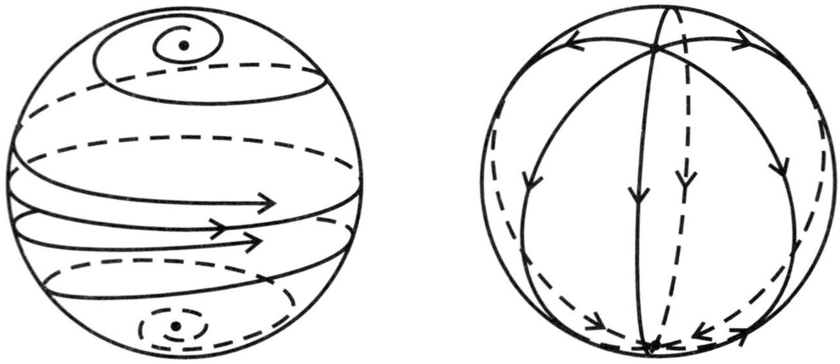

FIGURE 1.9. Examples of Morse–Smale systems on the 2-sphere.

REMARK. If the sum of dimensions of the two submanifolds is strictly smaller than the dimension of the ambient manifold (say, two curves in $\mathbb{R}^3$), then the transversality condition implies that their intersection is void.

Examples of transversally intersecting submanifolds are shown in Figure 1.8. Now everything is prepared to introduce the class of Morse–Smale systems.

1.7. Morse–Smale systems.

DEFINITION 1.13. A C^1 smooth vector field on a manifold or a diffeomorphism of this manifold is called a *Morse–Smale system* if the following three conditions hold:

(i) the set of nonwandering points consists of a finite number of singular points and periodic orbits (for a diffeomorphism a fixed point is a periodic point of period 1);
(ii) all singular points and periodic orbits are hyperbolic;
(iii) stable and unstable manifolds of hyperbolic singular points and periodic orbits intersect transversally.

Figure 1.9 shows two examples of Morse–Smale vector fields on the 2-sphere.

The fundamental property of Morse–Smale systems is their stability with respect to C^1-small perturbations.

THEOREM 1.3. *Morse–Smale systems on a compact manifold are structurally stable.*

Below we mention some results about Morse–Smale systems on closed surfaces. Morse–Smale vector fields on the 2-sphere admit a simple description, being characterized by the following two properties:

(i*) all singular points and periodic orbits of the field are hyperbolic;

(ii*) the field has no saddle connections; in other words, there are no separatrices common to two singularities, including loops through the same saddle point.

The equivalence of this particular definition to the general one for the case of the 2-sphere is not a very trivial fact. It follows from the Poincaré–Bendixson theorem, which is heavily based on Jordan's lemma (each Jordan curve divides the 2-sphere into two pieces). On the 2-torus and other orientable surfaces, conditions (i*) and (ii*) are not sufficient for a system to be of Morse–Smale type: irrational winding on the torus provides an easy counterexample; see Figure 1.1.

Andronov and Pontryagin were the first to define the concept of structural stability. They proved that conditions (i*) and (ii*) are necessary and sufficient for fields on the 2-sphere. In the early sixties Peixoto proved the following two statements.

THEOREM 1.4. *A vector field on a compact two-dimensional surface is structurally stable if and only if it is Morse–Smale.*

THEOREM 1.5. *For any natural r, Morse–Smale C^r smooth vector fields constitute an open dense subset of the space $\chi^r(M)$ for any compact orientable surface M of any genus and for any nonorientable surface of low ($\leqslant 3$) genus.*

Later Smale proved the converse statement for vector fields in the multidimensional case (in dynamical systems theory, "multidimensional" always means "of dimension greater than 2").

THEOREM 1.6. 1. *There exists a structurally stable system which is not Morse–Smale.*

2. *Structurally stable vector fields are not dense in $\chi^r(M)$ for* $\dim M \geqslant 3$.

Thus we see that the following challenging problem remains: *is it true that C^r-structurally stable vector fields on a nonorientable surface of high genus (> 3) are dense? The (positive) answer is known only for the case $r = 1$.*

Now we proceed with the description of the boundary of the set of Morse–Smale systems. For brevity we will refer to MS-sets and MS-boundaries.

1.8. Degeneracies occurring on the boundary of MS-sets. For simplicity (in order to avoid repetitious remarks) we restrict ourselves to the case of vector fields. On an MS-boundary the following phenomena may occur.

1. Systems with nonhyperbolic singular points.
2. Systems with nonhyperbolic cycles.
3. Systems with nontransversal intersections of stable and unstable invariant manifolds of singular points or cycles.
4. Systems with infinite number of nonwandering orbits.

We will study bifurcations occurring in generic one-parameter families crossing the MS-boundary. Genericity means that this crossing occurs at a generic point of the MS-boundary, therefore only one degeneration from the above list can materialize, and no additional equality-type obstructions arise (see below).

PROBLEM (Arnold). *Is it possible that a system with infinite nonwandering set may occur at the crossing of the boundary of an MS-set by a generic one-parameter family without the simultaneous occurrence of any system of type* 1–3 *above? In other words, may the explosion of the nonwandering set occur on the MS-boundary without the simultaneous occurrence of nonhyperbolic points or cycles or homoclinic tangency of stable and unstable manifolds of hyperbolic singular points or cycles?*

It is not known whether the first non-MS point on a generic curve passing through an MS-point towards MS-boundary in $\chi^r(\mathbb{R}^3)$ may be of type 4.

In all families studied below, the bifurcations are caused by degeneracies of types 1–3, and an infinite nonwandering set appears as a result of such bifurcations. Therefore we will study only phenomena relevant to the cases 1–3. The natural starting point is the investigation of *local* bifurcations.

§2. Equivalence and local bifurcations in generic one-parameter families

In this section we formulate the reduction principle of Shoshitaĭshvili. This principle gives the strongest possible assertion on the similarity of the germ of a vector field and its linear part. At the same time, it is the cornerstone for the foundation of local bifurcation theory.

2.1. Local and principal families. In local dynamics, the concept of *germ* is frequently used to refer to objects without specifying their domains explicitly. The precise definition follows.

DEFINITION 2.1. Two functions (vector fields, maps) defined in two neighborhoods of the same point are *equivalent* if they coincide on some smaller neighborhood of that point. The corresponding equivalence class is called the *germ* of the function (vector field, map). Any element of the equivalence class is a *representative* of the germ.

Evidently, all representatives of the same germ take the same value at the point p, so without ambiguity one may speak of the value of the germ at the reference point. In the same manner, derivatives of the germ at the reference point are well defined.

The notion of germ immediately finds an application. Let U be a subset of the Cartesian product $\mathbb{R}^n \times \mathbb{R}^p$ whose points are pairs (x, ε). A family of vector fields depending on the parameter $\varepsilon \in \mathbb{R}^p$ is a vector field defined in U and parallel to the phase space. In the coordinates (x, ε) this field gives rise to the equation

$$\begin{cases} \dot{x} = v(x, \varepsilon), \\ \dot{\varepsilon} = 0. \end{cases}$$

DEFINITION 2.2. A *local family* of vector fields is the germ of a family of vector fields considered as a single vector field on the Cartesian product of the phase space and the space of parameters at a certain point (x_0, ε_0). The latter is called the *reference point* of the local family, and ε_0 is the *initial value* of the parameters.

We denote the local family by $(v; x_0, \varepsilon_0)$. Sometimes we call it a *deformation* of the germ of the vector field $v_0 = v(\,\cdot\,, \varepsilon_0)$.

DEFINITION 2.3. Two local families of vector fields $(v; x_0, \varepsilon_0)$ and $(w; y_0, \eta_0)$ are (*topologically orbitally*) *equivalent* if there exists a germ of a homeomorphism H such that:

(i) the germ H has the reference point at (x_0, ε_0) and takes the value (y_0, η_0) at this point;

(ii) there exists a representative of the germ H which is fibered over the parameter spaces; this means that the representative has the form

$$H\colon (x, \varepsilon) \mapsto (y, \eta) = (H_1(x, \varepsilon), H_2(\varepsilon)); \tag{2.1}$$

(iii) for every ε, the map $H_1(\,\cdot\,, \varepsilon)$ is a homeomorphism taking phase curves of the field $v(\,\cdot\,, \varepsilon)$ to those of $w(\,\cdot\,, \eta)$, $\eta = H_2(\varepsilon)$ and preserving their orientation.

REMARK. We do not require that $H(x_0, \varepsilon) = y_0$ for $\varepsilon \neq \varepsilon_0$.

DEFINITION 2.4. We say that two germs of vector fields are *weakly* (*orbitally topologically*) *equivalent* if there exist a germ of a map H satisfying conditions (i)–(iii) above, except for the continuity assumption, which is relaxed in the following way: we do not require that H be continuously dependent on ε. More precisely, in (2.1) the map $H_1(\,\cdot\,, \varepsilon)$ must be continuous for every ε, but the dependence on ε may be discontinuous.

Sometimes it is natural to consider local families of vector fields depending on the same set of parameters. For such families we may avoid reparametrization by introducing the following definition.

DEFINITION 2.5. Two local families are *strongly equivalent* if they are equivalent and the corresponding homeomorphism H preserves the parameter:

$$H_2 = \mathrm{id}\,. \tag{2.2}$$

DEFINITION 2.6. The local family $(u; x_0, \mu_0)$ is *induced* from the local family $(v; x_0, \varepsilon_0)$ if there exists a germ of a continuous map from the parameter space of the first family to the parameter space of the second family, $\mu \mapsto \varepsilon = \phi(\mu)$, $\phi(\mu_0) = \varepsilon_0$, such that

$$u(x, \mu) = v(x, \phi(\mu)). \tag{2.3}$$

Now we describe local families that are in a sense maximal: they contain all possible deformations up to topological equivalence.

DEFINITION 2.7. A local family $(v; x_0, \varepsilon_0)$ is a *topologically orbitally versal* (in short, *versal*) deformation of the germ of the field $v_0 = v(\,\cdot\,, \varepsilon_0)$ if any other local family containing the same germ v_0 is strongly equivalent to a family induced from (v, x_0, ε_0).

If in Definition 2.7 we replace strong equivalence by weak equivalence, then the notion of a *weakly versal deformation* arises.

Now we define *principal families* as topological normal forms for unfoldings of a field having degenerate singular point. Fix some type D of degeneracy (say, a class of germs satisfying certain equality-type conditions imposed on lower order terms).

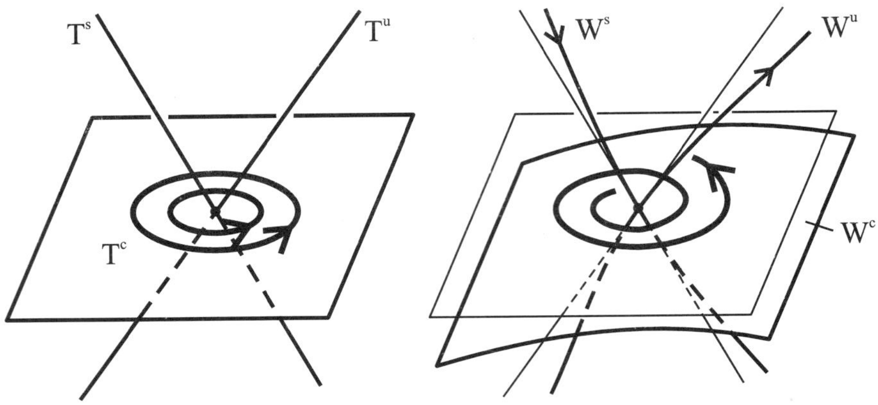

FIGURE 1.10. Stable, unstable and center manifolds.

DEFINITION 2.8. A *tuple of principal families for the degeneracy D* in codimension ν is a finite collection of ν-parameter local families with the following characteristic property: any generic ν-parameter family containing the degeneracy D for the initial value of the parameters is orbitally topologically equivalent to one of the families constituting the tuple.

Versal deformations and principal families contain, in very concentrated form, complete information about bifurcations occurring in local families unfolding degenerate singularities.

2.2. Center manifold theorems. Consider a linear operator $A: \mathbb{R}^n \to \mathbb{R}^n$. The linear space $\mathbb{R}^n$ is split into the direct sum of three A-invariant subspaces,

$$\mathbb{R}^n = T^s \oplus T^u \oplus T^c \tag{2.4}$$

(^{s} and ^{u} stand for stable and unstable respectively, ^{c} for center) in the following way: the restriction $A|_{T^s}$ possesses a spectrum contained in the open left half-plane, the spectrum of the restriction $A|_{T^u}$ lies to the right of the imaginary axis, and all eigenvalues of the restriction $A|_{T^c}$ have zero real parts. It turns out that a similar property holds for nonlinear vector fields. The next theorem is the strongest general result of that sort.

THEOREM 2.1 (the center manifold theorem for flows). *Let $v(x)$ be the germ at the origin of a C^{r+1} smooth vector field on $\mathbb{R}^n$ with $r < \infty$. Denote by A its linearization, so that $v(x) = Ax + \cdots$. Let T^s, T^u, and T^c be the invariant subspaces of the operator A defined in* (2.4).

Then the differential equation $\dot{x} = v(x)$ has three invariant manifolds W^s, W^u, and W^c of smoothness C^{r+1}, C^{r+1}, and C^r respectively, having the tangent spaces T^s, T^u, and T^c at the origin. The restriction of the flow on W^s (resp., on W^u) is exponentially contracting (resp., contracting in reverse time) exactly as in the assertion of Hadamard–Perron theorem; the behavior of the flow on W^c is determined by both linear and nonlinear terms of v.

Theorem 2.1 is represented by Figure 1.10.

REMARK. Under the same set of assumptions, the equation $\dot{x} = v(x)$ possesses two other invariant manifolds, the center-stable manifold $W^{sc} \supseteq W^s \cup W^c$ and the center-unstable manifold $W^{uc} \supseteq W^u \cup W^c$ tangent to $T^s \oplus T^c$ and $T^u \oplus T^c$ respectively and of the class C^r.

The manifolds W^s and W^u are referred to as the stable and unstable manifold as in the hyperbolic case; if the germ is C^∞ or C^ω smooth, then they are also of the same smoothness class. Unlike them, the manifolds W^c, W^{sc}, and W^{us} even for a C^∞ or C^ω smooth field are only of *finite differentiability*: for any $k \in \mathbb{N}$ there exists a neighborhood of the origin such that the intersection of the above invariant manifolds with this neighborhood is C^k smooth. With k increasing, the size of this neighborhood decreases and it shrinks to a singular point, so there can be no C^∞ smooth invariant manifold.

DEFINITION 2.9. The manifold W^c is called the *center manifold*. The plane $T^s \oplus T^u$ is the *plane of hyperbolic variables*.

2.3. Reduction principle. The dynamics of the restrictions of a flow to stable and unstable invariant manifolds was already described. It turns out that in general the topology of a flow is determined by its linear part and the restriction of this flow to the center manifold.

Consider a differential equation $\dot{x} = v(x)$ with C^r smooth right-hand side v, $r \geqslant 2$, having a singularity at the origin with the linearization A as in 2.2. Let $\mathbb{R}^n = T^s \oplus T^u \oplus T^c$ be the invariant splitting for A, and let W^s, W^u, and W^c be the corresponding invariant manifolds for the equation.

THEOREM 2.2. *In a sufficiently small neighborhood of the origin, the equation $\dot{x} = v(x)$ is topologically equivalent to the standard saddle suspension over its restriction to the center manifold*:

$$\begin{cases} \dot{x} = w(x), \\ \dot{y} = -y, \qquad x \in W^c,\ y \in T^s,\ z \in T^u. \\ \dot{z} = z, \end{cases} \tag{2.5}$$

This theorem has numerous applications. In fact, the entire modern theory of local stability of vector fields is based on this theorem, as well as the topological classification of germs of vector fields.

On the other hand, the theorem may be easily modified to cover the case of local families depending on parameters: to that end, we consider the extended system

$$\begin{cases} \dot{x} = v(x, \varepsilon), \\ \dot{\varepsilon} = 0, \end{cases} \qquad x \in \mathbb{R}^n,\ \varepsilon \in \mathbb{R}^k. \tag{2.6}$$

At the point $(0,0)$ the system (2.6) has a center manifold of dimension $k + \dim T^c$. The precise formulation of the reduction principle for local families follows.

DEFINITION 2.10. Let $u, s \geqslant 0$ be the dimensions of the stable and unstable manifolds. The *standard saddle suspension* over the local family

$$\dot{x} = w(x, \varepsilon) \tag{2.7}$$

is the family

$$\begin{cases} \dot{x} = w(x, \varepsilon), \\ \dot{y} = -y, & y \in \mathbb{R}^s, \ z \in \mathbb{R}^u. \\ \dot{z} = z, \end{cases} \tag{2.8}$$

DEFINITION 2.11. The center manifold of the local family $(v; 0, 0)$ satisfying $v(0, 0) = 0$ is the center manifold at the origin for the system (2.6).

SHOSHITAĬSHVILI REDUCTION PRINCIPLE. *An arbitrary local family of vector fields $(v; 0, 0)$ with $v(0) = 0$ is topologically equivalent to the saddle suspension over the restriction of the family to its center manifold.*

COROLLARY. *Let $(w; 0, 0)$ be the restriction of the family $(v; 0, 0)$ to the center manifold of the latter. If the former family is a versal deformation of the germ $w(\cdot, 0)$, then the family $(v; 0, 0)$ is a versal deformation of the germ $v(\cdot, 0)$.*

2.4. Local bifurcations of singular points in generic one-parameter families. In generic local one-parameter families of vector fields only two possible types of nonhyperbolic singularities may occur: exactly one zero eigenvalue (the *saddlenode case*) or exactly one pair of purely imaginary conjugate complex numbers (the *Andronov–Hopf case*). The restriction of the original local family to its center manifold will be called the *reduced family*. The Shoshitaĭshvili reduction principle asserts that without loss of generality only reduced families may be considered. In Table 1 below, we list all generic one-parameter reduced families. This table is organized as follows.

TABLE 1. Principal families in codimension 1.

Type of singularity	c	Typical germ	Genericity conditions	Principal families	Bifurcational diagram, phase portrait
One zero eigenvalue: the saddlenode case	1	$\dot{x} = ax^2$	$a \neq 0$	$\dot{x} = x^2 \pm \varepsilon$ (2.9$^\pm$)	Figure 1.11
One pair of imaginary eigenvalues: the Andronov–Hopf case	2	$\dot{z} = z(i\omega + A\rho)$	$\operatorname{Re} A \neq 0$	$\dot{z} = z(i + \varepsilon \pm \rho)$ (2.10$^\pm$)	Figure 1.12
Saddlenode maps (tangent to the identity)	1	$x \mapsto x + ax^2$	$a \neq 0$	$x \mapsto x + x^2 \pm \varepsilon$ (2.11$^\pm$)	Figure 1.13

In the first column we specify the type of degeneracy. The number c in the second column is the dimension of the center manifold. The third column gives the normalized jet of the degenerate vector field, and in column 4 the degenericity assumption (which rules out cases of deeper degeneracy) is given. Finally, we write principal families and make references to the figures showing the corresponding bifurcation diagrams.

THEOREM 2.3. *The set of all germs of vector fields at a nonhyperbolic singular point splits into the union of two open subsets and a complementary subset of codimension greater than one in the space of all germs of vector fields at the singular point. The first open subset corresponds to the saddlenode germs, the second consists of Andronov–Hopf germs. In both cases a generic germ from one of these two*

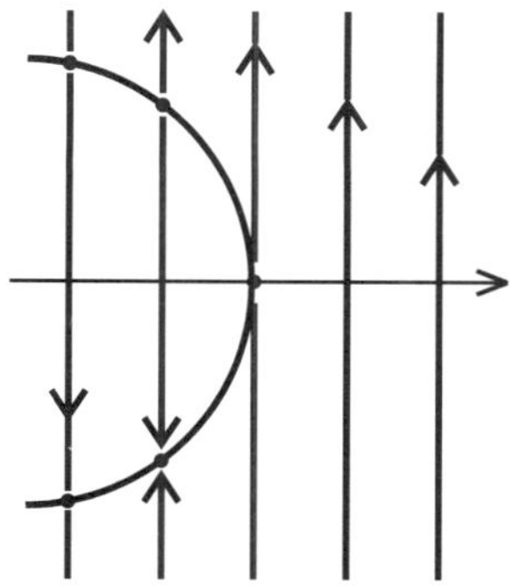

FIGURE 1.11. The bifurcation of a saddlenode singular point.

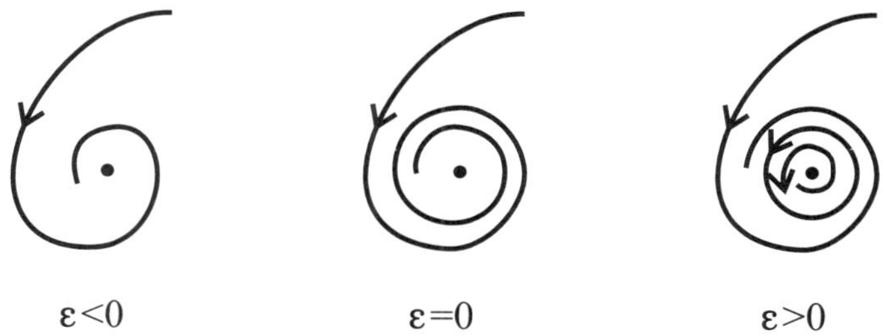

FIGURE 1.12. The Andronov–Hopf bifurcation for vector fields.

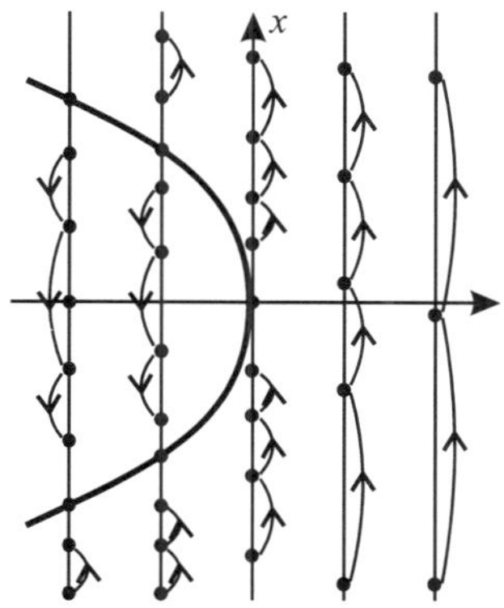

FIGURE 1.13. The bifurcation of a saddlenode fixed point.

subsets is orbitally topologically equivalent to its normalized jet given in Table 1. *Generic one-parameter deformations of such germs are stable equivalent* (*that is, become equivalent after an appropriate saddle suspension*).

The latter assertion means that the principal families shown in this table are stable versal deformations (i.e., become versal deformations after an appropriate saddle suspension).

The theorem is classical; the proof may be found, say, in [**A**].

The last line in the table corresponds to a so-called saddlenode local family of maps on the line. It is very simple in itself (see Figure 1.13). On the other hand, it is a key to nonlocal bifurcations of homoclinic surfaces of saddlenode cycles. These bifurcations are studied in Chapters 5 and 6.

We now describe the properties of the second principal family.

For the variable $\rho = z\bar{z}$, equations ($2.10^{\pm}$) (in Table 1) yield the so-called quotient system

$$\dot{\rho} = \rho(\varepsilon \pm \rho).$$

For the sign $+$, the equation has no nonzero singular points for $\varepsilon > 0$ and a unique nonzero singularity for $\varepsilon < 0$. The principal system ($2.10^{\pm}$) is itself invariant under rotations of the z-plane, and the nonzero singularity of the quotient system corresponds to the limit cycle of the principal family. Stability of this singular point for the quotient system coincides with that of the limit cycle of the principal system. This justifies Figure 1.12.

REMARKS. 1. The family (2.9^{+}) may be induced from the family (2.9^{-}) by reversing the parameter $\varepsilon \mapsto -\varepsilon$.

2. The families (2.10^{+}) and (2.10^{-}) are transformed into each other by the time reversal $t \mapsto -t$, the symmetry $z \mapsto \bar{z}$ and the parameter reversal. Still we distinguish between these two cases because the loss of stability occurs according to two different scenarios, called *soft* and *hard* stability loss. In the family (2.10^{-}) for $\varepsilon \leqslant 0$, the singular point at the origin is asymptotically stable; for $\varepsilon > 0$ it becomes unstable. Yet a small neighborhood of the singularity remains attracting for small positive ε: trajectories starting on the boundary of this neighborhood enter the neighborhood and stay in it forever, converging to the cycle of radius of order $\sqrt{\varepsilon}$ rather than to the singularity. This phenomenon is referred to as the soft stability loss in the physical slang.

In the family (2.10^{+}) for $\varepsilon < 0$ the singularity is stable, yet its basin of attraction shrinks to zero as $\varepsilon \to 0-$, and for $\varepsilon \geqslant 0$ the origin becomes unstable. So all trajectories (except for the singular point itself) must leave the neighborhood of the origin after sufficient time for any positive ε. This situation is called hard stability loss: when the parameter ε passes through the zero value, the system must jump to another stable regime. This new regime may be constant, periodic or more complicated, but in any case it must be far away from the original equilibrium point.

2.5. Local bifurcations of cycles in generic one-parameter families. Local bifurcations of cycles may be described in the same way as bifurcations of singular points. But this description is much more complicated and essentially lies beyond the scope of this book.

There are three possible types of nonhyperbolic cycles which may occur in generic one-parameter cycles: *saddlenode, flip*, and *Andronov–Hopf fixed point*; see 4.1 below. In all three cases the corresponding cycles may have homoclinic orbits, but only in the first case does the vector field belong to the boundary of the Morse–Smale set. This effect is briefly explained in §4 and in more detail in Chapter 2. Here we describe the local bifurcation of saddlenode cycles.

DEFINITION 2.12. A nonhyperbolic cycle of a vector field is of *saddlenode* type if exactly one of its multipliers is equal to $+1$, while the others are hyperbolic (not on the unit circle).

Now we need an analog of the local theory described in 2.1–2.4 for the case of maps rather than vector fields. This analog may be obtained by making the following replacements in all definitions and formulations:

1. germs of vector fields at singular points $\mapsto$ germs of maps at fixed points;
2. eigenvalues outside the imaginary axis $\mapsto$ eigenvalues outside the unit circle;
3. saddle suspensions of vector fields $\mapsto$ maps of the form

$$(x, y, z) \mapsto (x', y', z'), \qquad x' = w(x, \varepsilon), \quad y' = Ay, \quad z' = Bz,$$
$$y \in \mathbb{R}^s, \ z \in \mathbb{R}^u, \ \|A\| < 1, \ \|B^{-1}\| < 1.$$

In particular, the Shoshitaĭshvili reduction principle also holds for the case of maps and allows one to consider bifurcations occurring in a saddlenode family of maps in dimension 1. The last line in Table 1 gives the topological principal family.

REMARK 3. The families $(2.11^{\pm})$ from Table 1 may be transformed into each other by parameter reversal.

§3. Homoclinic trajectories of nonhyperbolic singular points

A *homoclinic trajectory* of a singular point is the trajectory which tends to this point both in direct and in reverse time. Only two types of nonhyperbolic singularities (see above) may appear in generic one-parameter families.

1. Saddlenodes (with exactly one zero eigenvalue).
2. Andronov–Hopf points (with only one pair of nonzero purely imaginary eigenvalues).

All other eigenvalues have nonzero real parts (and are thus hyperbolic). In the first case the center manifold is one-dimensional; in the second the dimension of the center manifold is two. No other degeneracies are allowed for singularities occurring in generic one-parameter families. Therefore the germ of the restriction of the vector field to the center manifold can be topologically normalized according to Table 1, §2. Hence a saddlenode generically occurring in such a family is topologically equivalent to the vector field

$$\dot{x} = x^2, \quad \dot{y} = -y, \quad \dot{z} = z, \qquad (x, y, z) \in (\mathbb{R}^1 \times \mathbb{R}^u \times \mathbb{R}^s, 0). \tag{3.1}$$

In the Andronov–Hopf case, the topological normal form is

$$\dot{z} = iz \pm z^2\overline{z}, \quad \dot{y} = -y, \quad \dot{p} = p,$$
$$z \in (\mathbb{C}, 0) \simeq (\mathbb{R}^2, 0), \quad (y, p) \in (\mathbb{R}^u \times \mathbb{R}^s, 0). \tag{3.2$^{\pm}$}$$

DEFINITION 3.1. A *positive* (*negative*) *semitrajectory* of a point is the part of its trajectory corresponding to nonnegative (resp., nonpositive) time values. The *stable* (*unstable*) *set* of a nonhyperbolic singular point of a vector field is the union of all positive (resp., negative) semitrajectories tending to this point (resp., in reverse time).

The stable and unstable sets of a nonhyperbolic singularity are denoted by S^s and S^u; if we want to stress the fact that a point rather than a cycle is considered, then we use the notation S_0^s, S_0^u. Actually we deal with germs of those sets rather

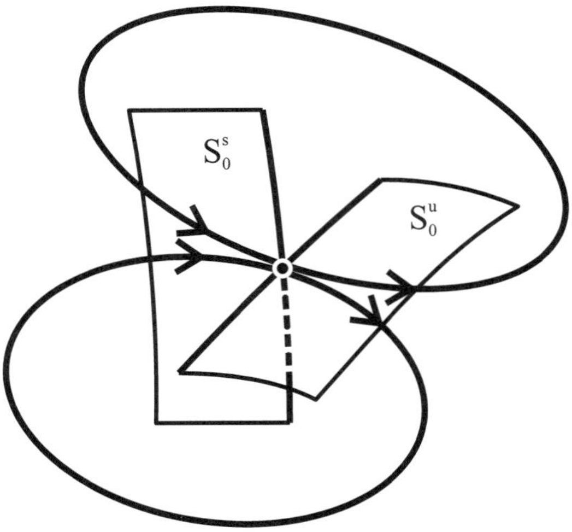

FIGURE 1.14. Homoclinic orbits of a saddlenode singular point.

than with the sets themselves. (A germ of a set at a point a is an equivalence class of sets with the following equivalence relation: two sets are equivalent if they coincide in some neighborhood of a.)

In the saddlenode case the (germs of) stable and unstable sets are homeomorphic to the (germ of the) closed half-space of dimension $s+1$ and $u+1$ respectively. Let n be the dimension of the total phase space. Then

$$n = s + u + 1,$$

since $\dim W^c = 1$. Therefore

$$\dim S_0^s + \dim S_0^u = n + 1.$$

These two manifolds with boundary may intersect in a one-dimensional manifold, which is either a single connected component or the union of several, perhaps an infinite number of, connected components (phase curves); see Figure 1.14.

For the Andronov–Hopf case ($3.2^{\pm}$), we distinguish between two subcases. If the sign is $+$, then the germs of the sets S_0^s and S_0^u are homeomorphic to T^s and $T^u \times W^c$ respectively, and $\dim W^c = 2$. Therefore

$$\dim S^s + \dim S^u = n.$$

When transversal, these two manifolds are disjoint (since the intersection must be at least one-dimensional, being invariant under the flow) except for one point of intersection at the origin. The other case (3.2^{-}) is treated similarly. Recall that the existence of an Andronov–Hopf point is a degeneracy itself. In generic one-parameter families no other degeneracies occur. Thus we conclude that no homoclinic trajectories are possible for an Andronov–Hopf point in codimension 1. So we need to study only homoclinic trajectories of saddlenodes. This is the subject of Chapter 4.

§4. Homoclinic trajectories of nonhyperbolic cycles

This section is parallel to the previous one but contains more pictures. A homoclinic trajectory of a cycle is the trajectory tending to the cycle both in direct and in reverse time. As usual, we operate with the Poincaré map P of a cycle and its iterate rather than with trajectories of the vector field itself. For example, a homoclinic trajectory intersects the global transversal section (if such a section exists) in an orbit $\{x_t\}|_{t=\dots,-1,0,1,\dots}$ of the Poincaré map P, $x_k = P^k(x_0)$, $k \in \mathbb{Z}$, such that

$$\lim_{k\to\pm\infty} x_k = 0.$$

In this section we also formulate the Birkhoff–Smale theorem, which gives the key argument for recognizing systems that do not belong to the MS-boundary. Several new modifications of this theorem are presented.

4.1. Types of nonhyperbolic fixed points generic in codimension 1. The three types of degeneracies possibly occurring in generic one-parameter families of diffeomorphisms are the following:

1. Saddlenode (one multiplier equal to $+1$).
2. Flip (one multiplier equal to -1).
3. Andronov–Hopf point: a pair of nonreal multipliers on the unit circle, $\exp(\pm i\phi)$, $\phi \in \mathbb{R}$, $\phi \notin \pi \cdot \mathbb{Z}$.

We call the corresponding cycle saddlenode, flip, or Andronov–Hopf cycle, respectively. The stable and unstable sets for fixed points of maps or for limit cycles of vector fields are defined as in §3 for singular points of vector fields.

4.2. Saddlenode fixed points and cycles. The Poincaré map for a nonhyperbolic saddlenode is topologically equivalent to the map

$$\begin{gathered}(x,y,z) \mapsto (x',y',z'),\\ x' = x + x^2, \qquad x \in (\mathbb{R}^1, 0),\\ y' = Ay, \quad z' = Bz, \qquad y \in \mathbb{R}^s,\ z \in \mathbb{R}^u,\\ \|A\| < 1,\ \|B^{-1}\| < 1.\end{gathered} \tag{4.1}$$

Hence, the "hyperbolic part" (y', z') in (4.1) has one of the four normal forms (2.9), where A is a linear contraction, and B a linear dilatation. In fact, there are only two topological normal forms for each operator A or B, depending on the preservation or reversal of orientation.

Thus the stable and unstable sets of the fixed point O for the map (4.1) are homeomorphic to the closed half-spaces of dimensions $s+1$ and $u+1$ respectively. The intersection of their interiors is generically a one-dimension manifold having up to an infinite number of connected components, which are no longer phase curves of a flow, but rather invariant curves of the diffeomorphism. The corresponding vector field has invariant surfaces. Some possibilities are shown in Figure 1.15.

Now consider a vector field with a saddlenode cycle. The invariant curves of the Poincaré map correspond to invariant surfaces of the vector field. These surfaces are filled with trajectories homoclinic to the same cycle. These surfaces may be homeomorphic to the 2-torus and the Klein bottle (as in case (a), Figure 1.15). Any number of surfaces may occur simultaneously in case (b). The union of all homoclinic trajectories and cycles may be nonarcwise connected in case (c). It may have a complicated topology, case (d). The homoclinic surface may be not smooth,

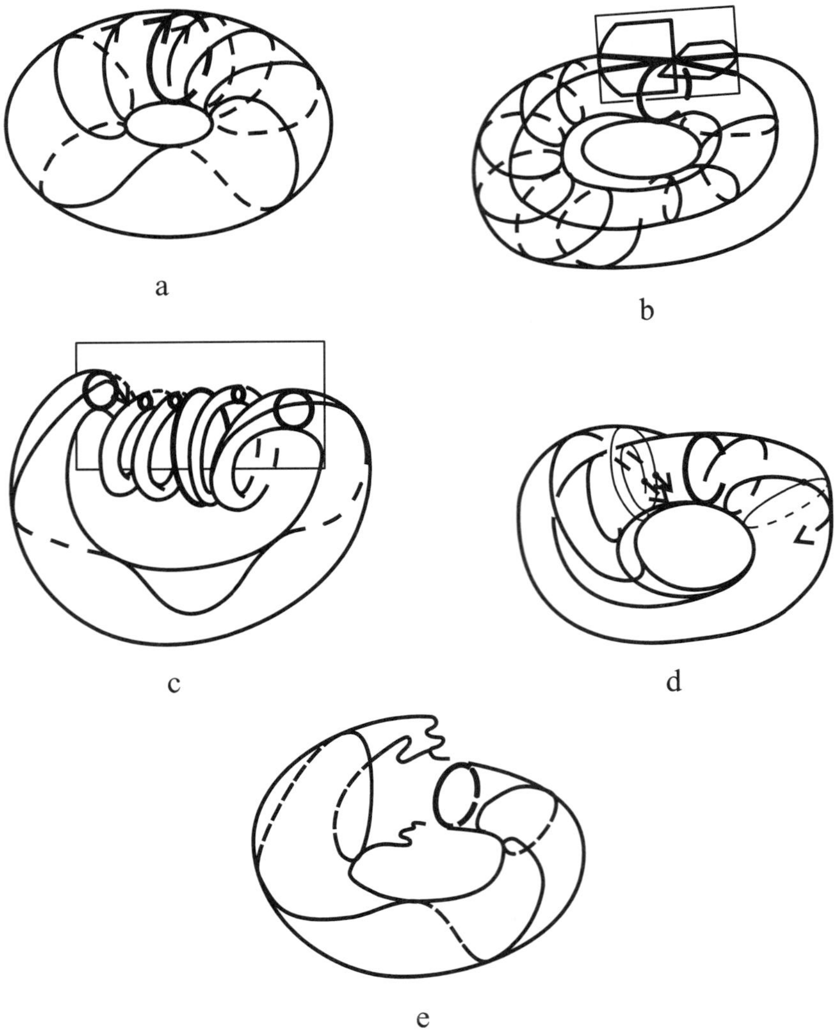

FIGURE 1.15. Homoclinic surfaces of the saddlenode cycle.
(a) A homoclinic torus. (b) Two homoclinic tori. (c) A snake.
(d) A twisted homoclinic torus. (e) Nonsmooth homoclinic torus.

case (e). The bifurcations in cases (a), (b), (d) are presented in Chapter 5. Case (e) is studied in Chapter 6. In case (c), a blue sky catastrophe may occur; see [**ST**].

4.3. Homoclinic intersections and the Birkhoff–Smale theorem. In this subsection we formulate and comment on the famous theorem due to Birkhoff and Smale, which shows that the set of Morse–Smale systems is not dense in the space of all vector fields on manifolds of dimension greater than 2.

THEOREM 4.1 (Birkhoff–Smale theorem). *Let* $f\colon \mathbb{R}^n \to \mathbb{R}^n$, $n \geqslant 2$, *be a diffeomorphism such that the origin is a hyperbolic fixed point and there exists a point*

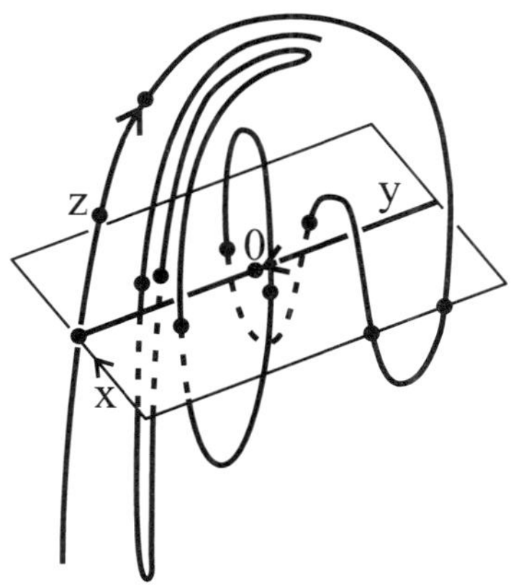

FIGURE 1.16. Intersection of the stable and unstable sets of a fixed point in the flip case.

$p \neq 0$ of transversal intersection of the stable and unstable manifolds of the origin. Then f has an infinite number of hyperbolic periodic orbits.

We derive this statement from a more general theorem of hyperbolic theory in Chapter 8 under slightly different assumption that the map f near the origin is C^1-equivalent to its linear part.

4.4. Flip maps and cycles. The topological normal form in the flip case is

$$(x, y, z) \mapsto (x', y', z'),$$
$$x' = -x \pm x^3, \quad x \in (\mathbb{R}^1, 0), \qquad y' = Ay, \quad z' = By, \tag{4.2$^{\pm}$}$$

where y, z, A, and B have the same meaning as in (4.1). For the minus sign the germs of stable and unstable sets of the origin O are homeomorphic to s- and $(u + 1)$-dimensional open balls, hence generically they intersect outside the origin in a set consisting of isolated points (see Figure 1.16). The case of the other sign exhibits a similar behavior, the center manifold now being part of the unstable rather than the stable set. Afraimovich (1985) conjectured that a vector field with a homogeneous orbir of a flip cycle cannot occur on the boundary of the MS-set. The main reason for that is in the similarity of this case with the case covered by the Birkhoff–Smale theorem. The only difference with the hyperbolic case is in the much slower rate of convergence of a point from the stable set to the fixed point. In the hyperbolic case, the Birkhoff–Smale theorem guarantees the existence of an infinite nonwandering set for the system under consideration and, at the same time, for all sufficiently close systems. In the flip case, similar arguments show that the same effect holds true for flip systems with homoclinic intersection of stable and unstable sets. This implies that such a system cannot occur on the MS-boundary.

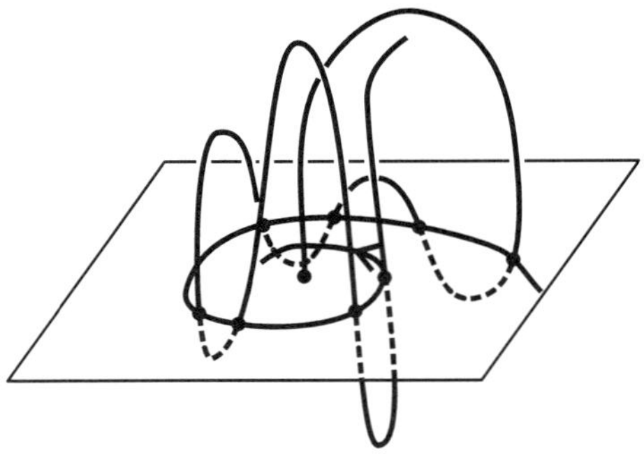

FIGURE 1.17. Intersection of the stable and unstable sets of a fixed point in the Andronov–Hopf case.

4.5. Andronov–Hopf points and cycles. The topological normal form in this case is

$$\begin{gathered}(z, y, p) \mapsto (z', y', p'),\\ z' = \exp(i\phi)z \pm z|z|^2, \qquad z \in (\mathbb{C}^1, 0),\\ y' = Ay, \quad p' = Bp, \qquad (y, p) \in \mathbb{R}^s \times \mathbb{R}^u,\\ \|A\| < 1, \quad \|B^{-1}\| < 1,\end{gathered} \tag{4.3$^\pm$}$$

with the same properties of the hyperbolic part. For the minus sign case, the germs of stable and unstable manifolds are open balls of dimensions $s + 2$ and u. The dimensions are complementary, hence in the generic case the intersection consists of isolated points (see Figure 1.17). We expect that the same arguments as in 4.4 will show that such type of behavior never occurs on the boundary of the set of Morse–Smale systems (Afraimovich, 1985). As far as we know, the detailed proof of this statement and the previous one has not been published yet. We have presented these conjectures here to explain why the homoclinic orbits of flip cycles and Andronov–Hopf cycles are beyond the scope of this book.

§5. Homoclinic loops of hyperbolic fixed points and other contours

The study of bifurcations of homoclinic loops of hyperbolic saddles in $\mathbb{R}^3$ signaled the starting point of multidimensional bifurcation theory.

For generic vector fields, the stable and unstable manifolds of a hyperbolic fixed point in $\mathbb{R}^3$ do not intersect. Indeed, they have supplementary dimensions s and u, $s + u = n$, and are foliated by the phase curves. If these manifolds intersect, then their projections along the phase curves onto a suitable cross-section intersect as well. These intersections have dimensions $s - 1$ and $u - 1$, $(s - 1) + (u - 1) = n - 2$, while the dimension of the cross-section is $n - 1$. Two transversal submanifolds of a domain in $\mathbb{R}^{n-1}$ of total dimension $n - 2$ do not intersect at all.

5.1. Two kinds of hyperbolic fixed points. There are two kinds of generic hyperbolic fixed points. Vector fields with loops homoclinic to fixed points of the first kind may be found on the boundary of the Morse–Smale set. Unfoldings of these loops generate a unique cycle.

Vector fields with homoclinic loops of points of the second kind never appear on the boundary of the Morse–Smale set. They have infinitely many periodic orbits, accumulating to the homoclinic loop. All the neighboring vector fields have an infinite number of periodic orbits as well and, generically, no homoclinic loop.

Let us now describe the two kinds of hyperbolic singular points. Consider the restriction of the linearization of a vector field at the hyperbolic singular points to its stable manifold. For a generic vector field, this restriction has exactly one maximum real eigenvalue λ or exactly two complex conjugate eigenvalues with maximum real part, which we also denote by λ. Obviously, $\lambda < 0$. The corresponding real line in the first case or real plane in the second are called *stable real* or *complex leading directions*. In the same way, the *unstable real* and *complex leading directions* are defined. The corresponding real eigenvalue or the real part of the complex eigenvalue is denoted by μ, where $\mu > 0$. Generically, $\lambda + \mu \neq 0$. We call the stable leading direction *subordinate* if $\lambda + \mu > 0$, and the unstable leading direction *subordinate* if $\lambda + \mu < 0$.

Vector fields of the first kind are those for which the subordinate leading direction is real. *Vector fields of the second kind* have subordinate complex leading direction.

We give a complete description of the bifurcation of the homoclinic loops for the singular points of the first kind. The 3-dimensional and n-dimensional cases are treated in §§7.1 and 7.3 respectively. For singular points of the second kind, we only study the 3-dimensional case, in §7.2. Formally, this study is beyond the scope of this book, because the vector fields under consideration do not belong to the boundary of the Morse–Smale set.

5.2. Definitions and examples of contours. A homoclinic trajectory of a singular point or a cycle is a particular case of so-called *contours*. In this section we give a general definition and describe contours that may occur on the MS-boundary.

TERMINOLOGICAL REMARK. *Contours* are often called *cycles* in other sources. We reserve the term "cycle" for a periodic orbit of a vector field for the multidimensional case as well as for the theory of vector fields on the plane, where it is of common usage. Instead, we use a different term, "contour", for spatial analogs of planar separatrix polygons.

A separatrix polygon on the plane is the finite union of cyclically enumerated singular points and separatrices connecting them in the prescribed order (see Figure 1.18). A contour is a multidimensional analog of this concept with an additional flexibility: some singularities may be replaced by cycles.

DEFINITION 5.1. A *contour* is a finite union of cyclically ordered singular points and cycles (together referred to as *elements*) such that there exist trajectories of the field connecting them in the prescribed order: for any two elements Q_k and Q_{k+1} there exists a trajectory tending to Q_k as $t \to -\infty$ and to Q_{k+1} as $t \to +\infty$. The total number of elements is called the (*combinatorial*) *length* of the contour.

The simplest example of a contour is the homoclinic orbit of a hyperbolic singular point, which will be studied in Chapters 3 and 7. Some different examples of contours of length 1 were presented in §§3 and 4 (in this case the trajectories are homoclinic orbits of a nonhyperbolic singular point or a cycle). Here we discuss contours of length $\geqslant 2$.

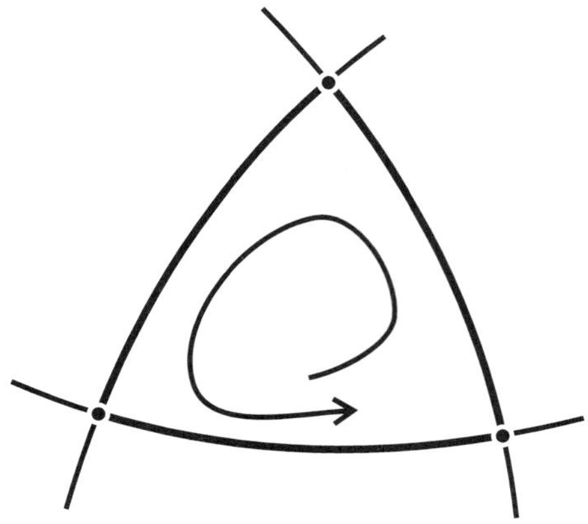

FIGURE 1.18. Separatrix polygon in the plane.

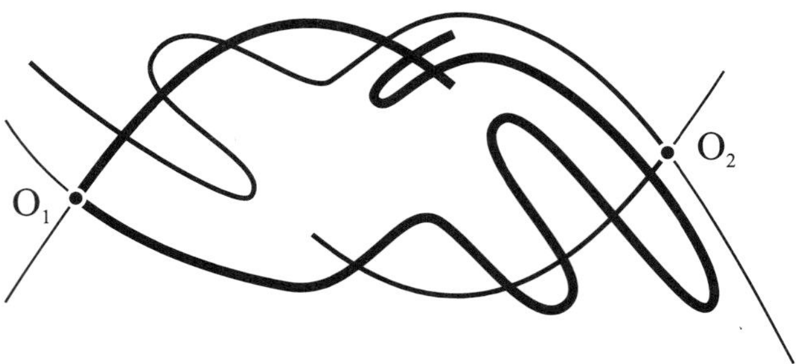

FIGURE 1.19. Contour of length one near a contour of greater length.

5.3. Contours in generic systems. The simplest example of a contour is a homoclinic curve of a hyperbolic cycle. The Birkhoff–Smale theorem cited in 4.3 applies to the corresponding Poincaré map, so such contours cannot occur on the MS-boundary. The same is true for similar contours of any length.

A vector field is said to be of *Kupka–Smale type* if and only if it has hyperbolic singular points and periodic orbits only, and their stable and unstable manifolds intersect transversally.

THEOREM 5.1 (Birkhoff–Smale). *Suppose that a contour of length more than one with only hyperbolic cycles appears for a vector field of Kupka–Smale type. Then this vector field has a homoclinic orbit of a hyperbolic cycle.*

COROLLARY 5.1. *The vector field in Theorem* 5.1 *has an infinite number of periodic orbits, and the same is true for all* C^1*-close vector fields.*

SKETCH OF THE PROOF OF THEOREM 5.1. The existence of a contour of length one near a contour of length more than one is clarified by Figure 1.19 for the case of two hyperbolic cycles replaced by two hyperbolic fixed points of the map. Denote these points by O_1, O_2.

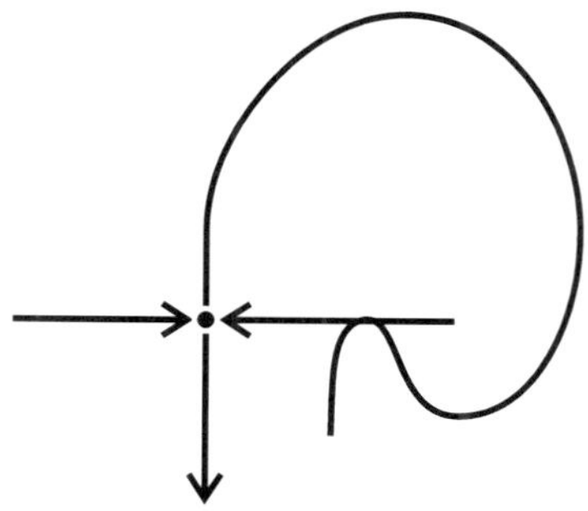

FIGURE 1.20. Homoclinic tangency of a saddle fixed point.

Take a small ball on the unstable manifold of O_1 containing the point of the heteroclinic orbit going from O_1 to O_2. The hyperbolicity of the point O_2 implies that the iterates of this ball accumulate to the unstable manifold of O_2. They stretch wider and wider along this manifold until they cross the stable manifold of O_1. The intersection point belongs to the required homoclinic orbit of O_1.

For contours of greater length the reasoning is the same.

REMARK. The same construction gives a countable number of intersections of $W^u_{O_1}$ and $W^s_{O_1}$. In general, they belong to different orbits of the map. Hence, the existence of a long contour formed by periodic orbits generically implies the existence of a countable set of homoclinic orbits for any of these cycles.

5.4. Contours with hyperbolic elements and nontransversal intersections of invariant manifolds. Contours of such type may appear in generic one-parameter families of vector fields only if:

1. All elements of the contour are hyperbolic cycles (not singular points).
2. There is only one hyperbolic singular point, and all the rest are hyperbolic cycles.

The first case has been the subject of intensive study in the last three decades. Recently the book by J. Palis and F. Takens [**PT**] devoted to this subject has appeared. Here we mention only several phenomena relevant to that case.

Ω-explosion. The term "explosion" means a sudden increase of the size of the nonwandering set triggered by a small perturbation of a system.

A contour exhibiting tangency between invariant manifolds in $\mathbb{R}^3$ may appear on the MS-boundary. A small perturbation produces a vector field with transversal intersection of the stable and unstable manifolds of the hyperbolic cycle. By the Birkhoff–Smale theorem, this field has an infinite nonwandering set.

Countable number of stable periodic orbits. Let f_ε be a generic one-parameter family of diffeomorphisms of $\mathbb{R}^2$ exhibiting a *homoclinic tangency* of a dissipative saddle fixed point for $\varepsilon = 0$: for $\varepsilon = 0$ one has a hyperbolic fixed point p with $|\det Df(p)| < 1$ and with tangent stable and unstable invariant curves as in Figure 1.20.

If we perturb a vector field having a Poincaré map of the above form, unfolding it in a generic one-parameter family $v_\varepsilon(\cdot)$, then a surprising phenomenon occurs. Arbitrarily close to the value $\varepsilon = 0$ on the parameter axis, there exist intervals carrying residual subsets (i.e., countable intersections of open dense sets) of parameters ε for which the field v_ε has an infinite number of periodic attracting orbits. This difficult result was established by S. Newhouse and is described in detail in the book [**PT**].

As far as we know, for the second type of contours only the three-dimensional case was studied in full detail [**BLMP**]. In this example, an Ω-explosion may also occur.

5.5. Nonhyperbolic contours. A contour occurring in a generic one-parameter family may have no more than one nonhyperbolic element, and when this happens, no other degeneracies are allowed. Therefore, the stable and unstable sets of all singular points and cycles intersect transversally. As was claimed in 5.2, the existence of a hyperbolic contour of combinatorial length $\geqslant 2$ implies the existence of a hyperbolic homoclinic loop (a contour of length 1). A similar effect can be observed in the nonhyperbolic case.

THEOREM 5.2. *Suppose that a nonhyperbolic contour of length $\geqslant 2$ occurs for a vector field belonging to the intersection of a generic one-parameter family with the MS-boundary. Then*:

(i) *If the nonhyperbolic element is a singular point, then it is a saddlenode, a saddle with respect to the hyperbolic variables, and it has a homoclinic curve.*
(ii) *If the nonhyperbolic element is a cycle, than it is a saddlenode cycle and has a homoclinic trajectory.*

SKETCH OF THE PROOF OF THEOREM 5.2. The existence of a homoclinic orbit for the nonhyperbolic element in both cases is proved in the same way as in the previous theorem. If this element is a nonhyperbolic singular point, and the contour corresponds to a generic point on the MS-boundary, then this singular point is a saddlenode; see §3. If this element is a nonhyperbolic cycle, then, under the same assumption, it is a saddlenode cycle; see 4.4, 4.5.

This result reduces the investigation of nonhyperbolic cycles to the case of loops (cycles of length 1) in the sense that any behavior occurring in bifurcations of the latter will be also present in the former general case. For example, the result from Chapter 4 implies that Ω-explosion will typically occur in unfoldings of type (i). However, this reduction is one-way: some special effects eventually may be observed for long contours only. However, this study is at its very beginning, so we will focus on the case of cycles of length 1, which were already described in §§3, 4, and 5.1 above.

§6. Summary of results

The central role in this book is played by the middle chapters. In this section we summarize the contents of these chapters and briefly describe the subject matter of the others.

6.1. Architecture of the book. Nonlocal spatial bifurcations are described in Chapters 4–7. The corresponding results are summarized in the table of 6.3. Chapter 3 deals with planar bifurcations. The technique of local normal forms turns these theorems into simple exercises. Chapters 8–10 form the foundation of our study. The theory of normal forms for local families is presented in Chapters 9 and 10. This theory allows one to replace the cumbersome asymptotic analysis

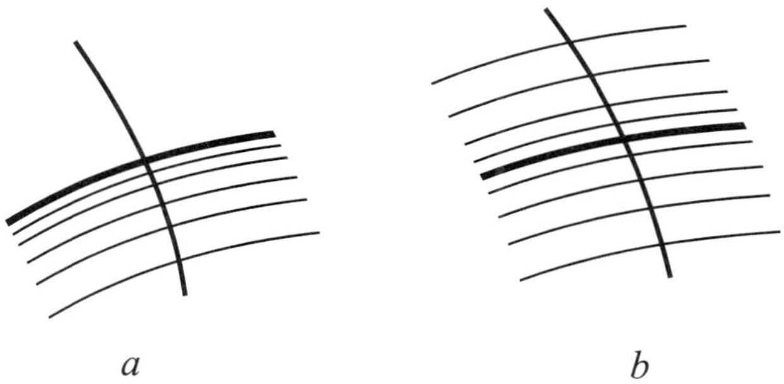

FIGURE 1.21. Accessible and nonaccessible boundary points of the MS-set.

used in earlier investigations by explicit formulas. These formulas describe the behavior of phase curves near singular points. The principal characteristic of this behavior is the so-called correspondence map, which may be explicitly calculated in the normalizing chart. The result allows for a heuristic description of bifurcations of the corresponding homoclinic curves.

The correspondence map has strong hyperbolic properties. Some basic theorems of hyperbolic theory allow us to use these properties to confirm the heuristic description of bifurcations. Chapter 8 contains the detailed proofs of these theorems.

Chapter 2 contains the summary of results of the last three chapters together with some heuristic explanations. It makes it possible to use the principal tools of the last chapters before the corresponding results are rigorously proved. Moreover, Chapter 2 contains an elementary description of the Smale horseshoe, a study of the fractal dimension of attractors and the concept of prevalence, i.e., the notion of "almost everywhere in function spaces". The last two topics may be used for the future study of nonlocal bifurcations.

In order to describe the table below, we need several definitions.

6.2. Accessible boundary classes and one-parameter families leading out of the Morse–Smale set. Both properties named in the title characterize the structure of the set of vector fields of Morse–Smale type (Morse–Smale sets for brevity) near a generic boundary point of this set. The characteristics is given in terms of generic one-parameter families passing through the boundary point.

DEFINITION 6.1. A class of boundary points of a Morse–Smale set is called *accessible from one side* (*respectively, from two sides*) if for any generic one-parameter family crossing this class, there exists a parametrization such that the intersection point of the family with the class corresponds to the zero value of the parameter, and all sufficiently small negative (respectively, nonzero) values of the parameter correspond to Morse–Smale systems. In the opposite case, the class is called *nonaccessible* (see Figure 1.21).

DEFINITION 6.2. A one-parameter family is *not leading out of the Morse–Smale set* if there exist a parametrization such that the zero value of the parameter

THE MAIN TABLE

Class	Subclass	Acc.	Out	Max. No.	Hyperbolic set	Str. attr.	Ref.
Nonhyperbolic singular point	One homoclinic orbit	++	+	1	–	–	4.1 4.3
	More than one homoclinic orbit	+–	–	∞	Ω	–	4.4
Nonhyperbolic cycle, compact noncritical case	One homoclinic surface of type T^2 or K^2	+–	+	∞	–	–	5.4 5.5
	More than one homoclinic surface of type T^2 or K^2	+–	–	∞	Partially hyperbolic	–	5.7
	Twisted homoclinic surface	+–	+	∞	Ω	+	5.8
Nonhyperbolic cycle, compact semicritical case	One homoclinic surface of type T^2	+–	–	∞	Ω	+	6.1
Hyperbolic singular point with a homoclinic loop	A loop in $\mathbb{R}^3$	++	+	1	–	–	7.1
	A loop in $\mathbb{R}^n$	++	+	1	–	–	7.3
	Subordinate complex leading direction in $\mathbb{R}^3$	*	*	∞	Ω	–	7.2

corresponds to a boundary point of the Morse–Smale set, and an open dense set in some neighborhood of zero on the parameter axes such that any point of this set corresponds to a Morse–Smale system. In the opposite case, the point is *leading out of the Morse–Smale set.*

6.3. Description of the table. The concepts of genericity and attractors used below may be specified in many different ways. This specification is discussed in §§1, 2 of Chapter 2. Hyperbolic sets are defined in §4 of Chapter 2. The heuristic meaning of genericity and attractors is clear anyway. The main example of a hyperbolic set is the nonwandering set of the Smale horseshoe map, the simplest example of which is given in §3 of Chapter 2.

The table consists of 8 columns. The first column, called *Class*, gives the name of the degeneration. The *Subclass* column distinguishes the subsets of the corresponding class with similar properties of unfoldings. In column 3, the sign ++ means that the subclass is accessible from two sides, and +– that it is accessible from one side in the sense of Definition 6.1. The sign + in column 4 means that the generic one-parameter family crossing the subclass is not leading out of the Morse–Smale set, and the sign – means that it is. The column *Maximum number of cycles* contains information on the periodic orbits generated by bifurcations of the generic equations from the corresponding subclass. The integer in the column gives the explicit number of cycles generated in typical one-parameter families crossing the subclass. The symbol ∞ means either that an infinite number of cycles may be generated in time or that an arbitrary large, though finite, number may be generated. The symbol Ω in column 6 means that a nontrivial hyperbolic set is generated, and the sign $-$ means that it is not. The signs $+$ and $-$ in column 7 mean the same for the strange attractor. *Strange attractor* here means an attractor which is not a finite union of smooth manifolds; see Chapter 2 for details. In the column *References*, theorems formalizing the brief description summarized in the preceding part of the same line are indicated. The first digit in the reference means the Chapter, and the second the section.

CHAPTER 2

Preliminaries

§1. Prevalence

In this section traditional and new concepts of the notion "almost everywhere" are discussed. This discussion leads to problems that are only stated, but not solved, in this book.

1.1. What does "almost everywhere" in a function space mean? There are different viewpoints on the subject. The first, dominating in singularity theory, is *categoric* (*in the sense of Baire*) or *topological.*

DEFINITION 1.1. A property is called *topologically typical* for a class of maps if it holds for all maps that belong to some set of second category, or, for brevity, to a thick set in an appropriate function space. A *thick* subset of a space is the countable intersection of open dense subsets of this space.

REMARK. The definition makes sense for dynamical systems, because vector fields are, at the same time, maps from the phase space to its tangent bundle.

EXAMPLE 1. Morse–Smale systems on oriented two-dimensional surfaces are typical in the sense of the previous definition in the space of C^k smooth vector fields for any $k \geqslant 1$.

Another point of view is the metric one.

DEFINITION 1.2. A property is called *metrically typical* or *prevalent* for maps from some smooth finite-dimensional family if the set of parameter values corresponding to those maps for which this property is violated has zero Lebesgue measure.

EXAMPLE 2. The following map in a prevalent subset of the family

$$z \mapsto e^{i\varphi} z + f(z), \qquad \varphi \in \mathbb{R},$$

of holomorphic transformations with $f(0) = f'(0) = 0$ is analytically equivalent near zero to its linear part at zero.

The condition on φ sufficient for this equivalence is the following: there exist positive numbers σ and C such that for any rational irreducible p/q we have

$$|\varphi - p/q| \geqslant Cq^{-\sigma}.$$

Any number φ satisfying this condition is called *Diophantine.* Any other real number is called *Liouvillian.*

REMARK. The set of all Liouvillian numbers has zero Lebesgue measure. This explains the statement of Example 2.

Now note that the set L of all Liouvillian numbers is typical in the sense of Definition 1.1. Indeed, for any fixed σ, C consider the set

$$L_{\sigma,C} = \{\varphi \in \mathbb{R} \mid \exists p/q \text{ such that } |\varphi - p/q| < Cq^{-\sigma}\}.$$

Then $L = \bigcap L_{\sigma,C}$. The set $L_{\sigma,C}$ is open by definition. It is dense, because it contains all rational numbers. Therefore, the set L is typical in the sense of Definition 1.1, while its complement is prevalent in the sense of Definition 1.2! In general, *what is negligible in the sense of KAM theory, is typical in the sense of singularity theory.* This contradiction (on the heuristic, not on the mathematical level) was known long ago.

1.2. Kolmogorov's concept of prevalence. In his plenary talk at the International Congress of Mathematicians (Amsterdam, 1954) Kolmogorov said:

"In order to get negative results showing that some effect has exclusive, negligible character, we will use the following somewhat handicraft tool. Suppose that a finite number of functionals

$$F_1(f), \ldots, F_r(f)$$

is defined on a class K of functions $f(x)$. Let these functionals take "in general" arbitrary values

$$F_1(f) = C_1, \ \ldots, \ F_r(f) = C_r$$

ranging in some domain of r-dimensional space. Then we will consider as "shy, negligible" any effect that takes place for a set of C having zero Lebesgue measure."

This concept was generalized in different directions.

We present a version due to Arnold, and its further development.

1.3. Prevalence in function spaces. The main idea in the following definitions of "negligible, shy" sets is to use finite-parameter families.

We consider two kinds of function spaces: Banach spaces, e.g., spaces of vector fields and their families, and nonlinear function spaces, e.g., spaces of maps of one manifold to another. The linear case is treated in [**HSY**]. Definitions of this section are applicable to the linear case as well as to the nonlinear.

Roughly speaking, an l-parameter family in a function space $\mathcal{F}$ is a smooth map of an l-dimensional ball B^l to $\mathcal{F}$. The notion of smoothness requires special definition.

DEFINITION 1.3. A map $F\colon B^l \to \mathcal{F}$ to a Banach space $\mathcal{F}$ is C^k *smooth* if for any linear continuous functional $\varphi\colon \mathcal{F} \to \mathbb{R}$, the function $\varphi \circ F$ is of class $C^k(B^l)$. This map is said to be a C^k*-family* in a Banach space. The image $F(\varepsilon)$ is denoted by f_ε.

The space of maps of one manifold to another has no linear structure. Therefore, a special definition of a family in such a space is required.

DEFINITION 1.4. Let M and N be two smooth manifolds, and $\mathcal{F} = C^k(M,N)$ the space of C^k-maps of M to N, B^l being the same as above. A C^k-map $F\colon B^l \times M \to N$ is said to be a C^k*-family of maps* $M \to N$. The restriction of F to $\{\varepsilon\} \times M$ is denoted by f_ε.

In what follows, we consider both kinds of families simultaneously. The following definition was proposed by Arnold (modulo some technical details).

DEFINITION 1.5. A subset S of a function space $\mathcal{F}$ is *shy* for a family F if the intersection of the family with S corresponds to a set of parameter values having Lebesgue measure zero, that is

$$\operatorname{mes}\{\varepsilon \in B^l \mid f_\varepsilon \in S\} = 0.$$

A subset $S \subset \mathcal{F}$ is *strongly l-shy* if there exists an open dense subset Σ in the space of C^k smooth families for some k such that S is shy for any family of this subset. A subset $S \subset \mathcal{F}$ is called *shy* if it is a countable intersection of strongly l-shy subsets for some l.

This definition is historically closely related to that of Kolmogorov; see 1.2.

Let us check whether this definition agrees with our intuition. To do this, in Definition 1.5 take the Banach space $\mathcal{F}$ to be finite-dimensional, with dimension greater than the number of parameters of the families. Then check whether a shy set in $\mathcal{F}$ has Lebesgue measure zero.

The answer turns out to be negative, as the following example shows. Let $\mathcal{F} = \mathbb{R}^2$, $l = 1$. Take the set $\mathcal{R}$ of $\mathbb{Q}$-polynomial curves in $\mathbb{R}^2$, that is, the union of all images of maps $P\colon [0,1] \to \mathbb{R}^2$, where P is a vector polynomial with rational coefficients. Any map $F\colon [0,1] \to \mathbb{R}^2$ is a one-dimensional family (of points) in $\mathbb{R}^2$; the family P above is a rational one.

The set $\mathcal{R}$ is of Lebesgue measure zero. For any ε there exists an open neighborhood U_ε of $\mathcal{R}$ with $\operatorname{mes} U_\varepsilon < \varepsilon$. Let S be the complement of U_ε. This set is shy in the sense of Definition 1.5. Indeed, the set Σ of one-parameter families in $\mathbb{R}^2$ for which S is shy contains all rational families (the intersection of any such family with S is void). On the other hand, any one-parameter family may be approximated by a rational one. Hence, Σ is dense. A small perturbation of a rational family still keeps its image in U_ε. Hence, Σ is open. Therefore, S is shy. On the other hand, its Lebesgue measure in any unit square is greater than $1 - \varepsilon$.

The following modification of the definition of shy sets, due to Kaloshin, forbids "bad" examples of this type.

DEFINITION 1.6 (modified Arnold's definition). A set S is *strongly l-shy* if it is strongly l-shy in the sense of Definition 1.5 and, moreover, for any point of $\mathcal{F}$ there exists a family F passing through this point such that $\mathcal{F}$ is shy for F.

PROPOSITION 1.1. *Let $S \subset \mathbb{R}^n$ be a strongly l-shy set in sense of Definition 1.6, with $l < n$. Then it has Lebesgue measure zero.*

PROOF. It is sufficient to prove that any point $x \in \mathbb{R}^n$ has a neighborhood whose intersection with S has Lebesgue measure zero. Take an l-parameter family through x for which S is shy. Then take a nearby family $\widetilde{F}$ such that $\tilde{f}_0 = x$ and $\operatorname{rank} d\widetilde{F}(0) = l$. This family may be chosen so that the subset S will be shy with respect to it, because the set Σ in Definition 1.5 is open dense. The image G of $\widetilde{F}$ is a submanifold near x. Hence, there exists a diffeomorphism of $\mathbb{R}^n$ to itself that rectifies G in some neighborhood U of x, that is, brings it to an l-dimensional disk D centered at x. Smooth maps preserve sets of measure zero. The image of S under this map is still denoted by S. Any l-disk parallel to D may be included into the image of some family close to $\widetilde{F}$. Any of these disks intersects S in a set of Lebesgue measure zero, by Definition 1.6. By the Fubini theorem, $\operatorname{mes}(U \cap S) = 0$. This proves the proposition.

DEFINITION 1.7. Any countable union of strongly l-shy sets is called *l-shy. A shy set* is a set that is l-shy for some l.

REMARK. l-shy sets in $\mathbb{R}^n$ have Lebesgue measure zero even if $l < n$.

DEFINITION 1.8. A set in a function space is *prevalent* if and only if its complement is shy.

EXAMPLE 3. (a) The set of all vector fields on the two-sphere generating Morse–Smale systems is prevalent.

(b) The analogous set of vector fields on the two-torus is neither prevalent, nor shy.

(c) The same is true if the two-torus is replaced by any manifold of dimension higher than two.

Statement (a) is very simple, although not written up anywhere. Statement (b) is a consequence of KAM theory (applied to diffeomorphisms of the circle). Statement (c) is a discovery of Smale.

1.4. Revision of classical results. The basic fact lying at the very foundation of singularity and bifurcation theory is Thom's big transversality theorem. It has a "prevalent counterpart".

THEOREM 1.1 [**K**]. *Let M and N be smooth manifolds and K a submanifold in the jet space $J^k(M, N)$. Then the set of maps $M \to N$ whose k-jet extensions are transversal to K is prevalent in an appropriate function space.*

COROLLARY 1.1 [**AAIS**]. *The set of vector fields on a smooth compact manifold having hyperbolic fixed points only is prevalent.*

COROLLARY 1.2 [**HSY**]. *The set of one-parameter families of vector fields in which the standard Andronov–Hopf bifurcation occurs is prevalent.*

REMARK. For the description of the standard Andronov–Hopf bifurcation, see 1.2.5.

Statement (a) of Example 3 is a simple corollary of Theorem 1.1.

These corollaries are given just as illustrations. We shall neither use, nor prove them. The main point of this section is the following generalization of Corollary 1.2:

GENERAL CONJECTURE. *The genericity assumptions in all theorems of Chapters 3–7 determine prevalent sets in appropriate function spaces.*

We shall not discuss this conjecture. In what follows we prove (or even only claim) that the genericity assumptions required for the theorems below are typical in the traditional categoric sense. Yet it is important to realize (and to prove in the future) that the sets of families studied below are "more residual" than it is claimed in this book.

§2. Attractors, their dimensions and projections

A general concept, going back to Kolmogorov, Landau, Hopf, Arnold, Ruelle, and Takens, presents the following heuristic description of a typical dynamical system. After finite time, a positive semitrajectory enters the neighborhood of some attracting set or simply *attractor*. If this set is large enough, that is, distinct from the finite union of fixed points and periodic orbits, then the behavior of the phase curve on the attractor and near it has chaotic features.

This concept is neither proved nor even formalized. There are different non-equivalent definitions of attractors. Below we describe the so-called maximal attractors. Whatever other concept of attractor is used, it defines a subset of a maximal attractor.

Our goal here will be to give some kind of upper estimate for the attractor. The characteristic to be estimated is dimension. If the dimension of a maximal attractor is much lower than the dimension of the phase space, this implies that the actual number of parameters describing the behavior of the dynamical system is much lower than the order of the system. A serious difficulty in this approach is that the attractor is not necessarily a manifold and can have a complex set-theoretical structure. Hence, it can have no dimension at all in the sense of smooth analysis. Therefore, some special definition of dimension is required. The so-called Hausdorff and box counting dimensions appear to be most useful in this context.

From the physical point of view, the dimension of a set is the number of parameters distinguishing points of this set. For a set of exotic structure, this concept may be formalized as the dimension of a typical plane onto which the set admits a one-to-one orthogonal projection. Below we discuss the upper estimate of the Hausdorff and box counting dimensions of maximal attractors, and one-to-one projections of these attractors onto generic planes of rather small dimension. Moreover, we study the properties of box counting and Hausdorff dimensions related to projections onto planes.

2.1. Maximal attractors. In this section we discuss so-called dissipative systems.

DEFINITION 2.1. A vector field is called *dissipative* in a compact manifold with boundary B if its phase flow shifts corresponding to positive time values map B strictly into itself. A diffeomorphism $f\colon B \to B$ that maps B strictly into its interior is also called *dissipative.*

The next definitions will be given for maps. Their counterparts for vector fields can be obtained by evident modifications.

DEFINITION 2.2. The *maximal attractor* of a dissipative diffeomorphism $f\colon B \to B$ is the intersection

$$A_{\max} = \bigcap_{n>0} f^n B.$$

This set is nonempty, because the sets $f^n B$ are nested and compact. In what follows, f will be at least C^2 smooth. Hence the nested compact sets above are manifolds with boundaries. Yet their intersection is not necessarily a manifold at all.

Maximal attractors characterize limit behavior of the orbits of dissipative maps in the sense of the following propositions.

PROPOSITION 2.1. *The maximal attractor is invariant under f:*

$$fA_{\max} = A_{\max}.$$

PROOF. By definition,

$$fA_{\max} = \bigcap_{n>0} f^{n+1}B = \bigcap_{n>1} f^n B = A_{\max},$$

because the intersection of nested sets does not change after elimination of the first set.

DEFINITION 2.3. A set $\{x_n \mid n \in \mathbb{Z}\}$ is called a *complete orbit* of a diffeomorphism f if $x_{n+1} = f(x_n)$ for any $n \in \mathbb{Z}$.

PROPOSITION 2.2. *The maximal attractor of a dissipative diffeomorphism contains all complete orbits.*

PROOF. Any point of the maximal attractor belongs to some complete orbit. Indeed,

$$fA_{\max} = A_{\max}, \qquad f^{-1}A_{\max} = A_{\max}$$

by Proposition 2.1. On the other hand, any complete orbit belongs to all the images $f^n B$, and hence to their intersection.

The next proposition shows that the maximal attractor is Lyapunov stable.

PROPOSITION 2.3. *For any neighborhood U of a maximal attractor there exists a number n such that $f^n B \subset U$.*

PROOF. To get a contradiction, suppose that the converse is true. Hence there exists a neighborhood U of A such that for any n there exists an x such that $x_n = f^n(x) \notin U$. Since B is compact, one may choose a convergent subsequence of the previous sequence. We denote it by the same symbol. Let y be its limit. By assumption, $y \notin U$. We will prove, nevertheless, that the complete orbit of y is well defined. Therefore, by Proposition 2.2, y belongs to the maximal attractor of f, a contradiction. We must prove that for any k the point $f^{-k}y$ is well defined. Once more, suppose that the converse is true. Then there exists a $k \geqslant 0$ such that $f^{-k}y$ exists and $f^{-k-1}y$ does not. That is, $f^{-k}y \in B \setminus f(B)$. For sufficiently large n, the points $f^{-k}x_n$ are well defined and tend to $f^{-k}y$ as $n \to \infty$. Since $B \setminus f(B)$ is open in B, all these points belong to $B \setminus f(B)$. On the other hand, $f^{-k-1}x_n$ is well defined for n sufficiently large. Hence, $f^{-k}x_n \in f(B)$. This contradiction proves the proposition.

2.2. Hausdorff and box counting dimensions. The idea of the definition of Hausdorff dimension is illustrated by the following example. Somebody is investigating a set in $\mathbb{R}^3$, knowing that it is either a curve, a surface, or a body. The set itself is nonaccessible, but one can obtain an answer to the question: "What is its length, area and volume?" Obviously, the answer: "Length ∞, area 1, volume 0" means that the unknown set is a surface.

Now let us proceed to the precise definition. A covering of a compact set by balls (for short, a covering of the set K) is a collection of balls whose union contains K. We denote by $\mathfrak{U}_\varepsilon(K)$ the collection of all the coverings of K by balls whose radius does not exceed ε. Denote by $\mathfrak{B}_\varepsilon(K)$ the set of all coverings from $\mathfrak{U}_\varepsilon(K)$ consisting of equal balls. The d-dimensional volume of a covering U of the set K by balls Q_j of radii r_j is the quantity

$$V_d(U) = \sum_j r_j^d.$$

DEFINITION 2.4. The *Hausdorff dimension* of K is the infimum of those d for which there exists a covering of the class $\mathfrak{U}_\varepsilon(K)$ with arbitrarily small d-dimensional volume:

$$\dim_H(K) = \inf\{d \mid \forall\,\delta\ \exists\,\varepsilon,\ U \in \mathfrak{U}_\varepsilon(K) : V_d(U) < \delta\}.$$

The *box counting dimension* of K is defined in the very same way, only the class $\mathfrak{U}_\varepsilon(K)$ is replaced by $\mathfrak{B}_\varepsilon(k)$:

$$\dim_B(K) = \inf\{d \mid \forall\,\delta\ \exists\,\varepsilon,\ U \in \mathfrak{B}_\varepsilon(K) : V_d(U) < \delta\}.$$

REMARK. Hausdorff and box counting dimensions are invariant under diffeomorphisms. Moreover, they do not increase under smooth maps.

2.3. Attractors of k-contracting systems. The result of this subsection is based on the following simple idea. A diffeomorphism that contracts k-dimensional volume cannot have an invariant k-dimensional compact manifold. Indeed, a manifold invariant under the action of a diffeomorphism preserves its volume.

Denote by Π^k a k-dimensional parallelepiped in a linear space and by $V(\Pi^k)$ its k-dimensional volume.

DEFINITION 2.5. A diffeomorphism $f\colon B \to B$ is called *k-contracting* if there exists a $q \in (0,1)$ such that for any k-dimensional parallelepiped Π^k in any tangent space to B we have

$$V(df\Pi^k) \leqslant qV(\Pi^k). \tag{2.1}$$

The map f in (2.1) is also called *k-contracting with constant q.*

THEOREM 2.1. *The Hausdorff dimension of the maximal attractor of a k-contracting diffeomorphism is no greater than k.*

PROOF. Let $K = A_{\max}$ be the maximal attractor of f. We will use only the fact that $A_{\max}$ is invariant under f; see Proposition 2.1. For any covering of K by sufficiently small balls, we will construct a new covering by balls no greater than the given ones such that the k-dimensional volume of this new covering equals half of that of the original covering. Replacing this covering in a similar way and using induction, one may construct a covering of K with arbitrary small k-dimensional volume, thus proving the theorem.

Take a covering U of K by balls Q_j. The union of the sets fQ_j still covers K because of the invariance of K. If the ball is small enough, then its image is similar to the ellipsoid. In more detail, if the ball Q_j centered at x_j is small, then

$$fQ_j \subset f(x_j) + LQ_j, \qquad L = 2df(x_j).$$

The operator L is k-contracting with coefficient $2q$. This coefficient may be considered to be smaller than any fixed value, say, than a constant depending on k only. Indeed, replacing f by its iterate, say f^n, we replace q by q^n, while K is still f^n-invariant.

LEMMA 2.1. *For any k there exists a constant $q(k)$ with the following property. Let L be a nonsingular linear operator that is k-contracting with coefficient $q < q(k)$, and Q a unit ball. Then there exists a covering $U(L)$ of the ellipsoid $E = 2L(Q)$ by balls of radius no greater than $q^{1/k}\sqrt{k} < 1/2$, having k-dimensional volume no greater than $1/2$: $V_k(U_0) < 1/2$. Here the constant $q(k)$ may be taken equal to $(k+1)^{-k/2}4^{-k-1}$.*

PROOF. We begin with the following

PROPOSITION 2.4. *Let L be a nonsingular k-contracting linear operator with coefficient q. Let the ellipsoid LQ have the semiaxes $a_1 \geqslant \cdots \geqslant a_n$. Then*

$$a_1 \cdots a_k \leqslant q.$$

PROOF. Consider the plane spanned by the k larger axes of the ellipsoid LQ and its inverse image P under L. The L-image of the unit ball in this plane is the ellipsoid with semiaxes $a_1 \geqslant \cdots \geqslant a_k$. Let V and W be k-dimensional volumes of this ball and ellipsoid respectively. Then

$$a_1 \cdots a_k = \frac{W}{V} \leqslant q.$$

This proves the proposition.

The key idea of the proof of Lemma 2.1 is that the ellipsoid $2LQ$ can be covered by cubes of width equal to the kth axis of the ellipsoid, with edges parallel to its axes, and centers located in the plane spanned by the k larger axes. The number of these cubes is no greater than

$$N = a_1 \cdots a_k \frac{2^k}{a_k^k} \leqslant q \frac{2^k}{a_k^k}.$$

Indeed, the number of segments of length a covering a segment of length b is no greater than $2b/a$. This is trivial for $a \geqslant b$ and wrong for $a \leqslant b$.

Now replace any cube of this covering by a ball with the same center and radius $2\sqrt{k+1}\, a_k$. Denote by E_k the section of the ellipsoid $2LQ$ by the plane spanned by its k larger axes. Let Q_k be the ball in the orthogonal $(n-k)$-plane centered at 0 with radius $2a_k$. Than $2LQ \subset E_k \times Q_k$. By the Pythagoras theorem, if E_k is covered by k-dimensional cubes with edge $4a_k$, then $E_k \times Q_k$ is covered by n-dimensional balls with the same centers and radii $2\sqrt{k+1}\, a_k$.

Now let U_0 be the covering of $2LQ$ by these balls. Then

$$V_k(U_0) \leqslant N(2\sqrt{k+1}\, a_k)^k \leqslant q(4\sqrt{k+1})^k. \tag{2.2}$$

For sufficiently small q, $q < q(k)$ (see Lemma 1.2), we have

$$V_k(U_0) < 1/2.$$

By Proposition 2.4, the radii of the balls of the covering are no greater than $q^{1/k}\sqrt{k} < 1/2$ for $q < q(k)$. This proves the lemma.

The theorem is now easily derived from Lemma 2.1. Let U be an arbitrary covering of K by small balls Q_j with centers x_j and radii r_j. Let $U(L_j)$ be the covering from Lemma 2.1 for the ellipsoid $2L_jQ$, $L_j = df(x_j)$. Then the image U_j of $U(L_j)$ under the affine map $x \mapsto x_j + r_j x$ consists of balls of radii no greater than $r_j/2$ and covers the image fQ_j. Its k-dimensional volume is at most half of r_j^k. The union of the coverings U_j forms the desired covering $F(U)$ of K (F for *filial*). This proves Theorem 2.1.

REMARK. The proof exploits the possibility of using different balls constituting the coverings in the definition of Hausdorff dimension. Indeed, even if the initial covering U_0 consists of equal balls, then the ellipsoids obtained as the images of these balls under the action of the derivative of f taken at their centers have different axes a_k. Hence, the new covering U_1 consists of different balls. This prevents the generalization of Theorem 2.1 from Hausdorff to box counting dimension.

APPENDIX TO THEOREM 2.1. *The Hausdorff dimension of the maximal attractor of a k-contracting diffeomorphism is smaller than k.*

PROOF. Suppose that in Lemma 2.1, the kth axis of the ellipsoid LQ is greater than some positive α: $a_k \geqslant \alpha$. Then Lemma 2.1 may be improved:

$$V_{k-\varepsilon}(U_0) < 1/2,$$

provided that $\alpha^{-\varepsilon} < 2$. Indeed, for the same U_0 as in the lemma, we have the following counterpart of (2.2):

$$V_{k-\varepsilon}(U_0) \leqslant N(2\sqrt{k+1}\,a_k)^{k-\varepsilon} \leqslant q2^{2k-\varepsilon}k^{(k-\varepsilon)/2}\alpha^{-\varepsilon}.$$

For $q < q(k)$, we get

$$V_{k-\varepsilon}(U_0) \leqslant \alpha^{-\varepsilon}/4 \leqslant 1/2.$$

Now let $F(U)$ be the same covering of K as in the proof of Theorem 2.1. Let $a_k(j)$ be the kth axis of the ellipsoid L_jQ, and α the minimum of these axes. Take $\varepsilon > 0$ such that $\alpha^{-\varepsilon} < 2$. Then, by the improved version of Lemma 2.1,

$$V_{k-\varepsilon}(F(U)) < \frac{1}{2}V_{k-\varepsilon}(U).$$

This proves the appendix.

2.4. Box counting dimension of attractors. Nevertheless the generalization mentioned above holds true.

THEOREM 2.2 (Hunt). *The box counting dimension of the maximal attractor of a k-contracting diffeomorphism is no greater than k.*

This theorem will be proved in three steps.

Filial covering. Here we describe more formally than above the covering constructed in the proof of Theorem 2.1.

Let f be a k-contracting diffeomorphism, and $L(x) = 2df(x)$ and let $U(L(x))$ be the covering from Lemma 2.1 consisting of balls denoted by $Q_j(L)$. For any ball $\mathbf{Q}$ with center y and radius r in the absorbing domain of f, denote by $U(\mathbf{Q}, f)$ the image of the covering $U(L(x))$ under the affine map $x \mapsto f(y) + rx$:

$$U(\mathbf{Q}, f) = \{f(y) + rQ_j(L) \mid Q_j(L) \in U(L(x))\}.$$

All the balls of this covering will be called *filial* to the ball $\mathbf{Q}$, which in turn is called the *ancestor* of these balls.

Let K be a compact set invariant under the k-contracting diffeomorphism f. Consider a covering U of the compact set K. Let $\mathbf{B}_j$ be the balls of this covering. Let $U(\mathbf{B}_j, f)$ be the set of the balls filial to $\mathbf{B}_j$. Then the set of balls

$$F(U) = \bigcup U(\mathbf{B}_j, f)$$

covers K. This covering is called *filial* for U.

Define the rank of the ball $\mathbf{B}$ of radius $\mathbf{R}$ as $r(\mathbf{B}) = -\log_2 \mathbf{R}$. The smaller the ball, the larger is its rank.

Let $L(x)$ be the same as above. Denote by $a_k(x)$ the kth semiaxes of the ellipsoid $L(x)Q$. Let

$$\omega(x) = -\log_2(\sqrt{k}\, a_k(x)).$$

Note that the radius of the ball used in Lemma 2.1 stands under the logarithm. This is a continuous function on the absorbing domain of f, hence, it is bounded:

$$\mu < \omega < \nu. \tag{2.3}$$

Let U be a covering of K by small equal balls of rank ρ. We will construct the covering W of K by smaller equal balls such that the ratio of k-dimensional volumes of these coverings is less than any prescribed constant. This will prove the theorem.

The generation of filial coverings and heuristic proof of the theorem. Let ρ be a large constant, to be specified below, and U a covering of K by equal balls of radius greater than ρ. Define the sequence of filial coverings inductively:

$$U_0 = U, \qquad U_{m+1} = F(U_m).$$

The index m will be called the *age* of the covering U_m and of the corresponding balls. Then the rank of the balls of age m lies between $\rho + \mu m$ and $\rho + \nu m$. Note that

$$V_k(U_m) \leqslant 2^{-m} V_k(U). \tag{2.4}$$

On the heuristic level, the proof goes on in the following way.

For any sufficiently large R the compact set K will be covered by some balls of the coverings U_j with ranks approximately equal to $\rho + R$. Denote the covering thus obtained by $\mathbf{W}$ (see Figure 2.1). The age n of the coverings U_j used in this construction takes approximately cR subsequent values between R/μ and R/ν. Any of these coverings has k-dimensional volume no greater than $2^{R/\mu} V_k(U)$. The k-dimensional volume of the union of these coverings is no greater than $cR\,2^{-n} V_k(U)$. The same is true for $V_k(\mathbf{W})$. The oscillation of the ranks of the balls of the covering $\mathbf{W}$ is small as compared to R. Hence, the following construction works. Replace all the balls of the covering $\mathbf{W}$ by balls with the same centers and radius equal to the largest radius of the balls from $\mathbf{W}$. The k-dimensional volume of the covering W thus obtained will be much smaller than that of U. This will prove Hunt's theorem.

Final covering by equal balls. Let us pass to the detailed exposition.

LEMMA 2.2. *Let K be a compact set invariant under a k-contracting diffeomorphism f of a compact manifold with boundary into itself, let μ and ν be defined by f as above; see* (2.3). *Then for any ε there exists a ρ depending on ε and f such that for any covering U of the set K by balls of rank ρ and for any positive R there exists a covering W of K with the following properties*:

1. *W consists of equal balls.*
2. *For some positive c depending only on f we have*

$$\frac{\log_2 V_k(U)}{V_k(W)} \geqslant \frac{R}{\nu} - \log_2 cR - 2k\varepsilon\,\frac{R+\nu}{\mu} - \mu.$$

The lemma immediately implies Hunt's theorem, because the right-hand side of the previous inequality tends to infinity together with R.

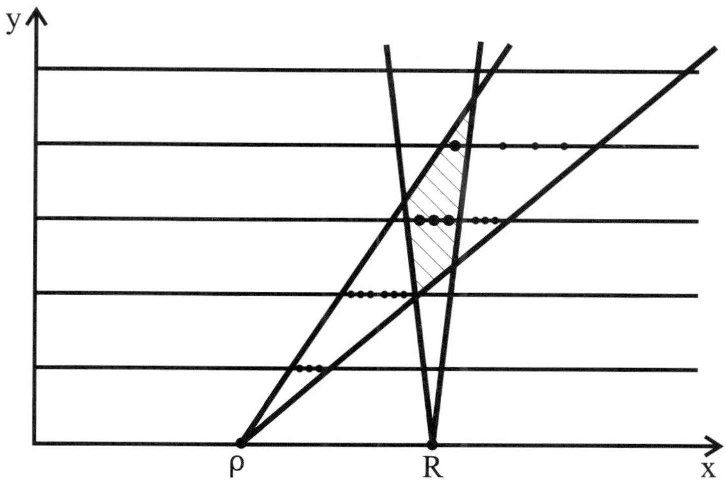

FIGURE 2.1. Gathering the balls for the covering **W**. The horizontal axis shows the rank of the balls, and the vertical one is for the age of the covering.

PROOF OF LEMMA 2.2. Let $\{U_j\}$ be a series of filial coverings with $U = U_0$. Let us construct an auxiliary covering **W** formed by the balls of the filial coverings U_j with rank approximately equal to $R + \rho$. To do this, define an integer-valued "age" function $n(x)$ on K in the following way. Take a backward orbit of x:

$$x_0 = x, \qquad x_{l+1} = f^{-1}x_l.$$

It is well defined, due to the invariance of K under f. Define $n = n(x)$ by the inequalities:

$$R \leqslant \sum_1^n \omega(x_l) < R + \mu. \tag{2.5}$$

Such an n is well defined because of (2.3). Take a sequence of balls $B_l(x)$ of the covering U_l defined in the following way: $B_0(x)$ is an arbitrary ball of the covering U_0 containing x_n; $B_l(x)$ is an arbitrary ball filial to $B_{l-1}(x)$ and containing x_l. Let $B(x) = B_n(x)$. This will be a ball of the covering U_n.

Denote by **W** the covering formed by all the balls $B(x)$:

$$\mathbf{W} = \{B(x) \mid x \in K\}.$$

Let r be the minimal rank of the balls of this covering. Recall that the smaller is the rank, the greater is the radius of the ball. Replace all balls of the covering **W** by balls with the same centers and ranks equal to r. This is the desired covering.

Let us now estimate the k-dimensional volume of this covering. The age $n(x)$ of the ball $B(x)$ can be estimated as

$$\frac{R}{\nu} \leqslant n(x) < \frac{R+\nu}{\mu}.$$

This follows from (2.3) and (2.5). For any $x \in K$ and $n = n(x)$, we have

$$\log_2 \frac{V_k(U)}{V_k(U_n)} \geqslant n \geqslant \frac{R}{\nu}. \tag{2.6}$$

This follows from the previous inequality and (2.4). The function $n(x)$ takes no more than cR values with c depending on μ, ν only, where one may take

$$c = \frac{1}{\mu} - \frac{1}{\nu} + 1$$

for R large enough.

Let us now estimate the ranks of the balls $B(x)$. Let $B_l(x)$ be the same balls as above. Denote their centers by y_l. Their ranks are no less than ρ. Now take $\rho = r(B_0(x))$ so large that the oscillation of the function ω in the ball of rank ρ will be smaller than ε. Then

$$r(B_n(x)) - r(B_0(x)) = \sum_1^n \omega(y_l) \in (R - \varepsilon n, R + \varepsilon n + \mu). \tag{2.7}$$

The k-dimensional volume of the covering $\mathbf{W}$ is no more than cR times greater than the maximal k-dimensional volume of the covering V_n with $n \geqslant R/\nu$. Hence,

$$\log_2 \frac{V_k(U)}{V_k(\mathbf{W})} \geqslant \frac{R}{\nu} - \log_2(cR). \tag{2.8}$$

Recall that the rank of all the balls of the covering u_0, hence of $B_0(x)$, equals ρ. Then the oscillation of the ranks of the balls constituting the covering $\mathbf{W}$ is no greater than

$$2\varepsilon \max n(x) \leqslant 2\varepsilon \frac{R+\nu}{\mu} + \mu. \tag{2.9}$$

Hence,

$$\log_2 \frac{V_k(U)}{V_k(W)} \geqslant \log_2 \frac{V_k(U)}{V_k(\mathbf{W})} - 2\varepsilon k \max n(x) - \mu.$$

Together with (2.8), (2.9), this implies Lemma 2.2. The lemma, in turn, proves Theorem 2.2.

2.5. The easy Whitney theorem. In this and next two subsections we discuss one-to-one projections of attractors of dissipative systems. A sample is given by the theorem on projections of smooth manifolds.

THEOREM (Easy Whitney Theorem). *A compact k-dimensional submanifold of Euclidean space admits a one-to-one orthogonal projection onto a plane L in general position along the orthogonal complement to L, provided that* $\dim L > 2k$.

REMARK. We say that a general k-dimensional plane in an n-dimensional linear space has a certain property if the set of planes having this property is in a sense "residual" in the space of all k-dimensional planes in n-dimensional linear space. The traditional meaning of the word "residual" is "the intersection of open everywhere dense sets". The word residual in the easy Whitney theorem may be replaced by the term "prevalent". A subset of a finite-dimensional space is called *prevalent* if its complement is of zero Lebesgue measure; see Definition 1.2 in 1.1.

PROOF OF THE EASY WHITNEY THEOREM. We shall give the proof by induction on the codimension of the plane L. If the codimension is 0, then the projection is the identity and the theorem is trivial. The induction step exploits the following

LEMMA 2.3. *A compact k-dimensional submanifold of Euclidean space $\mathbb{R}^n$ with $n > 2k+1$ admits a one-to-one projection to a generic hyperplane.*

DERIVATION OF THE THEOREM FROM LEMMA 2.3. Suppose that for k-dimensional submanifolds of $\mathbb{R}^n$ the theorem is already proved. Then for k-dimensional submanifolds of $\mathbb{R}^{n+1}$ it follows immediately from the above lemma and the induction hypothesis.

PROOF OF LEMMA 2.3. Let k be the submanifold in question. Any oriented line in $\mathbb{R}$ defines a direction; that is, all parallel lines with the same orientation have the same direction. A direction is called *bad* if some line of this direction intersects K in more than one point. Any direction corresponds to a point on the unit sphere S^{n-1}. The set B of all bad directions in S^{n-1} is given by the formula

$$B = \{(x-y)/|x-y| \mid x \in K,\, y \in K,\, x \neq y\}.$$

Denote by Δ the diagonal

$$\Delta = \{(x,x) \in K \times K \mid x \in K\}.$$

The set B is the image of the following smooth map:

$$\pi\colon K \times K - \Delta \to S^{N-1}, \qquad (x,y) \mapsto \frac{x-y}{|x-y|}. \tag{2.10}$$

By assumption, the dimension of the target is greater than that of the source. Hence, all the points of the source are singular for f: the rank of the derivative df is smaller than the dimension of the target. Therefore, by the Sard lemma, the image of this map is of zero Lebesgue measure in S^{n-1}. Hence, the set of good (i.e., not bad) directions has complete Lebesgue measure and thus forms a prevalent subset of the sphere. This completes the proof of the lemma.

REMARK. Prevalence of the set of good directions in Lemma 2.3 implies prevalence of the set of planes admitting one-to-one projection of K in the Easy Whitney Theorem as a subset of the Grassmann manifold. We will not discuss this implication in detail.

2.6. Generalization to the box counting dimension.

THEOREM 2.3 (Generalized Easy Whitney Theorem). *The conclusion of the previous theorem holds if the compact submanifold of dimension k in its statement is replaced by a compact set of box counting dimension k (now k is not necessarily an integer).*

PROOF. The proof repeats the previous one verbatim with only one exception, the study of the image of the map (2.10). Without paying special attention, we claimed that

$$\dim K \times K = 2 \dim K$$

for compact smooth manifolds. This is trivial. Yet if manifolds are replaced by compact sets and the dimension in question is fractal, then this statement is no longer trivial. It is valid for box counting, but it is not true for Hausdorff dimension.

PROPOSITION 2.5. *Let $K \subset \mathbb{R}^n$ be an arbitrary compact set. Then*

$$\dim_B(K \times K) = 2\dim_B K. \tag{2.11}$$

This proposition will be proved below. Let $K_m \subset K \times K - \Delta$ be the compact set

$$K_m = \{(x, y) : |x - y| \geqslant 1/m\}, \qquad K_\infty = K \times K - \Delta\,.$$

By Proposition 2.5

$$\dim_B K_m \leqslant \dim_B(K \times K) = 2\dim_B K.$$

Let π be the same as in (2.11). Then $\dim_B \pi K_m \leqslant 2k$, because π is smooth on K_m; see the remark at the end of 2.2. Hence $\operatorname{mes}\pi K_m = 0$, where mes is $(n-1)$-Lebesgue measure in S^{n-1}. Hence $\operatorname{mes}\pi K_\infty = 0$ and the set of good directions is prevalent once more.

The end of the proof of Theorem 2.3 follows the same lines as the proof of the Easy Whitney Theorem.

PROOF OF PROPOSITION 2.5. The definition of box counting dimension is well suited for obtaining upper estimates of the dimension of the Cartesian square. Indeed, suppose that $d > k$ is any number. Then there exists a covering U of K by equal balls having an arbitrary small d-dimensional volume: for an arbitrary δ there exist ε and $U \in \mathfrak{B}_\varepsilon(K)$ such that $V_d(U) < \delta$. Let the balls of the covering U be centered at the points $x_1, \dots, x_N$ and have radius R. Then the balls in $\mathbb{R}^n \times \mathbb{R}^n$ of radius $\sqrt{2}R$ cover $K \times K$. Denote this covering by U'. We have $V_d(u) = NR^d$,

$$V_{2d}(U') = N^2(2R^2)^d = 2^d(V_d(U))^2 < 2^d\delta^2.$$

Hence, $\dim_B(K \times K) \leqslant 2K$. This implies (2.11).

THEOREM 2.4. *A compact set K of Hausdorff dimension no greater than k admits a one-to-one projection onto a generic plane of dimension greater than $2k$ provided that*

$$\dim_H(K \times K) \leqslant 2k. \tag{2.12}$$

PROOF. The proof repeats the previous one verbatim, only Proposition 2.5 is replaced by assumption (2.12).

Now the main result of the section follows.

THEOREM 2.5. *The maximal attractor of any k-contracting system has a one-to-one projection onto a generic $(2k+1)$-plane.*

FIRST PROOF. Denote the attractor by K as before. By Theorem 2.2, we have $\dim_B K \leqslant k$. Theorem 2.5 now follows from the generalized easy Whitney theorem (for box counting dimension).

SECOND PROOF. By Theorem 2.1, $\dim_B K \leqslant k$. If we check assumption (2.12), then Theorem 2.5 will follow from Theorem 2.4. In order to check (2.12), consider the Cartesian square of f:

$$F\colon B \times B \to B \times B, \qquad (x, y) \mapsto (f(x), f(y)).$$

The map F is dissipative and $2k$-contracting. Its attractor is $K \times K$. Inequality (2.12) now follows from Theorem 2.1.

The next section shows that the counterpart of the easy Whitney theorem for box counting dimension is false in general.

2.7. Counterexamples for Hausdorff dimension. Ittai Kan sets. There is a strong temptation to repeat the proof of Theorem 2.3 word for word, replacing the box counting dimension by its Hausdorff counterpart. It is evident that (2.12) must hold for any reasonable dimension! However, for Hausdorff dimension, relation (2.12) and the easy Whitney theorem are not true. Below we describe the counterexample to the Easy Whitney Theorem for Hausdorff dimension due to Ittai Kan.

Let us first construct the so-called Ittai Kan sets. Consider the splitting of the set of natural numbers given by the series $\Sigma = \sum b_l$, $b_l \in \mathbb{N}$:

$$\mathbb{N} = \bigcup \mathbb{N}_j, \qquad \mathbb{N}_j = \{n \in \mathbb{N} \mid s_{j-1} \leqslant n < s_j\}, \tag{2.13}$$

where $s_0 = 0$, s_j is a partial sum of Σ, $s_j = \sum_1^j b_l$. The splitting itself is denoted by Σ.

DEFINITION 2.6. A binary number ω is called *plus-subordinate* to the splitting (2.13) if

$$\omega = 0.a_1 \dots a_{n+1}, \qquad a_l = 0 \text{ for } l \in \mathbb{N}_{2j+1},$$

and the digits a_l for $l \in N_{2j}$ take arbitrary values 0 or 1. Next, ω is *minus-subordinate* to the same splitting, provided that the inverse holds:

$$a_l = 0 \text{ for } l \in N_{2j}, \qquad a_l \in \{0, 1\} \text{ is arbitrary for } l \in N_{2j+1}.$$

The subsets of $[0, 1]$ plus- and minus-subordinate to the splitting Σ are denoted by Σ^+ and Σ^- respectively.

PROPOSITION 2.6. *Let the splitting Σ satisfy*

$$b_{l+1}/b_l \to \infty \tag{2.14}$$

as $l \to \infty$. Then $\dim_H \Sigma^+ = \dim_H \Sigma^- = 0$.

This proposition will be proved at the end of this subsection.

LEMMA 2.4. *In the plane $\mathbb{R}^2$ there exists a set of Hausdorff dimension 0 that has no one-to-one projection to any line $L \subset \mathbb{R}^2$.*

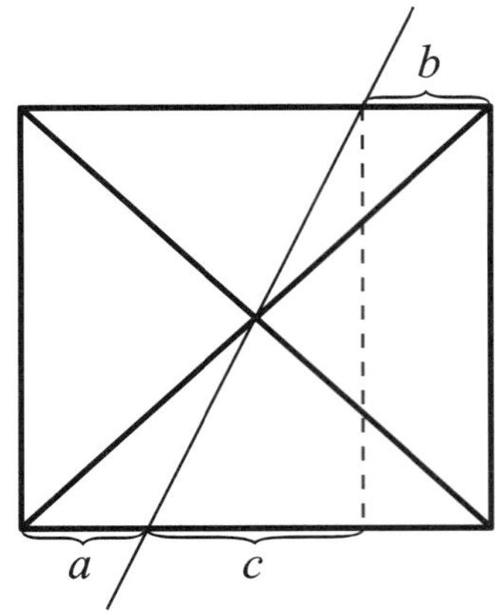

FIGURE 2.2. All the directions are bad for the set in the Ittai Kan counterexample.

PROOF. Take the arbitrary splitting Σ described in Proposition 2.6. Take the set $K' \subset [0,1]$:

$$K' = \Sigma^+ \cup (1 - \Sigma^-), \qquad (1 - \Sigma^-) = \{1 - x \mid x \in \Sigma^-\}.$$

Take a set K on the boundary of the square $[0,1] \times [0,1]$:

$$\begin{gathered} K = K_1 \cup K_2 \cup K_3 \cup K_4, \\ K_1 = K' \times \{0\}, \quad K_2 = K' \times \{1\}, \quad K_3 = \{0\} \times K', \quad K_4 = \{1\} \times K'. \end{gathered} \tag{2.15}$$

The set K is of Hausdorff dimension 0 by Proposition 2.6. We shall prove that any direction is bad for the set K. Let us prove that any line with polar angle in $[\pi/4, 3\pi/4]$ intersects $K_1 \cup K_2$ in two points. For any fixed direction, the difference between the x coordinates of the intersection points of the line having this direction with the lines $y = 0$ and $y = 1$ is constant. Denote this difference by c and let $\omega = 1 - |c|$.

Suppose that $c > 0$ (the case $c < 0$ is treated along the same lines (see Figure 2.2)). Now the definition of the Ittai Kan sets can be used: any number $\omega \in [0,1]$ may be split into the sum

$$\omega = a + b, \qquad a \in \Sigma^+,\ b \in \Sigma^-.$$

Indeed, for $\omega = 0.a_1^+ \ldots a_l^+ \ldots$ we define:

$$\begin{aligned} a &= 0.a_1^+ \ldots a_l^+ \ldots, \qquad & a_l^+ &= \begin{cases} 0 & \text{for } l \in \mathbb{N}_{2j+1}, \\ a_l & \text{for } l \in \mathbb{N}_{2j}, \end{cases} \\ b &= 0.a_1^- \ldots a_l^- \ldots, \qquad & a_l^- &= \begin{cases} a_l & \text{for } l \in \mathbb{N}_{2j}, \\ 0 & \text{for } l \in \mathbb{N}_{j+1}. \end{cases} \end{aligned}$$

This gives the desired splitting and proves that the direction given by the vector $(c, 1)$ with $|c| \leqslant 1$ is bad.

The same is true for the directions with polar angle from $[-\pi/4, \pi/4]$; only the horizontal edges in the above argument must be replaced by vertical ones. This proves the lemma.

LEMMA 2.5. *In any Euclidean space $\mathbb{R}^n$ there exists a set of Hausdorff dimension 0 that has no one-to-one projection to any hyperplane $L \subset \mathbb{R}^n$.*

SKETCH OF THE PROOF. The desired example is obtained as a subset of the boundary of the cube $[0,1]^n$. The sets on the faces are the appropriate Cartesian powers of the set (2.15).

Let us now prove Proposition 2.6. The proposition will be proved for Σ^+. The proof for Σ^- follows the same lines. Let A_j be a set of binary approximations to the elements of Σ^+ with the accuracy $2^{-s_{2j}}$. In more detail,

$$A_j = \{\omega^+ = 0.\omega_1 \dots \omega_s \mid \omega = 0.\omega_1 \dots \omega_s \dots \in \Sigma^+,\ s = s_{2j}\}.$$

For all $\omega \in \Sigma^+$ the truncation $\omega^+ \in A_j$ thus defined approximates ω with accuracy $2^{-(s_{2j}+b_{2j+1})}$. Indeed, for any such ω the b_{2j+1} digits following the sth digit are zeros by the definition of Σ^+. Hence the set

$$W_j = \bigcup_{\omega^+ \in A_j} [\omega^+, \omega^+ + 2^{-b_{2j+1}}]$$

is a covering of Σ^+. We shall prove that for any $\varepsilon > 0$ we have $V_\varepsilon(W_j) \to 0$ as $j \to \infty$. This will prove the proposition. By the definition of Σ^+, the set W_j consists of N segments of length R with

$$R = 2^{-(s_{2j}+b_{2j+1})} N = |A_j| = 2^{\sigma_j}, \qquad \sigma_j = \sum_{l=1}^{j} b.$$

Hence

$$\log_2 V_\varepsilon(W_j) = -\varepsilon(s_{2j} + b_{2j+1}) + \sigma_j.$$

It is sufficient to prove that $s_{2j}/b_{2j+1} \to 0$ as $j \to \infty$. This statement follows easily from (2.14). This proves the proposition.

2.8. Problems and applications. The most attractive problem in the field is how to include the maximal attractor into a smooth low-dimensional manifold. The desired result is provided by the following theorem.

Consider a system

$$\dot{x} = Ax + f(x) \tag{2.16}$$

in Euclidean space $\mathbb{R}^n$. Let A be a selfadjoint operator with eigenvalues $\lambda_1, \dots, \lambda_n$. Let $\mathbb{R}^k$ be the linear hull of the first k eigenvectors.

THEOREM 2.6. *Suppose* (2.16) *is a dissipative system with absorbing domain B. Let*

$$\|df\| < L$$

everywhere in B. Suppose the following spectral gap condition holds:

$$\lambda_k - \lambda_{k+1} > 2L.$$

Then the system (2.16) *restricted to B has a C^1 smooth k-dimensional invariant manifold having a one-to-one projection to $\mathbb{R}^k$. All the orbits of* (2.16) *tend to M exponentially as $t \to \infty$.*

The proof of this theorem is beyond the scope of this book.

REMARKS. 1. The manifold in the previous theorem is said to be an *inertial manifold.*

2. It may be proved that the maximal attractor in the previous theorem belongs to the inertial manifold. Hence, this manifold contains the "natural phase space" of the system.

Theorems 2.5, 2.6 have infinite-dimensional counterparts. These counterparts allow us to reduce some well-known problems of partial differential equations to ordinary ones. For the cases in which Theorem 2.6 is applicable, Theorem 2.5 is of no importance. Yet there are numerous problems (e.g., equations of reaction-diffusion type in space, or Navier–Stokes equations on surfaces) in which the first theorem cannot be applied, while the second one may be. Some of these applications are the subject matter for future research.

§3. Smale horseshoe for high school students

In this section an elementary treatment of the Smale horseshoe map is presented. At the same time, some basic concepts of symbolic dynamics are introduced.

3.1. An olympiad problem on integer parts. In the 1993 Moscow Mathematical Olympiad for high school students, the following problem was proposed.

For any pair of real numbers α and β consider the sequence

$$\omega_n = [2\{\alpha n + \beta\}].$$

*Any subsequence of this sequence of length k is called a k-*word. *Is it true that any ordered set of zeros and ones may be realized as a k-word for some α and β for* a) $k = 4$, b) $k = 5$?

Here $[c]$ and $\{c\}$ are the integer and fractional parts of c respectively.

This problem has a natural geometric interpretation. Consider the circle $S^1 = \mathbb{R}/\mathbb{Z}$ and the shift

$$f\colon S^1 \to S^1, \qquad x \mapsto x + \alpha. \tag{3.1}$$

Take an orbit $b_n = f^n(b)$, $n \in \mathbb{Z}$. Then the sequence ω_n characterizes the behavior of this orbit in the following way: $\omega_n = 0$ if b_n belongs to the upper half-circle $[0, 1/2)$ and $\omega_n = 1$ otherwise.

Now the answer to the problem may easily be obtained. If the sequence $\{\omega_n\}$ has many subsequent zeros, this means that the orbit spends a long time in the upper half-circle, and hence the angle of rotation is small. When the orbit finally enters the lower half-circle, it will spend a long time there also. Therefore, a sequence having a lot of zeros one after another, then "a sparkling", and zeros after that, can never be realized as a word.

For instance, this is true for the ordered set 00010. Indeed, three zeros may occur in succession for $|\alpha| < 1/4$ only. On the other hand, the combination 010 shows that the point of the orbit comes after one rotation from the upper half-circle to the lower one and after the next rotation returns to the upper half-circle. This implies that $|\alpha| > 1/4$. The contradiction thus obtained shows that for $k = 5$, the answer in the problem is NO.

The answer YES for $k = 4$ may be easily confirmed by concrete examples.

3.2. Concepts of symbolic dynamics. Symbolic dynamics studies orbits of maps from a very robust point of view. The phase space is divided into a finite number of pairwise disjoint subsets D_j; this splitting is said to be a *partition*. Any point x of the phase space now determines the sequence $\{\omega_n \mid n \in \mathbb{Z}\}$:

$$\omega_n = j \iff f^n x \in D_j.$$

Therefore, instead of watching the explicit formulas for the orbit, we remember only to what domain of the fixed partition the nth point of the orbit belongs. This sequence may be called the *fate of the point*. In the literature the term *itinerary* is also used. One of the principal problems studied in symbolic dynamics is the following.

What sequences may be realized as the fate of a point for a given map? How many points have the same fate?

REMARK. For irrational rotations of the circle ($\alpha \notin \mathbb{Q}$ in (3.1)) the fate determines the point in a unique way. This is not true for rational rotations.

EXAMPLE. The problem from 3.1 shows that many sequences can never be realized as fates of points for any rotation of the circle.

3.3. Definition of the horseshoe map. The map described here is not at all similar to the horseshoe. The origin of the name will be explained in 3.10.

Take the unit square $B = [0, 1] \times [0, 1]$ in the coordinate plane. Divide it into five equal horizontal rectangles and call the second and the fourth rectangles from the bottom D_0 and D_1 respectively. In a similar way, divide the same square into five vertical rectangles and call the second and the fourth rectangles from the left D'_0 and D'_1.

Consider a hyperbolic rotation f_0 of D_0 with respect to the point $(1/4, 1/4)$ which contracts 5 times in the horizontal direction, expands 5 times in the vertical direction and therefore maps D_0 onto D'_0. Consider a similar hyperbolic rotation with respect to $(3/4, 3/4)$,

$$f_1 \colon D_1 \to D'_1.$$

Let

$$f|_{D_0} = f_0, \qquad f|_{D_1} = f_1 \tag{3.2}$$

(see Figure 2.3). This is a piecewise linear map defined in a disconnected domain. It is the subject matter for future study. The domain of this map is denoted by D and called the *phase space* of the map. The iterates of f are not defined in the whole phase space. By the *orbit* of a point x we mean the naturally enumerated set of the iterates of x that are well defined. If the iterates are defined for all integer exponents, then the orbit is called *complete*. It turns out that the union of all complete orbits is very shy: it is the product of nowhere dense Cantor sets.

For any $x \in D$ and any integer n let

$$\omega(x) = \{\omega_n(x) | n \in \mathbb{Z}\}, \qquad \omega_n(x) = \begin{cases} 0 & \text{if } f^n x \in D_0, \\ 1 & \text{if } f^n x \in D_1, \\ * & \text{if } f^n \text{ is not defined at } x. \end{cases}$$

We are interested in the fates that contain no asterisks.

THEOREM 3.1. *Any two-sided sequence of zeros and ones may be realized as the fate of one and only one point $x \in D$ under the map f.*

The theorem will be proved in the next two subsections.

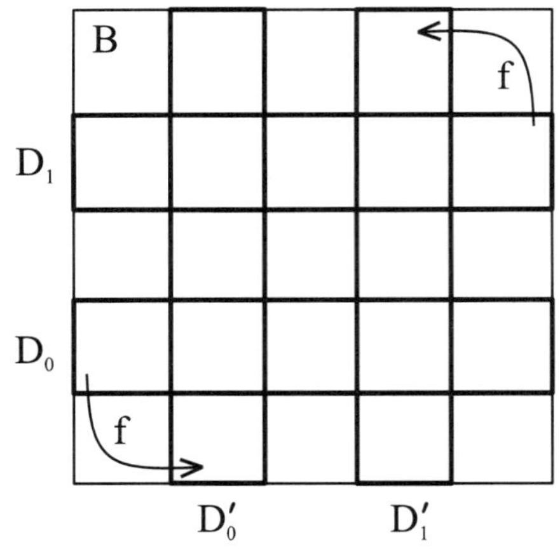

FIGURE 2.3. Smale horseshoe for high school students.

3.4. Points with prescribed future fate. Consider a finite sequence of zeros and ones,

$$\omega_n^+ = \omega_0 \dots \omega_{n-1}. \tag{3.3}$$

It will be called the *future fate of length* n. Consider the set

$$D_{\omega_n^+} = \{x \in D \mid \omega_j(x) = \omega_j,\ j = 0, \dots, n-1\}. \tag{3.4}$$

It consists of points with given future fate of length n equal to ω_n^+. We will describe this set in Lemma 3.1 below. Consider at first the cases of small n.

The case $n = 1$. The set $D(\omega_0^+)$ is the set of all points taken to D_{ω_0} by the map $f^0 = \mathrm{id}$. In more detail,

$$D(\omega_0^+) = \begin{cases} D_0 & \text{if } \omega_0 = 0, \\ D_1 & \text{if } \omega_0 = 1. \end{cases}$$

The case $n = 2$. The future fate of length 2 may take the following forms: 00, 01, 10, 11. The set $D(00)$ consists of those points that belong to the rectangle D_0 together with their image. Divide the rectangle D_0' into five equal squares. The second from the bottom, called Y, forms the intersection $D_0 \cap D_0'$. We look for the points which come to D_0' and still stay in D_0. Their image belongs to Y. The inverse image of Y is described as follows. Divide D_0 into five equal horizontal rectangles. The second from the bottom is the desired inverse image of Y.

In the same way, $D(01)$ is the fourth rectangle from the bottom of the previous division; $D(10)$ and $D(11)$ are the second and the fourth rectangles from the bottom of a similar division of D_1.

Now we can describe the set of points with prescribed future fate by means of a construction similar to that of a Cantor set. Let us define and enumerate by induction on n the segments of the nth rank in $I = [0, 1]$. The segments of the first rank are, by definition, $\sigma(0) = [1/5, 2/5]$ and $\sigma(1) = [3/5, 4/5]$. They have the numbers 0 and 1 respectively. The segments of rank n will be enumerated

FIGURE 2.4. Construction of the Cantor set C.

by the sequences (3.3) and denoted $\sigma(\omega_n^+)$. Let $\omega_{n+1} = \omega_n 0$ or $\omega_n 1$. In order to construct the segments $\sigma(\omega_n 0)$ and $\sigma(\omega_n 1)$, divide the segment $\sigma(\omega_n)$ into five equal segments and define the desired segments as the second and the fourth segments of this division respectively. Denote the union of all the segments of rank n by W_n (see Figure 2.4).

LEMMA 3.1. 1. *For any ω_n^+ of the form* (3.3), *the set of the points with the future fate ω_n^+ has the form*

$$D(\omega_n^+) = [0,1] \times \sigma(\omega_n^+). \tag{3.5}$$

Moreover, the set $Y_n = f^{n+1}(D(\omega_n^+))$ is a vertical rectangle of height 1.
2. *The set of all points for which the iterate f^n is well defined is*

$$D_n = [0,1] \times W_n.$$

PROOF. Statement 2 is an immediate consequence of Statement 1. Statement 1 is proved by induction on n. The base of induction, the case $n = 1$, has already been carried out. Now let us pass to the induction step (see Figure 2.5).

Suppose that (3.5) holds for n replaced by $n - 1$. Let

$$\omega_n^+ = \omega_{n-1} j^+, \qquad j \in \{0, 1\}.$$

By definition,

$$X_n = D(\omega_n^+) = \{x \in D(\omega_{n-1}^+) \mid f^n(x) \in D_j\}.$$

By the induction assumption, the set $X_{n-1} = D(\omega_{n-1}^+)$ is given by (3.5) with n replaced by $n - 1$, and the set

$$Y_{n-1} = f^n(X_{n-1})$$

is a vertical rectangle of height 1. The intersection $Z_n = Y_{n-1} \cap D_j$ is the f^n-image of the set we are looking for. This intersection is the ith rectangle from the bottom ($i = 2(j+1)$) obtained by the division of Y_{n-1} into five equal rectangles by horizontal lines. Therefore the desired set $X_n = f^{-n} Y_n$ is the ith rectangle from

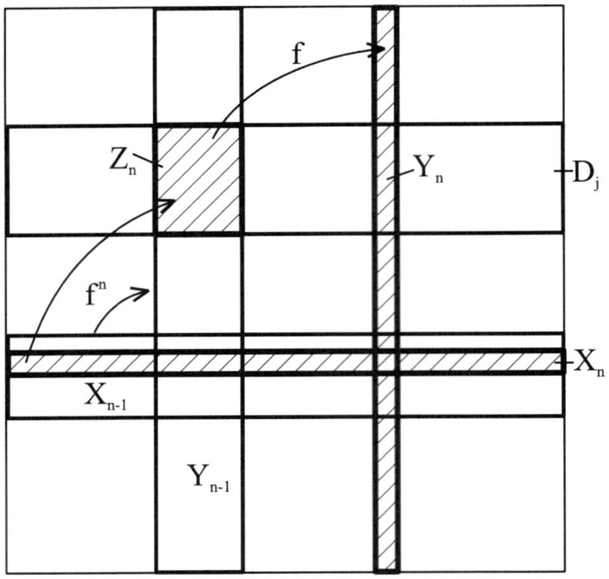

FIGURE 2.5. Points with prescribed finite future fate.

the bottom obtained by the division of X_{n-1} into five equal rectangles by horizontal lines. Therefore, it is given by (3.5).

Moreover, the horizontal edges of Z_n belong to those of D_j. Hence, $fZ_n = f^{n+1}X_n$ is a vertical rectangle of height 1. This proves Statement 1 and Lemma 3.1.

We can now find the set of all points with the prescribed infinite future fate

$$\omega^+ = \omega_0 \dots \omega_n \dots .$$

Let (3.3) be the n-term truncation of this sequence. Consider the Cantor set

$$C = \bigcap W_n.$$

Let $x(\omega^+)$ be a point of this set:

$$x(\omega^+) = \bigcap_{n \geqslant 0} \sigma(\omega_n^+). \tag{3.6}$$

Lemma 3.1 immediately implies the following

COROLLARY 3.1. *The set $D(\omega^+)$ of points with prescribed future fate ω^+ is given by the formula*

$$D(\omega^+) = [0,1] \times x(\omega^+). \tag{3.7}$$

3.5. Points with prescribed past fate and proof of Theorem 3.1. The past fate of a point under the map f is the future fate of this point under the inverse map, with some slight change of numeration discussed below. The map f^{-1} has the same properties as f, only the vertical and horizontal coordinates must change places. In more detail, define the *reversal* of the sequence enumerated by negative integers

$$\omega^- = \omega_{-1} \dots \omega_{-n} \dots$$

as the sequence enumerated by nonnegative integers

$$\alpha^+ = \alpha_0 \dots \alpha_{n-1} \dots, \qquad \alpha_n = \omega_{-n-1}.$$

Denote

$$\alpha^+ = \mathrm{rev}(\omega^-).$$

The above heuristic argument may be formalized as in the following lemma, which is a direct analog of Corollary 3.1.

LEMMA 3.2. *The set of all points $D(\omega^-)$ with prescribed past fate ω^- has the form*

$$D(\omega^-) = x(\alpha^+) \times [0,1], \tag{3.8}$$

where α^+ is the reversal of ω^-.

PROOF. The future fate of the point x under the map f^{-1} is defined as $\alpha^+ = \alpha_0\alpha_1 \dots \alpha_n \dots$, where $\alpha_n = j$ iff $f^{-n}x \in D'_j$. The last inclusion is equivalent to $f^{-(n+1)}x \in D_j$. Hence, if the past fate of the point x under f is ω^-, then the future fate of x under f^{-1} is $\alpha^+ = \mathrm{rev}\, \omega^-$.

Lemma 3.2 now follows from Lemma 3.1. For the reduction, vertical and horizontal directions must change places.

Corollary 3.1, together with Lemma 3.2, immediately implies the theorem. Indeed, the required point with fate ω may be constructed in the following way. Split ω into the future and past parts:

$$\omega = \omega^-\omega^+.$$

Then the point $x(\omega)$ with fate ω is the intersection

$$x(\omega) = D(\omega^+) \cap D(\omega^-). \tag{3.9}$$

This is the unique intersection point of the horizontal and vertical segments. Theorem 3.1 is proved.

3.6. Some calculations. Formula (3.9) deciphered by means of (3.6), (3.7), (3.8) allows us to express the coordinates x, y of the point $x(\omega)$ in terms of ω. Indeed, the Cantor set C in the above construction is the set of all points of the unit segment in whose expression in the pentadic scale only the digits 1 and 3 can occur. More precisely,

$$x = x(\omega^+) = \sum_0^\infty \frac{\gamma_j}{5^{j+1}}, \qquad y = x(\alpha^+) = \sum_{-1}^{-\infty} \frac{\gamma_j}{5^{-j}}\,. \tag{3.10}$$

Here $\gamma_j = 1$ if $\omega_j = 0$, $\gamma_j = 3$ if $\omega_j = 1$, $\alpha^+ = \mathrm{rev}(\omega^-)$.

3.7. Bernoulli shift. Consider the set Σ^2 of all the bi-infinite sequences of zeros and ones,

$$\omega = \ldots \omega_{-n} \ldots \omega_{-1}\omega_0 \ldots \omega_n \ldots ,$$

and consider the shift transformation:

$$\omega \mapsto \sigma\omega, \qquad \sigma\omega = \ldots \omega'_{-n} \ldots \omega'_{-1} \ldots \omega'_0 \ldots \omega'_n \ldots , \qquad \omega'_j = \omega_{j+1}.$$

It is called the *Bernoulli shift.*

The metric in the space Σ^2 is generated by the norm:

$$\|\omega\| = \sum_{-\infty}^{\infty} \frac{|\omega_n|}{2^n}. \tag{3.11}$$

REMARK. The Bernoulli shift is defined in the same way in the space Σ^N of all bi-infinite sequences of elements $\omega_n \in \{0, 1, \ldots, N-1\}$. Formula (3.10) motivates the definition of the metric in this space.

If the point x has the fate ω, then the point $f(x)$ has the fate $\sigma(\omega)$. The correspondence: point $\to$ fate,

$$\omega \colon x \mapsto \omega(x),$$

conjugates the map f restricted to $C \times C$ with the shift σ, i.e., the diagram

$$\begin{array}{ccc} C \times C & \xrightarrow{\ f\ } & C \times C \\ {\scriptstyle \omega}\big\downarrow & & \big\downarrow{\scriptstyle \omega} \\ \Sigma^2 & \xrightarrow[\ \sigma\ ]{} & \Sigma^2 \end{array}$$

is commutative. Moreover, the map ω is a homeomorphism in the sense of the topology induced by the Euclidean metric of $\mathbb{R}^2$ in $C \times C$ and by the metric induced by the norm (3.11) in Ω. This follows immediately from (3.10) and (3.11) and motivates the definition of the metric in the space of sequences Ω.

The conjugation to the Bernoulli shift is a powerful tool for describing maps by means of symbolic dynamics. It will be used below for numerous maps provided by nonlocal bifurcations. Some applications will be given in the next subsection.

3.8. Corollaries.

COROLLARY 3.2. *The map f has infinitely many periodic points.*

PROOF. A periodic point of the map is a point that returns to the same place after a finite number of iterations. A periodic fate corresponds to such a point. The theorem above implies the inverse statement: a point with periodic fate is periodic itself. Indeed, suppose that the fate of the point x is periodic with period N. This means that the points x and $f^N x$ have the same fate. By the uniqueness statement of the theorem, these points coincide.

The next corollary ensures the same property for homoclinic orbits.

COROLLARY 3.3. *The map f has infinitely many homoclinic orbits of any of the fixed points.*

PROOF. By definition, the orbit is homoclinic if it tends to the same fixed point for $n \to +\infty$ and for $n \to -\infty$. Therefore, the fate of the point generating the homoclinic orbit of the fixed point from D_0 consists of zeros only except for a finite number of elements. The above theorem implies that the inverse is true: any sequence consisting of zeros only (except for a finite number of elements) is the fate of a point generating the homoclinic orbit of the fixed point from D_0. Indeed, the positive shifts $\sigma^n\omega$ of such sequences tend to the zero sequence in Ω as $n \to \pm\infty$ in the sense of the metric (3.11). The same is true for negative shifts, $n \to -\infty$. The zero sequence corresponds to the fixed point in D_0. The correspondence ω: point $\to$ fate is continuous. Therefore, the sequence with a finite number of nonzero elements is really the fate of a homoclinic orbit. The same is true for sequences with all elements (except a finite number) equal to 1.

Recall that a heteroclinic orbit is an orbit tending to one fixed point as $n \to +\infty$ and to the other as $n \to -\infty$.

The following corollaries may be proved in the same way as the previous one.

COROLLARY 3.4. *The map f has infinitely many heteroclinic orbits.*

COROLLARY 3.5. *For any points $x, y \in C \times C$ there exists a point such that its orbit approximates the orbit of x as $n \to +\infty$ and the orbit of y as $n \to -\infty$.*

COROLLARY 3.6. *For any N the map f has only a finite number of periodic points with period N.*

The discovery of the Smale horseshoe was a sensation not only in dynamics. In order to understand that we should go back two centuries.

3.9. Dynamics and philosophy: determinism and chaos. About two hundred years ago in his treatise "Essay on the philosophy of probability theory", Laplace wrote: "The mind that would know all the forces animating nature at some fixed moment, and the relative position of all its components; if, moreover, it should be so voluminous as to make this data a subject of analysis, cover by one formula the motion of the greatest masses of Universe together with the slightest atoms; nothing would remain unknown to this mind, but the future, together with the past, would appear before its gaze."

A great philosophic discovery stands behind these words: the understanding that all the evolutionary processes in the Universe may be described by ordinary differential equations, apparently in a phase space of very high dimension.

This concept is based on the mathematical fact that, at the time, was located not even on, but beyond the frontline of science. That was the existence and uniqueness theorem for solutions of ordinary differential equations, which was proved later by Augustin Louis Cauchy.

For a long time the philosophy based on this theorem, which is now part of any mathematical undergraduate differential equations course, was regarded as giving an adequate description of reality. The Smale horseshoe map motivates quite another point of view. Suppose that we observe the process modeled by the map f described above, and study the orbits of this map. We are interested, as before, not in the orbit itself, but only in the fate of the points. Suppose that we deal twice

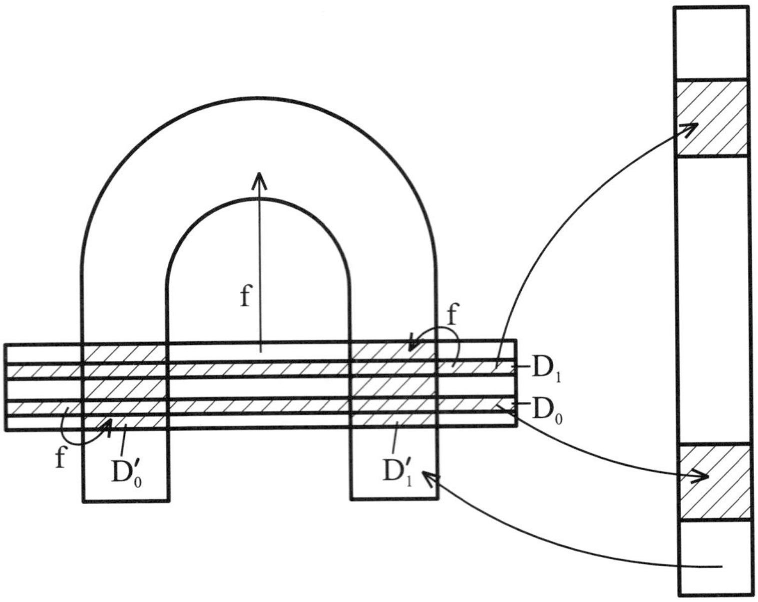

FIGURE 2.6. Actual horseshoe.

with the fate of the same point, beginning with the calculation of the coordinates of the initial point. The problem is that the calculation of the coordinates may be performed only approximately. If the data are of experimental nature, the results of two attempts can be slightly different. The smaller the difference, the longer will the fates coincide. But in due time they will always split.

Suppose that a complete fate is written for some point x. On the other hand, suppose that in parallel a coin is being tossed. The result of the nth toss is the nth term of the sequence: one is written for head and zero for tail. Then the two sequences are merged: up to some number, say 1000, the first sequence is taken, and after that the other, beginning with the same number, is added. By the realization theorem of 3.3, there exists a point whose fate coincides with this merged sequence. This point is pretty close to x; the distance is about 5^{-1000} and therefore no observer would be able to distinguish between these points. Yet their fates, beginning at some moment, would split and their difference would be of random character.

This effect occurs in different processes modeled by differential equations. It usually called the *nonreproducibility of the experiment.* One may repeat the same experiment and after some moment obtain quite different results. This is related to the effect of *exponential splitting of orbits*, which was described here for the horseshoe map.

In the last decades the investigation of the chaotic behavior of deterministic systems has become a subject of vivid interest. Many systems described by ordinary differential equations, even in phase spaces of low dimension, are deterministic only in theory. In practice the results of two experiments with data that seem to be identically the same, after some moment may differ in a chaotic manner.

Now we return to our example. Our goal is to explain why the map defined in 3.3 is known as the *horseshoe* map.

3.10. The actual horseshoe. The original Smale horseshoe map is constructed as follows. A long horizontal rectangle is strongly contracted in the horizontal direction and strongly expanded in the vertical one. Then it is bent to a horseshoe and after that placed over its preimage (see Figure 2.6). The composition of these transformations is the horseshoe map. At first it does not look like the map defined in 3.3 at all. Yet, after the restriction of its domain, one can easily find familiar details (see Figure 2.6).

Consider the intersection of the domain of the horseshoe map with its image. It consists of the two rectangles shadowed in the figure. Denote the left one by D_0', and the right one by D_1'. Suppose that the inverse map restricted to the union of these rectangles is linear; this assumption was part of Smale's original construction. Then the inverse images of D_0' and D_1' are the long horizontal rectangles. Denote them D_0 and D_1 respectively. The horseshoe map brings these horizontal rectangles to the vertical ones just like the map f studied before. It is a hyperbolic rotation on the lower rectangle and a hyperbolic rotation composed with a central symmetry on the upper rectangle. It has the same properties as the map f.

The examples of this section are piecewise linear and therefore highly degenerate. What are the properties of their nonlinear perturbations? It turns out that, from the point of view of symbolic dynamics, all the properties are preserved. The proofs make use of *hyperbolic theory*.

§4. Some results in hyperbolic theory

Hyperbolic theory studies the global behavior of dynamical systems which are similar in one way or another to the hyperbolic rotation $\mathbb{R}^2 \to \mathbb{R}^2$:

$$(x, y) \mapsto (x/2, 2y).$$

The family of horizontal lines as well as that of vertical ones is invariant under this map. The iterates of the map exponentially contract the horizontal lines and exponentially expand the vertical ones.

The generic hyperbolic map has two invariant families of manifolds, which are also exponentially contracting and expanding under the iterates of the map. The method for finding them is the so-called *graph transformation method.* In order to make it work, the *cone condition* is often used. Moreover, this condition is involved in one of the definitions of hyperbolic maps (see 4.2 below).

We now pass to the oldest fundamental fact of the theory.

4.1. Hadamard–Perron theorem for maps. We say that an invertible linear map $A\colon \mathbb{R}^n \to \mathbb{R}^n$ is *hyperbolic* if it has no eigenvalues on the unit circle. Consider a hyperbolic linear map and the corresponding splitting of $\mathbb{R}^n$:

$$\mathbb{R}^n = T^s \oplus T^u. \tag{4.1}$$

The spaces T^s and T^u are invariant for A, with eigenvalues lying inside and outside the unit circle respectively.

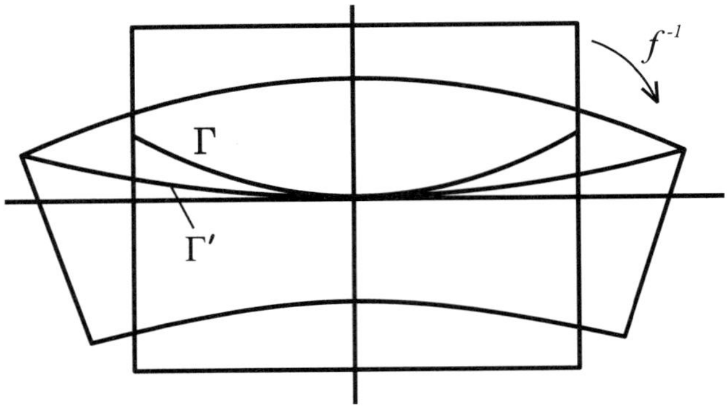

FIGURE 2.7. Construction of the graph transformation operator.

HADAMARD–PERRON THEOREM FOR MAPS. *Consider a germ of a map*

$$f\colon (\mathbb{R}^n, 0) \to (\mathbb{R}^n, 0), \qquad x \mapsto Ax + \cdots$$

with hyperbolic linear part A. Let (4.1) be the corresponding splitting of the phase space. Then the map f has two germs of invariant manifolds at zero: W^s and W^u with the tangent spaces T^s and T^u at zero respectively. These germs of manifolds are uniquely determined. If the map f is of class C^k ($k \geqslant 1$), C^∞, or C^ω (i.e., analytic), then the manifolds W^s and W^u are of the same class.

REMARK. A similar theorem for vector fields is stated in 1.1.5. It may be reduced to the previous one by considering the phase flow transformation.

SKETCH OF THE PROOF OF THE HADAMARD–PERRON THEOREM FOR MAPS. We restrict the domain of the map f to the product

$$D = D_h \times D_v$$

of two balls centered at zero in the spaces T^s and T^u respectively, and small enough for what follows. The first space is called *horizontal*, the second *vertical*, whence the notation. Consider a space H (H for *horizontal*) consisting of Lipschitz maps $D_h \to D_v$ with Lipschitz constant $L \leqslant 1/2$ and graph Γ_φ lying in D. The map f^{-1} takes D to a domain located as shown in Figure 2.7. For D small enough, the projection $\pi_h\colon \mathbb{R}^n \to T^s$ along T^u maps the set $\Gamma' = f^{-1}\Gamma_\varphi \cap D$ bijectively to D_h. Therefore, Γ' is the graph of a new map $\psi\colon D_h \to D_v$. This new map belongs to the space H. The operator

$$F\colon H \to H, \qquad \varphi \mapsto \psi,$$

is called the *graph transformation operator*. It is contracting in the C norm, because f^{-1} is contracting in the vertical and expanding in the horizontal direction (see Figure 2.7). Therefore, the graph transformation operator has a fixed point which corresponds to the expanding (unstable) invariant manifold of the map f^{-1} and hence to the contracting (stable) manifold of f. It is a graph of a Lipschitz map with constant $L \leqslant 1/2$. The existence of the Lipschitz unstable manifold of f is proved in the same way.

The same argument proves the Hadamard–Perron theorem in the analytic case. One must only pass to the complex domain. This means that $\mathbb{R}^n$ is replaced by $\mathbb{C}^n$, T^s and T^u are complex linear, and D_h and D_v are balls in the Hermitian metric. The space of holomorphic maps in the C-topology is complete. Therefore, all the previous arguments (together with Figure 2.7) are valid and imply the existence of holomorphic stable and unstable manifolds in the holomorphic case.

The intermediate cases of $f \in C^k$ require a C^k metric in the space H and verification of the fact that the graph transformation operator is contracting in the C^k norm. This involves additional technical difficulties.

The case $f \in C^\infty$ is C^k for all k. Hence, together with the uniqueness property of the invariant manifolds, the C^k version of Hadamard–Perron theorem implies the C^∞ version.

Therefore, the extreme cases of C^0 and C^ω invariant manifolds are extremely simple, while the intermediate cases C^k, C^∞ require more technical efforts.

4.2. Cone condition. There are maps with the hyperbolic properties for which "horizontal" and "vertical" directions are not so clearly distinguished as they were in the Hadamard–Perron theorem. In order to recognize the "hyperbolic situation", the *cone condition* is often used.

DEFINITION 4.1. Consider the splitting

$$\mathbb{R}^{n+m} = \mathbb{R}^n \oplus \mathbb{R}^m. \tag{4.2}$$

Let D_h and D_v be compact connected manifolds with boundary in $\mathbb{R}^n$ and $\mathbb{R}^m$, diffeomorphic to unit balls in these spaces respectively. A *standard rectangle* B is the Cartesian product of these domains:

$$B = D_h \times D_v.$$

Moreover,

$\partial_h B = D_h \times \partial D_v$ is called the *horizontal part* of the boundary of B,

$\partial_v B = \partial D_h \times D_v$ is called the *vertical part* of the boundary of B.

The splitting (4.2) induces the splitting of the tangent space at any point of $\mathbb{R}^{n+m}$. Tangent vectors will be denoted by ξ, η and so on; they split into components:

$$\xi = (\xi^-, \xi^+), \qquad \eta = (\eta^-, \eta^+).$$

For any two positive constants μ_h, μ_v with $\mu_h \mu_v < 1$ define two (μ_h, μ_v) *families of cones*:

$$K_p^+ = \{\xi \in T_p B \mid |\xi^-| \le \mu_v |\xi^+|\}, \qquad K_p^- = \{\xi \in T_p B \mid |\xi^+| \le \mu_h |\xi^-|\}.$$

DEFINITION 4.2. A map

$$f\colon D \to D' \subset \mathbb{R}^{n+m}$$

satisfies the (μ_h, μ_v) *cone condition* on D if there exists a $\lambda > 1$ such that

(1) $df K_p^+ \subset K_{f(p)}^+$;

(2) $(df)^{-1} K_{f(p)}^- \subset K_p^-$.

For any p in the domain of f and $\xi \in T_p\mathbb{R}^{n+m}$, let $\eta = df(p)\,\xi$. Then

(3) $|\eta^+| \geqslant \lambda |\xi^+|$ for any $\xi \in K_p^+$;

(4) $|\xi^-| \geqslant \lambda |\eta^-|$ for any $\eta \in K_p^-$.

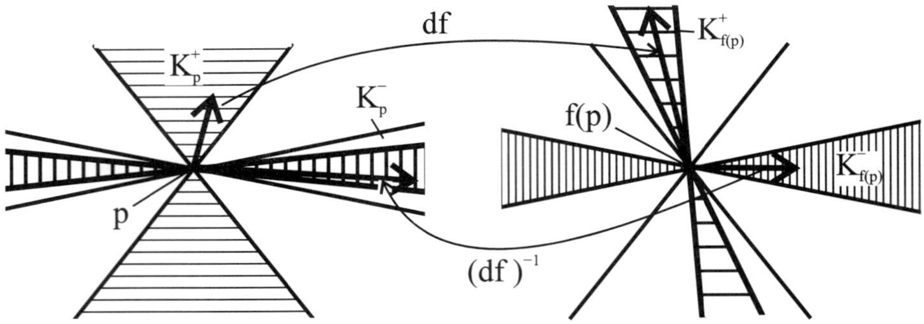

FIGURE 2.8. Cone condition.

REMARK. We have simplified the classical cone condition a little. In general, the splitting (4.2) is replaced by a splitting of the tangent space depending on the point where the space is attached. The simplification above is sufficient for our goals. Definition 4.2 is illustrated in Figure 2.8.

In what follows we deal with Lipschitz surfaces rather than smooth ones. The reason is that the space of Lipschitz maps with the same Lipschitz constant is complete in the C topology.

DEFINITION 4.3. Consider a standard rectangle $B = D_h \times D_v$ and two positive constants μ_h, μ_v, $\mu_h\mu_v < 1$. Call the graph H of a Lipschitz map $G \to D_v$ with the constant μ_h a *μ_h-horizontal surface*. Here G is a domain (with boundary) in D_h. The surface H is *horizontal* in B if $G = D_h$ in the previous definition. A *μ_v-vertical surface* (in B) is defined in the same way by taking D_v, μ_v instead of D_h, μ_h and vice versa.

REMARK. Recall that a Lipschitz map φ with constant L must satisfy, by definition, the inequality

$$|\varphi(x) - \varphi(y)| \leqslant L\,\mathrm{dist}(x, y)$$

for any x, y in the domain of φ. If this domain is convex, then the distance $\mathrm{dist}(x, y)$ is the same as in the ambient Euclidean space. If the domain is only connected, then the distance is intrinsic, that is, induced by the Riemannian metric of the ambient Euclidean space. It is equal to the infimum of the length of the curves that link x and y in the domain of φ.

EXAMPLE. The graph of a smooth map $f\colon D_h \to D_v$ is a μ_h-horizontal surface if and only if its tangent plane at each point belongs to the cone of the family $\{K^+\}$ at this point.

This is clear for the case in which the surface is smooth: the tangent plane to the horizontal surface belongs to the cone of the family $\{K^-\}$ and so does its image under the map f. In the Lipschitz case, the same arguments work; one should only consider tangent cones to the surface instead of tangent planes.

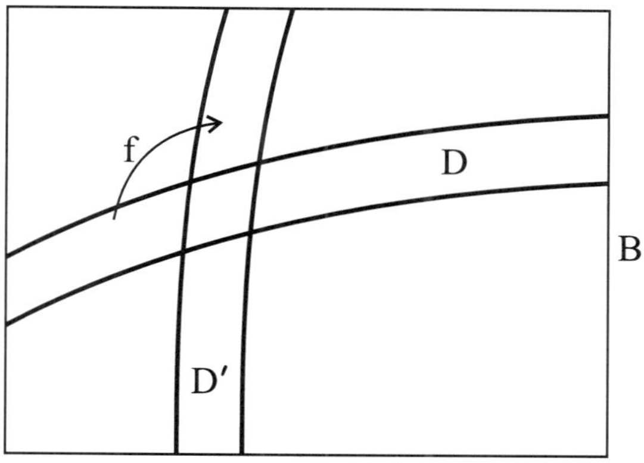

FIGURE 2.9. The map in the hyperbolic fixed point theorem.

This property gives rise to function spaces (of vertical and horizontal surfaces). To these spaces the graph transformation method may be applied.

4.3. Hyperbolic fixed point theorem. The theorem below gives a sufficient condition for the existence of a hyperbolic fixed point of a map. The crucial assumption is that the map satisfies the cone condition.

In order to give the statement of the hyperbolic fixed point theorem, we need the following

DEFINITION 4.4. A (μ_h, μ_v)-rectangle D in $\mathbb{R}^n \oplus \mathbb{R}^m$ is a domain of the form

$$D = F(B^n \times B^m),$$

where B^n and B^m are the unit balls in $\mathbb{R}^n$ and $\mathbb{R}^m$ respectively, F is a homeomorphism $B^n \times B^m \to \mathbb{R}^n \oplus \mathbb{R}^m$ and the surfaces $F(B^n \times \{y\})$ and $F(\{x\} \times B^m)$ are μ_h-horizontal and μ_v-vertical respectively for any $y \in B^m$, $x \in B^n$. The *horizontal* and *vertical* parts of the boundary of D are defined as the images under F of the horizontal and vertical parts of the boundary of $B^n \times B^m$ respectively:

$$\partial_h D = F(B^n \times \partial B^m), \qquad \partial_v D = F(\partial B^n \times B^m).$$

DEFINITION 4.5. A (μ_h, μ_v)-rectangle is called *vertically cylindric* (*v-cylindric*) in a standard rectangle B if

$$D \subset B, \quad \partial_v D \in \partial_v B.$$

It is called *horizontally cylindric* (*h-cylindric*) if the same holds for h and v interchanged, i.e.,

$$D \subset B, \quad \partial_h D \in \partial_h B.$$

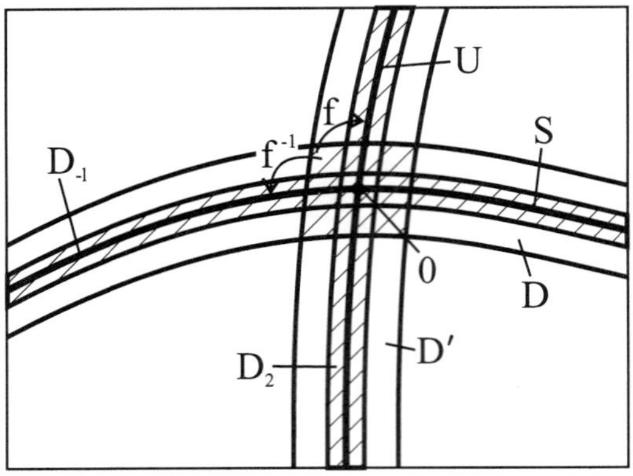

FIGURE 2.10. Proof of the hyperbolic fixed point theorem.

HYPERBOLIC FIXED POINT THEOREM. *Let $B = D_h \times D_v$ be a standard rectangle. Let $D \subset B$ and $D' \subset B$ be v-cylindric and h-cylindric (μ_h, μ_v)-rectangles in B (with projections D_h and D_v respectively). Let $f\colon D \to D'$ be a map satisfying the (μ_h, μ_v) cone condition. Then f has a unique fixed point O in D:*

$$O = \bigcap_{i=-\infty}^{\infty} f^i D.$$

The theorem is illustrated in Figure 2.9.

SKETCH OF THE PROOF (a detailed proof is given in §8.3). The plan of the proof is in a way inverse to that for the Hadamard–Perron theorem. Namely, we first find the invariant manifolds of the fixed point and after that find the fixed point itself as their intersection.

The invariant manifolds are constructed as successive images of the rectangle D under the iterates of f. Namely, let

$$D_1 = D', \qquad D_k = f(D_{k-1} \cap D), \quad k \geqslant 2$$

(see Figure 2.10).

These domains form a nested sequence of "curvilinear rectangles" with the "width in the horizontal direction" tending exponentially to zero. Their intersection is the expanding vertical invariant surface we are looking for:

$$U = \bigcap_1^{\infty} D_k.$$

In a similar way one finds the contracting surface

$$S = \bigcap_0^{\infty} D_{-k}, \quad \text{where } D_0 = D,\ D_{-k} = f^{-1}(D_{1-k} \cap D')$$

(see Figure 2.10). The inequality $\mu_h\mu_v < 1$ implies that these manifolds have a unique intersection point:

$$O = U \cap S.$$

This proves the existence of the hyperbolic fixed point O with the stable and unstable manifolds S and U respectively.

4.4. Smale Horseshoe Existence Theorem. A similar technique allows us to find a complicated invariant set for a nonlinear map similar to the one described in §3. The following theorem deals with the map which amalgamates the features shown in Figures 2.3 and 2.9.

SMALE HORSESHOE EXISTENCE THEOREM. *Let μ_h, μ_v with $\mu_h\mu_v < 1$ be two positive constants, and B a standard rectangle. Let*

$$D_i \subset \mathbb{R}^n \oplus \mathbb{R}^m, \qquad i = 1, \dots, N,\ N > 1,$$

be N pairwise disjoint (μ_h, μ_v)-rectangles. Let

$$f\colon D = \bigcup_{i=1}^{N} D_i \to f(D) \in B$$

be a diffeomorphism of the domain D onto its image satisfying the following assumptions for $i, j = 1, \dots, N$:

1) *the map f satisfies the (μ_h, μ_v) cone condition in D;*
2) *the domains D_i are vertically cylindric in B, while the domains $D_i' = f(D_i)$ are horizontally cylindric in B.*

Then the set

$$\Lambda = \bigcap_{-\infty}^{\infty} f^k D$$

is a hyperbolic invariant Cantor set. The restriction of f to Λ is topologically conjugated to the Bernoulli shift σ on the space Σ^N of sequences of the elements taking N different values $1, \dots, N$. This means that there exists a homeomorphism ϕ such that the diagram

$$\begin{array}{ccc} \Lambda & \xrightarrow{\ f\ } & \Lambda \\ {\scriptstyle\phi}\downarrow & & \downarrow{\scriptstyle\phi} \\ \Sigma^N & \xrightarrow{\ \sigma\ } & \Sigma^N \end{array}$$

is commutative.

The theorem is illustrated in Figure 2.11.

REMARKS. 1. In the case $N = 2$, $n = m = 1$, the theorem of §3 is a particular case of the previous one.

2. The map Φ is the map point $\to$ fate defined in the same way as in §3.

3. The set Λ turns out to be hyperbolic in the sense of the definition given later in §8.1. This property implies the structural stability of the map in some neighborhood of the invariant set (see §8.1).

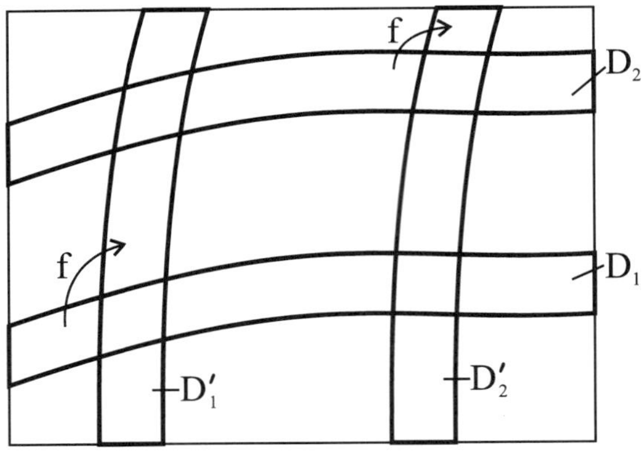

FIGURE 2.11. Maps in Smale horseshoe existence theorem.

4. All the conditions of the theorem are open in the sense that they hold for an open set of maps with the C^1 topology.

The idea of the proof of the theorem is to merge the arguments of §3 and of 4.3.

The theorem is applied in Chapters 4 and 7 in order to find infinite hyperbolic invariant sets for vector fields or their perturbations.

The complete proof of the last two theorems will be given in Chapter 8.

§5. Normal forms for local families

In this section the next principal tool of our investigation is described: normal forms for local families of vector fields and diffeomorphisms. These normal forms turn out to be polynomial and integrable, or linear suspensions over integrable systems. They exist for unfoldings of relatively simple germs only, namely, for germs in general position and for some germs appearing in typical one-parameter families. For some kinds of germs appearing in such families, the smooth classification has functional moduli. The list of nice normal forms is given in 5.3.

5.1. Choice of classification. The greater part of the theory of normal forms deals with the similarity of a germ of a vector field or a map to its linear part at the fixed point. Similarity means the possibility of transforming the germ to its linear part by a coordinate change. Different branches of the theory occur for different classes of these changes: topological, smooth, analytical. The topological theory involves the most significant simplifications: by the Grobman–Hartman theorem, the hyperbolic germ of a vector field or a map is topologically equivalent to its linear part. Nevertheless it turns out that for the answers to nonlocal topological questions, local topological normal forms are insufficient, and smooth ones are required.

This is shown by the following example. Consider a planar vector field with a hyperbolic saddle having a separatrix loop. The question is whether this loop is stable from the inside, that is to say, whether the phase curves wind to or off this loop. The Grobman–Hartman theorem implies that the vector field near the saddle is topologically equivalent to

$$\dot{x} = x, \qquad \dot{y} = -y.$$

This gives no information on the subject, as shown below.

The smooth theory gives more precise information. For the case in which the ratio of eigenvalues of the saddle is irrational, the vector field near this saddle is C^∞ equivalent to its linear part:

$$\dot{x} = \lambda_1 x, \qquad \dot{y} = -\lambda_2 y.$$

The coordinates taking the field to its normal form are called *normalizing*. Take two half-intervals, Γ^+, Γ^-, with the vertices on the incoming and outcoming separatrices respectively. If they are sufficiently small, the *correspondence map* $\Delta^{\text{sing}}\colon \Gamma^+ \to \Gamma^-$ along the phase curves of the field is well defined. If the coordinates on the half-transversals Γ^+ and Γ^- are chosen as the restrictions of the normalizing coordinates x and y respectively, then the correspondence map takes the form

$$\Delta^{\text{sing}}\colon \Gamma^+ \to \Gamma^-, \quad x \mapsto y = x^\lambda, \quad \lambda = \lambda_2/\lambda_1.$$

On the other hand, a map of Γ^- to some segment $\Gamma \supset \Gamma^+$ along the phase curves close to the regular part of the loop (containing no singular points) is well defined and smooth up to the vertex. It is therefore Lipschitz. Since $\lambda \notin \mathbb{Q}$, we have $\lambda \neq 1$. Suppose $\lambda > 1$. In this case the map Δ^{sing} is strongly contracting: its Lipschitz constant tends to zero together with the length of Γ^+. Therefore, the Poincaré map of the loop, $\Delta = \Delta^{\text{reg}} \circ \Delta^{\text{sing}}$, is contracting. Hence, in the case $\lambda > 1$, the loop is stable. In the opposite case the loop is unstable.

This example shows that the classification we need is at least finitely smooth. In the next subsection we show that it can be neither C^∞ nor analytic.

5.2. Resonances and the final choice of classification. In what follows *smooth* means *infinitely smooth*. The obstruction to smooth equivalence of a germ of a map or a vector field at a fixed point to its linear part is formed by the so-called *resonance relations* on the eigenvalues of the linear part.

DEFINITION 5.1. A tuple $\lambda \in \mathbb{C}^n$ is *resonant* if the following relation holds:

$$\lambda_j = (\lambda, k), \quad k \in \mathbb{Z}_+^n, \quad |k| \equiv k_1 + \cdots + k_n \geqslant 2.$$

Here $\mathbb{Z}_+$ is the set of nonnegative integers, and the number $|k|$ is called the *order of the resonance*.

DEFINITION 5.2. A tuple $\lambda \in \mathbb{C}^n$ is *multiplicatively resonant* if we have $\lambda_j = \lambda^k$, k being the same as in the previous definition; the order of the resonance is defined in the same way.

A linear vector field or a map is called *resonant* if its spectrum is resonant (for vector fields), or multiplicatively resonant (for maps). A germ of a vector field or a map at a fixed point is called *resonant* if its linear part is resonant.

EXAMPLES. 1. A germ of a vector field with one zero eigenvalue $\lambda_1 = 0$ is resonant. The corresponding resonance relations are

$$\lambda_j = \lambda_j + k_1\lambda_1, \qquad k_1 \in \mathbb{N}.$$

2. A germ of a vector field with two nonzero imaginary conjugate eigenvalues $\lambda_{1,2} = \pm i\omega$ is resonant with the relations

$$\lambda_j = \lambda_j + k_1(\lambda_1 + \lambda_2), \qquad k_1 \in \mathbb{N}.$$

Consequently, any nonresonant germ of a real vector field is hyperbolic. The same is true for real hyperbolic germs of maps.

STERNBERG'S THEOREM. *A nonresonant hyperbolic germ of a vector field or a map at its fixed point is smoothly equivalent to its linear part.*

Germs of vector fields or maps at a fixed point with resonant linear parts are, in general, not smoothly equivalent to their linear parts.

It is now easy to understand why local families of germs cannot in general be linearized by a smooth coordinate change. Consider, for instance, a planar vector field with a saddle fixed point having eigenvalues with irrational ratio. Its linear part is therefore nonresonant, and, by Sternberg's theorem, it is smoothly equivalent to its linear part. Now consider a one-parameter perturbation of this vector field. When the parameter begins to change, resonances immediately occur. Namely, the resonance relation holds at any time when the ratio of the eigenvalues is rational, and rational numbers are dense in the line. Therefore, for a generic family, the resonant saddles correspond to a dense set of parameter values. Hence, for arbitrarily small parameter values, there are vector fields of the family that are not smoothly equivalent to their linear parts.

The key idea for overcoming this difficulty is suggested by the finitely smooth version of Sternberg's theorem.

THEOREM. *For any tuple $\lambda \in \mathbb{C}^n$ and for any k there exists number $N = N(k, \lambda)$ such that any germ of a vector field with the eigenvalues λ is C^k equivalent to its linear part provided that the tuple λ satisfies the resonant relations of order greater or equal to $N(k)$.*

A similar theorem is true for germs of maps.

Now consider the previous example. The order of resonances generated by rational numbers approximating an irrational number tend to infinity as the approximations tend to the irrational limit. Therefore, the smaller the neighborhood of the zero value of the parameter, the greater the order of the resonances for the corresponding vector fields, and the greater the rate of smoothness of the linearizing coordinates.

This gives rise to the following definition.

DEFINITION 5.3. Two smooth local families of vector fields or maps at a fixed point are *finitely smoothly equivalent* if for any natural k, C^k equivalent representatives of the families exist.

REMARK. The rate of smoothness of the corresponding coordinate change increases when the base of the families decreases.

This equivalence relation will be used in what follows.

5.3. Finitely smooth normal forms: main results.

THEOREM 5.1. *If the eigenvalues of a germ of a vector field at a fixed point form a nonresonant tuple, then the local family obtained as the perturbation of this germ is finitely smoothly equivalent to the linear one*:

$$\dot{x} = A(\varepsilon)x. \tag{5.1}$$

REMARK. Any nonresonant tuple is automatically hyperbolic. The same is true for the germ in the previous theorem.

A similar theorem is true for maps.

THEOREM 5.2. *If the eigenvalues of the linearization of a germ of a map at a fixed point form a multiplicatively nonresonant tuple, then the local family obtained as the perturbation of this germ is finitely smoothly equivalent to the linear one*:

$$x \mapsto A(\varepsilon)x. \tag{5.2}$$

These results are true for multiparameter deformations. The next theorem is stated for one-parameter families. It deals with the deformations of so-called *saddlenode vector fields*. Recall that a *multidimensional saddlenode* is a germ of a vector field at a singular point with one zero eigenvalue of its linearization, all the other eigenvalues being hyperbolic (i.e., with nonzero real parts).

THEOREM 5.3. *In generic one-parameter families of vector fields only those germs of saddlenode vector fields occur whose unfoldings in these families are finitely smoothly equivalent to the following normal form*:

$$\begin{aligned} \dot{x} &= (x^2 + \varepsilon)(1 + a(\varepsilon)x)^{-1}, \\ \dot{y} &= A(x, \varepsilon)y, \\ \dot{z} &= B(x, \varepsilon)z. \end{aligned} \tag{5.3}$$

The matrices $A(0,0)$ and $B(0,0)$ have their eigenvalues in the open left and right half-planes respectively.

REMARK. The function $a(\varepsilon)$ is the invariant of this classification: if two local families with normal form (5.3) are finitely smoothly equivalent, and the conjugating map preserves the parameter, then the corresponding functions a coincide.

Consider the saddlenode germ of the map, that is, the germ with one multiplier equal to 1, other multipliers being hyperbolic, i.e., not lying on the unit circle. The main example of such a map is the time one shift along the phase curves of a saddlenode vector field. A map that may be represented as the time one shift along the phase curves of some vector field is called *embeddable*. A family of these maps is also called embeddable. It turns out that a generic unfolding of a saddlenode germ of a map is not embeddable. However, it is *partially embeddable* in the following sense.

THEOREM 5.4. *In generic one-parameter families of diffeomorphisms, only those germs of saddlenode maps occur whose perturbation in the subfamily parametrized by an appropriate half-interval of the parameter axis with vertex at the critical value, is finitely smoothly equivalent to the family*

$$F(x, y, z, \varepsilon) = (x', y', z'),$$
$$x' = f(x, \varepsilon), \quad y' = A(x, \varepsilon)y, \quad z' = B(x, \varepsilon)z, \qquad \varepsilon \geqslant 0.$$

Here f is the time one shift along the phase curves of the vector field w_ε from the first formula of (5.3):

$$w_\varepsilon = (x^2 + \varepsilon)(1 + a(\varepsilon)x)^{-1}.$$

Moreover,

$$\|A\| \leqslant \lambda < 1, \qquad \|B^{-1}\| \leqslant \lambda < 1.$$

REMARK. The theorem means that the restriction of the map of the family to its central manifold is embeddable for the critical parameter value and for those noncritical values that correspond to maps without fixed points. The corresponding vector field is smooth with respect to the phase variable and the parameter and may be smoothly extended to the zero parameter value.

In one situation in Chapter 3 we will use the following modification of the previous theorem.

THEOREM 5.5. *In generic two-parameter families of maps only those germs of saddlenode maps with multiplicity two occur whose perturbation in the subfamily parametrized by an appropriate half-interval of the parameter axis with vertex at the critical value is finitely smoothly equivalent to the time one shift along the phase curves of the equation*

$$\dot{x} = \omega_\varepsilon, \qquad w_\varepsilon = (x^2 + \varepsilon)(1 + a(\varepsilon, \mu)x)^{-1}.$$

These are the only theorems on normal forms for local families that will be used in the main part of this book. They will be proved in Chapters 9 and 10. These chapters contain more general results that will not be used in the book. Yet they are included there for two reasons. First, they have applications in nonlocal bifurcation theory for multiparameter families. Second, they are proved in exactly the same way as the previous theorems. Their proofs are also presented in the last two chapters.

CHAPTER 3

Bifurcations in the Plane

In this chapter we study nonlocal bifurcations occurring in typical one-parameter families of planar vector fields. This study demonstrates the basic ideas and techniques that will be applied below to the investigation of nonlocal spatial bifurcations. From now on we assume that the families under consideration are of class C^∞, unless explicitly stated otherwise.

§1. Bifurcations of homoclinic loops of planar saddles

In this section we study bifurcations of homoclinic orbits of hyperbolic saddles.

1.1. Statement of the theorem and genericity assumptions.

ANDRONOV–LEONTOVICH THEOREM. *In a generic one-parameter family of vector fields in the plane, only one limit cycle can be created from a saddle loop.*

We will now describe what the word "generic" means in the theorem above.

Suppose that a vector field contained in a one-parameter family X_ε has a homoclinic loop γ of a hyperbolic saddle O. The corresponding parameter value, say $\varepsilon = 0$, will be called *critical*. Suppose that the family X_ε satisfies the following *genericity assumptions*:

1. Let $\lambda < 0 < \mu$ be the eigenvalues of the saddle O of the vector field X_0. Then λ/μ is irrational.
2. In the family X_ε, the homoclinic loop occurs and breaks in a transversal manner as ε passes through zero. More precisely, let Γ be an oriented segment transversal to the field X_0 at a point of γ. Let $p_s(\varepsilon)$, $p_u(\varepsilon)$ be the first intersection points of the stable and unstable manifolds of O with Γ respectively. We denote the oriented distance between $p_s(\varepsilon)$ and $p_u(\varepsilon)$ by $\rho(\varepsilon)$. The genericity assumption is

$$\left.\frac{d\rho(\varepsilon)}{d\varepsilon}\right|_{\varepsilon=0} \neq 0$$

(see Figure 3.1).

Recall the definition of the *Hausdorff distance* between compact sets. Let A, B be two compact subsets of $\mathbb{R}^n$. We define the *Hausdorff distance* between A and B by the following equality:

$$\rho_H(A, B) = \max\left\{ \max_{x\in A}\min_{y\in B} |x-y|,\ \max_{x\in B}\min_{y\in A} |x-y| \right\}.$$

From now on, when we say that a compact set A_ε depending on a parameter ε tends to a compact set B as $\varepsilon \to \varepsilon_0$, we mean that the Hausdorff distance $\rho_H(A_\varepsilon, B)$

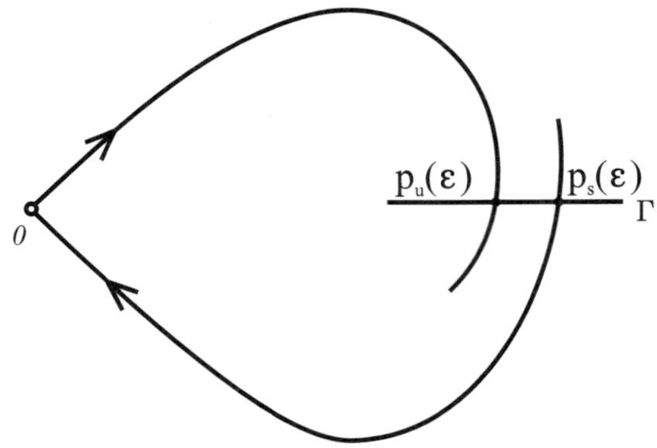

FIGURE 3.1. Characterization of separatrix splitting.

tends to zero as $\varepsilon \to \varepsilon_0$. We can now give a more detailed statement of the previous theorem. Recall that the sum of eigenvalues of a hyperbolic saddle is called the saddle value and is denoted by σ.

THEOREM 1.1. *Let the family X_ε satisfy assumptions* 1 *and* 2. *Then there is a neighborhood U of γ and a neighborhood V of $\varepsilon = 0$ satisfying the following conditions.*

1) *If $\sigma = \lambda + \mu < 0$, then for all $\varepsilon \in V$ lying on one side of zero, there exists a unique stable periodic orbit in U tending to γ as $\varepsilon \to 0$. For all ε on the opposite side of zero, X_ε has no periodic orbits in U.*

2) *If $\sigma > 0$, the same conclusions hold except that the periodic orbit is unstable.*

The second statement is reduced to the first by time reversal. Hence we will prove the first statement only.

The conclusions of the above theorem are shown in Figure 3.2.

1.2. Outline of the proof. In what follows we will prove Theorem 1.1 by constructing a Poincaré map near γ. The Poincaré map will be decomposed into the product of two mappings: a "singular" and a "regular" one. To do this, let us take two segments transversal to the field X_0 at two points of γ close to the singular point O. Denote them by T^+ and T^-; the first is called the "entrance" segment; it intersects the stable manifold of the saddle at a point p. The second, the "exit" segment, intersects the unstable manifold at a point q (see Figure 3.3). For a planar hyperbolic saddle, a *correspondence map* is defined for any hyperbolic sector of the saddle. It brings the half-interval Γ^+ with vertex on the stable manifold transversal to this manifold, to a similar half-interval Γ^- with vertex on the unstable manifold. This is a map along the phase curves of the field: the initial point of the curve from the first section comes to the final one on the second section. The curve and the sections belong to the hyperbolic sector chosen above. If the vector field depends on the parameter, the correspondence map also depends on the parameter. We denote the parameter-dependent correspondence map by

$$\Delta_\varepsilon^{\text{sing}} \colon \Gamma^+ \to \Gamma^-.$$

It is also called the *singular map* because it is not well defined in the full neighborhood of zero in the product of the parameter and coordinate spaces, and its

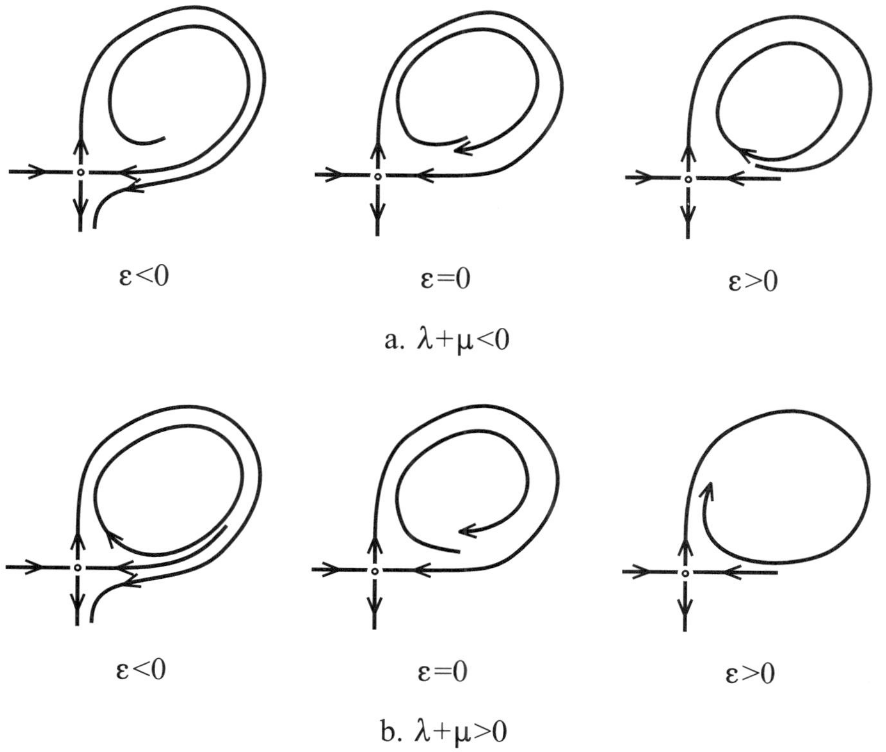

FIGURE 3.2. Bifurcations of a saddle loop.

derivatives may tend to infinity at some points of the boundary of its domain. This will be shown later.

By the theorem on the differentiability of solutions of ordinary differential equations with respect to initial conditions and parameters, for sufficiently small $|\varepsilon|$, the positive half-orbit starting at a point of T^- near q intersects T^+ for the first time at some point near p. The point on T^+ thus obtained is the image of the initial point. This map is denoted by $\Delta_\varepsilon^{\text{reg}}$ and called the *regular map*. It is a diffeomorphism depending smoothly on the parameter ε. The Poincaré map $\Delta_\varepsilon \colon \Gamma^+ \to T^+$ is defined by the equality

$$\Delta_\varepsilon = \Delta_\varepsilon^{\text{reg}} \circ \Delta_\varepsilon^{\text{sing}}.$$

We will study the bifurcation of the homoclinic orbit by considering the fixed point of the map Δ_ε.

The left factor in the above decomposition for the Poincaré map is a diffeomorphism with Lipschitz constant bounded away from zero and from infinity for all sufficiently small ε. The right factor, namely, the correspondence map, is contracting, if the saddle value is negative. The contraction coefficient tends to zero as the domain of the contraction tends to p. Hence, the composition is a contracting map for small ε. Therefore, it has a unique stable fixed point, provided that the Poincaré map takes its domain into itself.

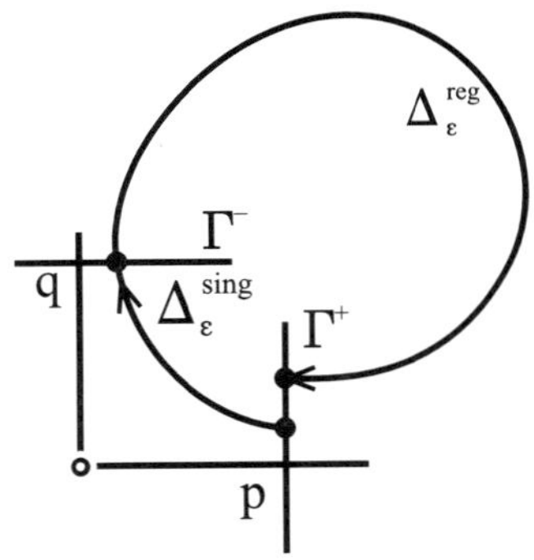

FIGURE 3.3. Definitions of regular and singular maps.

The correspondence map is studied by means of the theory of normal forms.

1.3. Normal forms for perturbations of planar hyperbolic saddles and correspondence maps. In order to calculate the map $\Delta_\varepsilon^{\text{sing}}$, we need to choose "good" coordinates near O. According to the theory of finitely smooth normal forms for local families (see §5 of Chapter 2), and by virtue of the genericity assumption 1, there exists a finitely smooth family of charts in some neighborhood D of O transforming the initial system to a linear one, namely

$$\begin{cases} \dot{x} = \lambda(\varepsilon)x, \\ \dot{y} = \mu(\varepsilon)y, \end{cases} \tag{1.1}$$

where $\lambda(0) = \lambda < 0 < \mu = \mu(0)$. By rescaling the picture, if necessary, we can assume that

$$\{|x|, |y| \leqslant 1\} \subset D.$$

We call (x, y) a *normalizing chart.* Choose T^+, T^-, and Γ^+ as follows:

$$T^+ = \{(x, y) \in D \mid x = 1\}, \qquad T^- = \{(x, y) \in D \mid y = 1\},$$
$$\Gamma^+ = \{(x, y) \in D \mid x = 1,\ y > 0\}.$$

The flow defined by (1.1) is given by

$$x(t) = x_0 \exp(\lambda(\varepsilon)t), \qquad y(t) = y_0 \exp(\mu(\varepsilon)t).$$

Thus, the correspondence map $\Delta_\varepsilon^{\text{sing}} \colon \Gamma^+ \to T^-$ is given by

$$\Delta_\varepsilon^{\text{sing}}(1, y) \to (y^{\alpha(\varepsilon)}, 1), \quad \text{where } \alpha(\varepsilon) = -\lambda(\varepsilon)/\mu(\varepsilon) > 1. \tag{1.2}$$

1.4. The Poincaré map and its fixed point. The map $\Delta_\varepsilon^{\text{reg}} \colon T^- \to T^+$ is given in the normalizing chart by the formula

$$\Delta_\varepsilon^{\text{reg}}(x, 1) \to (1, f_\varepsilon(x)).$$

By the genericity assumption 2, we have $\partial f_\varepsilon(0)/\partial\varepsilon|_{\varepsilon=0} \neq 0$. Without loss of generality, we assume

$$\left.\frac{\partial f_\varepsilon(0)}{\partial\varepsilon}\right|_{\varepsilon=0} > 0. \tag{1.3}$$

Otherwise we change the sign of the parameter. Now the Poincaré map has the form

$$\Delta_\varepsilon(y) = f_\varepsilon(y^{\alpha(\varepsilon)}).$$

Moreover, the Lipschitz constant L of the correspondence map restricted to the segment $[0, \delta]$ can be estimated in the following way:

$$L < \alpha(\varepsilon)\,\delta^{\alpha(\varepsilon)-1}.$$

Hence L tends to 0 as $\delta \to 0$ uniformly with respect to ε. On the other hand, the Lipschitz constant of f_ε is bounded. So we can take δ so small that $\operatorname{Lip}\Delta_\varepsilon < 1/2$. Since $\Delta_0(0) = f_0(0) = 0$, we have

$$\Delta_0(\delta) < \operatorname{Lip}(\Delta_0)\,\delta < \delta/2. \tag{1.4}$$

Therefore, for $\varepsilon > 0$ and small enough, we have $\Delta_\varepsilon(0) = f_\varepsilon(0) > 0$ by (1.3), and $\Delta_\varepsilon(\delta) < \delta/2$ by (1.4). Hence,

$$\Delta_\varepsilon([0,\delta]) \subset [0,\delta].$$

Thus, for small positive ε the Poincaré map is contracting and takes its domain into itself. Therefore, it has a unique stable fixed point.

For negative ε, the Poincaré map diminishes the y coordinate, and hence has no fixed points.

The proof of Theorem 1.1 is complete.

§2. Homoclinic orbit of a saddlenode

In this section we study bifurcations of homoclinic orbits of saddlenodes in the plane.

2.1. Statement of the theorem and genericity assumptions.

THEOREM (Andronov). *In a generic one-parameter family of vector fields in the plane, only one limit cycle can be created from a homoclinic orbit of a saddlenode.*

Let us give explicit genericity assumptions for the family in the previous theorem. Let X_ε be a one-parameter family of vector fields in the plane. The vector field X_0 is subject to the first two of the following genericity assumptions, and the family itself is subject to the third one.

1. The origin O is a saddlenode of multiplicity two. This means that the singular point O has two eigenvalues: $\lambda \neq 0$, 0, thus having a one-dimensional center manifold. Moreover, the restriction of X_0 to the local center manifold of O has the form

$$(ax^2 + \cdots)\frac{\partial}{\partial x}, \qquad a \neq 0,\ x \in (\mathbb{R}, 0),$$

where the dots denote higher order terms.

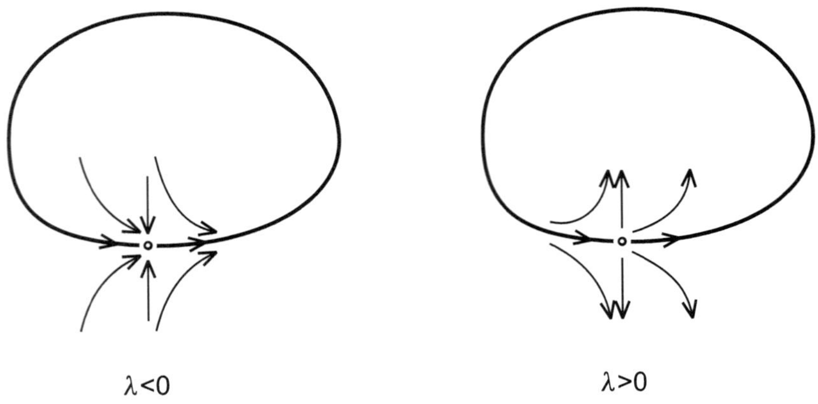

FIGURE 3.4. Homoclinic curves of saddlenodes.

2. The system X_0 has a homoclinic orbit γ which enters O as t tends to infinity and is tangent to the center manifold.

The phase portrait of a vector field X_0 satisfying assumptions 1 and 2 is shown in Figure 3.4.

The third assumption deals with the family X_ε itself.

3. The family X_ε, at $\varepsilon = 0$, is transversal to the hypersurface consisting of vector fields with a degenerate singular point near O. More precisely, if the restriction of X_ε to the local center manifold of O has the form

$$\dot{x} = f(x, \varepsilon), \qquad (x, \varepsilon) \in (\mathbb{R}^2, 0),$$

then the third assumption is equivalent to the inequality

$$\left. \frac{\partial f(0, \varepsilon)}{\partial \varepsilon} \right|_{\varepsilon=0} \neq 0.$$

We now give a detailed version of the previous theorem.

THEOREM 2.1. *Suppose that a one-parameter family X_ε of vector fields in the plane satisfies the above genericity assumptions 1–3. Then there exists a neighborhood U of $O \cup \gamma$ and a neighborhood V of $\varepsilon = 0$ such that the following holds.*

Case 1: $\lambda < 0$. *For all $\varepsilon \in V$ on one side of zero, the nonwandering set in U consists of a unique stable limit cycle that tends to $\gamma \cup O$ as $\varepsilon \to 0$. For all ε on the other side of zero, the nonwandering set consists of two hyperbolic singular points that tend to O as $\varepsilon \to 0$; one of them is a saddle, and the other is a stable node.*

Case 2: $\lambda > 0$. *The same conclusions hold as in case* 1, *except that the word "stable" is replaced by "unstable".*

2.2. Normal forms for perturbations of saddlenodes and correspondence maps. Without loss of generality, we assume $\lambda < 0$ (otherwise perform time reversal). According to the theory of finitely smooth normal forms for local families, the initial family near zero is finitely smoothly equivalent to

$$\begin{cases} \dot{x} = (x^2 + \varepsilon)(1 + a(\varepsilon)x)^{-1}, \\ \dot{y} = \lambda(x, \varepsilon)\, y, \end{cases} \tag{2.1}$$

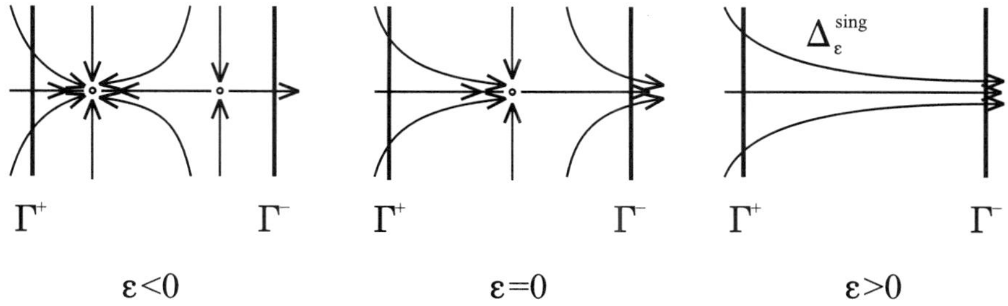

FIGURE 3.5. Bifurcation of a saddlenode.

where $\lambda(0,0) = \lambda < 0$. The expression (2.1) implies the following description of local bifurcations (see Figure 3.5).

1) When $\varepsilon > 0$, the system (2.1) has no singular points.

2) When $\varepsilon < 0$, the system has two singular points $(\pm\sqrt{-\varepsilon}, 0)$, one is a stable node, and the other is a saddle.

In what follows we study the nonlocal bifurcation that occurs when the singular point disappears ($\varepsilon > 0$). We construct a parameter-dependent Poincaré map as in Theorem 1.1. Let

$$\Gamma^+ = \{(x,y) \mid x = -\delta,\ |y| \leqslant \delta\}, \qquad \Gamma^- = \{(x,y) \mid x = \delta,\ |y| \leqslant \delta\}$$

be the entrance and exit transversal segments: the phase curves of the family (2.1) enter the neighborhood of the singular point through Γ^+ and exit through Γ^-. Here δ is a small constant to be chosen later.

The normal form (2.1) is a system with separated variables. It is linear with respect to y. The orbits of the system pass from the entrance segment to the exit one in the case $\varepsilon > 0$. For $\varepsilon \leqslant 0$, the invariant curves $x = \pm\sqrt{-\varepsilon}$ form an obstacle for passing from Γ^+ to Γ^-. Let T_ε be the time taken by the solutions of the system (2.1) to pass from Γ^+ to Γ^-:

$$T_\varepsilon = \int_{-\delta}^{\delta} \frac{1 + a(\varepsilon)x}{x^2 + \varepsilon}\, dx = (\pi + o(1))\frac{1}{\sqrt{\varepsilon}}\,.$$

In the same time the y-coordinate changes along the solution in the following way:

$$y \mapsto \Lambda(\varepsilon)y. \tag{2.2}$$

Choose a δ so that

$$\lambda(x,\varepsilon) < \lambda/2 \tag{2.3}$$

for $|x| \leqslant \delta$. Then

$$\Lambda(\varepsilon) = \exp\left(\int_0^{T_\varepsilon} \lambda(x_\varepsilon(t),\varepsilon)\, dt\right) < \exp\left(\frac{\lambda T_\varepsilon}{2}\right) \to 0 \tag{2.4}$$

as $\varepsilon \to 0$. Formulas (2.2), (2.4) give the expression for the correspondence map

$$\Delta_\varepsilon^{\text{sing}} \colon \Gamma^+ \to \Gamma^-.$$

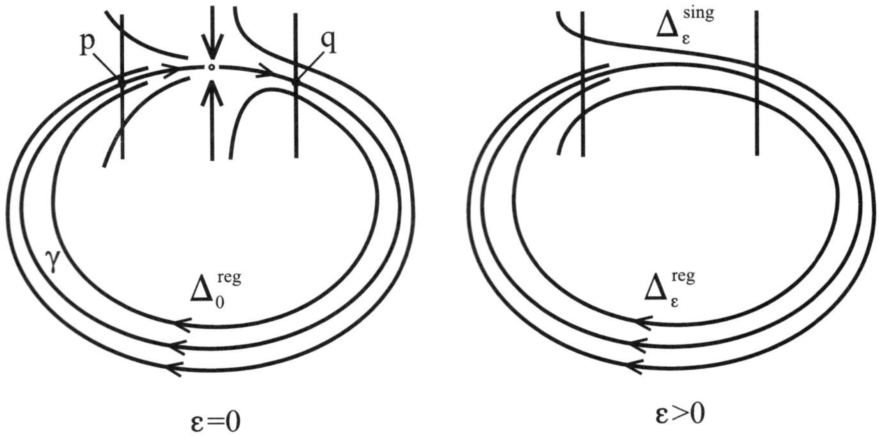

FIGURE 3.6. Generation of a limit cycle from the homoclinic orbit of a saddlenode.

Note that in the normalizing chart this map is a linear contraction with the coefficient shrinking as $\varepsilon \to 0$.

2.3. The Poincaré map and its fixed point. The genericity assumption 2 from 2.1 implies that, for δ small enough, the homoclinic orbit γ intersects the transversal segment Γ^+ at its inner point p. Denote by q the intersection point of γ with the exit segment Γ^- (see Figure 3.6). Our further analysis follows the same lines as in the case of the separatrix loop in §1.

By the theorem on the differentiability of solutions with respect to initial conditions and parameters, for sufficiently small $|\varepsilon|$ the positive half-orbit starting at a point of Γ^- near q intersects Γ^+ for the first time at some point near p. The correspondence thus defined will be denoted by $\Delta_\varepsilon^{\text{reg}}$ and called the *regular map*. It is a diffeomorphism depending smoothly on the parameter ε. The Poincaré map $\Delta_\varepsilon : \Gamma^+ \to \Gamma^+$ is defined by the equality

$$\Delta_\varepsilon = \Delta_\varepsilon^{\text{reg}} \circ \Delta_\varepsilon^{\text{sing}}. \tag{2.5}$$

We will study the bifurcation of the homoclinic orbit of the saddlenode by considering the fixed point of this map.

The left factor in the above decomposition for the Poincaré map is a diffeomorphism with the Lipschitz constant bounded away from zero and from infinity for all sufficiently small positive ε. The right factor, namely, the correspondence map, is linear and contracting, with the coefficient tending to zero as $\varepsilon \to 0$. Hence the composition is a contracting map. Note that the regular map takes a sufficiently small neighborhood U^+ of q into Γ^+ for sufficiently small values of ε. Moreover, the singular map takes the entire segment Γ^+ into U^+ for sufficiently small ε. Hence, the composition (2.5) takes the segment Γ^+ into itself, and is contracting. Therefore, it has a unique stable fixed point. This proves Theorem 2.1.

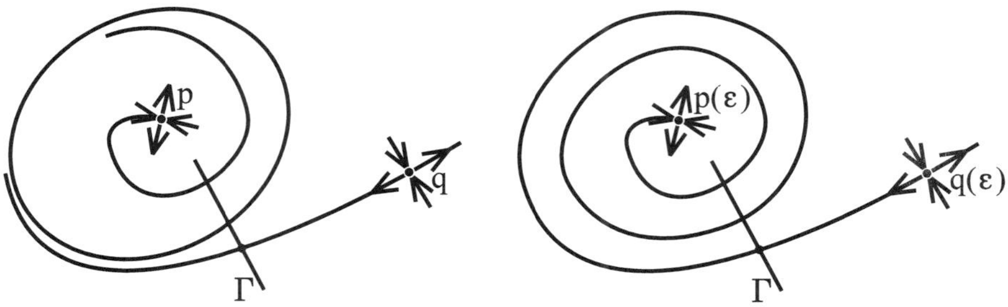

FIGURE 3.7. Semistable limit cycle, winding separatrices, and saddle connections.

§3. Semistable cycles breaking saddle connections

In this section we discuss global bifurcations produced by the disappearance of a semistable limit cycle.

3.1. Sparkling saddle connections. We begin with the picture formed by a semistable limit cycle and two hyperbolic saddles.

THEOREM 3.1. *Suppose that a vector field* X *in the plane contained in a generic one-parameter family* X_ε $(X_0 = X)$ *has a semistable limit cycle* L. *Let this field have two hyperbolic saddles, one inside, and the other outside the cycle. Suppose that the separatrix of one saddle winds to the cycle as* $t \to +\infty$ *and the separatrix of the other one does the same as* $t \to -\infty$ *(see Figure* 3.7*). Then on one side of* $\varepsilon = 0$, *there exist two limit cycles that tend to* L *as* $\varepsilon \to 0$, *and one is stable, the other is unstable. For* ε *on the other side of* $\varepsilon = 0$, *there exist no limit cycles near* L. *Moreover, there exists a sequence of parameter values of the form*

$$\varepsilon_n = \frac{1}{n^2}(c + o(1)), \qquad c \neq 0,$$

such that the field X_{ε_n} *has a saddle connection for any* n *large enough.*

The bifurcation described in the theorem is shown in Figure 3.7.

Before proving the theorem, we specify, as usual, the explicit genericity assumption on the family X_ε.

Let Γ be a segment transversal to X at a point O of L. Denote by Δ_ε the parameter-dependent Poincaré map that corresponds to the field X_ε, cycle L, and segment Γ. Let x be any smooth chart on Γ with zero value at the intersection point of the transversal segment and the cycle.

The genericity assumption in Theorem 3.1 is the following:

The diffeomorphisms Δ_ε form a saddlenode family, i.e.

$$\Delta_0(0) = 0, \qquad \left.\frac{\partial \Delta_0(x)}{\partial x}\right|_{x=0} = 1, \qquad \left.\frac{\partial^2 \Delta_0(x)}{\partial x^2}\right|_{x=0} \neq 0, \qquad \left.\frac{\partial \Delta_\varepsilon(0)}{\partial \varepsilon}\right|_{\varepsilon=0} \neq 0. \tag{3.1}$$

PROOF OF THEOREM 3.1. All results in the theorem except the description of the bifurcations of saddle connections may be derived from the theory of local bifurcations. Hence we will only consider the bifurcation of saddle connections. Without loss of generality, we can assume that for $\varepsilon > 0$ there are no cycles near L.

By the one-dimensional partial embedding theorem (see Theorem 2.5.5), there exists a C^∞ smooth family of charts x on Γ taking the family Δ_ε to the family of time one shifts along the phase curves of the following equations:

$$\dot{x} = (x^2 + \varepsilon)(1 + a(\varepsilon)x)^{-1}, \qquad \varepsilon \geqslant 0. \tag{3.2}$$

By abuse of notation, we write x instead of x_ε, although the chart x depends on the parameter. Let p be the saddle of the unperturbed field lying inside the cycle L and let q be the outside one. These saddles depend smoothly on the parameter of the family together with their separatrices. Let $x^+(\varepsilon) > 0 > x^-(\varepsilon)$ be the intersection points of the separatrices of the saddles p and q respectively winding to the cycle L with Γ. These intersections depend smoothly on ε.

Equation (3.2) defines a *time function* t on the section Γ. It is equal to the time taken by the point 0 to get to the position x:

$$t_\varepsilon(x) \stackrel{\text{def}}{=} \int_0^x \frac{1 + a(\varepsilon)\xi}{\xi^2 + \varepsilon}\, d\xi. \tag{3.3}$$

The time one shift of equation (3.2) is equal to the Poincaré map Δ_ε. The field X_ε ($\varepsilon > 0$) has an orbit connecting p and q if and only if the points $x^+(\varepsilon)$ and $x^-(\varepsilon)$ belong to the same orbit of the Poincaré map Δ_ε. Equivalent statement: the values of the time function t_ε at these points differ by an integer.

This motivates the following definition:

$$T(\varepsilon) = t_\varepsilon(x^+(\varepsilon)) - t_\varepsilon(x^-(\varepsilon)).$$

We are interested in the integer values of the function T_ε. Namely, we must prove that the equations $T(\varepsilon) = n$ have the solutions ε_n mentioned in Theorem 3.1. To do this, it is sufficient to prove the following

PROPOSITION 3.1. *$T(\varepsilon)$ is a monotonic function of the form*

$$T(\varepsilon) = \frac{1}{\sqrt{\varepsilon}}(\pi + o(1)).$$

This proposition immediately implies Theorem 3.1.

PROOF OF PROPOSITION 3.1. The function T_ε may be explicitly expressed through the functions $x^\pm(\varepsilon)$:

$$T(\varepsilon) = F_\varepsilon(x^+(\varepsilon)) - F_\varepsilon(x^-(\varepsilon)), \qquad F_\varepsilon(x) = \frac{1}{\sqrt{\varepsilon}} \arctan \frac{x}{\sqrt{\varepsilon}} + \frac{a}{2}\log(x^2 + \varepsilon). \tag{3.4}$$

The second term is smooth when ε ranges in some neighborhood of zero and x ranges in some segment that does not contain zero. Hence the function (3.4) has the form

$$T(\varepsilon) = \frac{c(\varepsilon)}{\sqrt{\varepsilon}} + b(\varepsilon).$$

Here b is a smooth function near zero, c is smooth for small positive ε, and $c \to \pi$ as $\varepsilon \to 0$.

The explicit formula for c is the following:

$$c(\varepsilon) = c^+(\varepsilon) - c^-(\varepsilon), \qquad c^\pm = \arctan x^\pm(\varepsilon)/\sqrt{\varepsilon}.$$

Now an elementary calculation shows that $T'(\varepsilon) > 0$ for small ε. This proves Proposition 3.1.

At the same time, Theorem 3.1 is proved.

3.2. No simultaneous saddle connections in the case of many saddles. Now we consider the semistable cycle with more than one separatrix of hyperbolic saddles winding to it from the inside and more than one winding from the outside. A natural question arises: when the cycle L disappears, is the simultaneous occurrence of two saddle connections possible?

THEOREM 3.2 [**MP**]. *In a generic one-parameter family of vector fields X_ε, the simultaneous occurrence of two saddle connections is impossible.*

The genericity assumptions in this theorem are the following. Let Γ, Δ_ε, x be the same transversal section, Poincaré map, and normalizing chart as before. Let p_i be the saddles lying inside L, and q_i the outside ones. Choose an arbitrary intersection point of the separatrix of the saddle p_i winding to L, with Γ, such that this intersection continuously, hence smoothly, depends on the parameter $\varepsilon \in (\mathbb{R}, 0)$. Denote it by $x_i^-(\varepsilon)$. Let $x_j^+(\varepsilon)$ be a similar intersection for the saddle q_j.

Let t_ε be the time function (3.3). We define the time intervals τ_{ij}^+ and τ_{ij}^- by the relations:

$$\tau_{ij}^+(\varepsilon) = t_\varepsilon(x_i^+(\varepsilon)) - t_\varepsilon(x_j^+(\varepsilon)), \qquad \tau_{ij}^-(\varepsilon) = t_\varepsilon(x_i^-(\varepsilon)) - t_\varepsilon(x_j^-(\varepsilon)). \tag{3.5}$$

Now the genericity assumptions in Theorem 3.2 are the following:

1. The genericity assumption of Theorem 3.1 holds.
2. For any i, j, k, l the following inequality holds:

$$\tau_{ij}^+(0) \neq \tau_{kl}^-(0) \mod \mathbb{Z}.$$

REMARKS. 1. The last inequality is called the *Malta–Palis condition.*

2. It may be proved that the vector field generating the Poincaré map Δ_0 is uniquely determined. Hence the time function t_0 is well defined. Therefore, the Malta–Palis condition is invariant under coordinate changes.

PROOF OF THEOREM 3.2. If the conclusion of the theorem is not true, then there exists a sequence of parameter values $\varepsilon_n \to 0$ such that X_{ε_n} has two heteroclinic curves, say, for example, connecting the points $x_1^+(\varepsilon_n)$ and $x_1^-(\varepsilon_n)$, $x_2^+(\varepsilon_n)$ and $x_2^-(\varepsilon_n)$. Define the time functions

$$T_{11}(\varepsilon) = t_\varepsilon(x_1^+(\varepsilon)) - t_\varepsilon(x_1^-(\varepsilon)), \qquad T_{22}(\varepsilon) = t_\varepsilon(x_2^+(\varepsilon)) - t_\varepsilon(x_2^-(\varepsilon)).$$

The evident identity holds:

$$T_{11}(\varepsilon) - T_{22}(\varepsilon) = \tau_{12}^+(\varepsilon) - \tau_{12}^-(\varepsilon). \tag{3.6}$$

The above assumption implies

$$T_{11}(\varepsilon_n) \in \mathbb{Z}, \qquad T_{22}(\varepsilon_n) \in \mathbb{Z}.$$

Hence

$$\tau_{12}^{+}(\varepsilon_n) = \tau_{12}^{-}(\varepsilon_n) \mod \mathbb{Z}.$$

Passing to the limit as $n \to \infty$, we get

$$\tau_{12}^{+}(0) = \tau_{12}^{-}(0) \mod \mathbb{Z}.$$

This contradicts the Malta–Palis condition.

The proof of Theorem 3.2 is complete.

3.3. Violating the Malta–Palis condition. The following theorem implies that the genericity assumption (2) in 3.2 is in a sense a necessary and sufficient condition for the coexistence of saddle connections. Namely, if the Malta–Palis condition is violated for a vector field X, then in any neighborhood of this field in the function space there are vector fields having at least two saddle connections simultaneously. The vector field X is the subject of two independent hypotheses: existence of a semistable cycle and violation of the Malta–Palis condition. Hence, this field cannot appear in a generic one-parameter family, but can be in a two-parameter one. The following theorem deals with two-parameter families.

THEOREM 3.3 [**IY2**]. *If the Malta–Palis condition is violated for a vector field X, then in a generic two-parameter deformation of X, there are countably many values of the parameters accumulating to the zero value, for which two saddle connections occur simultaneously.*

PROOF. Consider a two-parameter family of vector fields $X_{\varepsilon,\mu}$ with $X_{0,0} = X$. Let the transversal section Γ be the same as before. The Poincaré map Δ corresponding to this section and to the cycle L depends on two parameters of the family and on the coordinate x on the section. Suppose that this map restricted to the line $\mu = 0$ satisfies the genericity assumption (3.1). Then, by Theorem 2.5.5, the family Δ is finitely smoothly equivalent to the time one shift along the phase curves of the family of differential equations:

$$\dot{x} = v(x, \varepsilon, \mu), \qquad v(x, \varepsilon, \mu) = (\varepsilon + x^2)(1 + a(\varepsilon, \mu)x)^{-1}, \ (x, \mu) \in (\mathbb{R}^2, 0), \ \varepsilon \geqslant 0.$$

From now on in this subsection x is the normalizing chart on Γ. The curve

$$\theta = \{(\varepsilon, \mu) \in (\mathbb{R}^2, 0) \mid \varepsilon = 0\}$$

for small μ corresponds to equations with a semistable limit cycle close to L. Let p_1, p_2 be the saddles lying inside L and q_1, q_2 the outside ones. Denote by $x_i^{+}(\varepsilon, \mu)$ some intersection point of the separatrix of $q_i(\varepsilon, \mu)$ winding to L, with Γ, $i = 1, 2$. Denote by $x_i^{-}(\varepsilon, \mu)$ a similar point for $p_i(\varepsilon, \mu)$. Choose these points to depend smoothly on the parameters. Define the time function and time intervals as in the two previous subsections. The only difference is that the parameter ε is everywhere replaced by the pair ε, μ:

$$T_{11}(\varepsilon, \mu) = t(x_1^{+}(\varepsilon, \mu)) - t(x_1^{-}(\varepsilon, \mu)), \qquad T_{22}(\varepsilon, \mu) = t(x_2^{+}(\varepsilon, \mu)) - t(x_2^{-}(\varepsilon, \mu)).$$

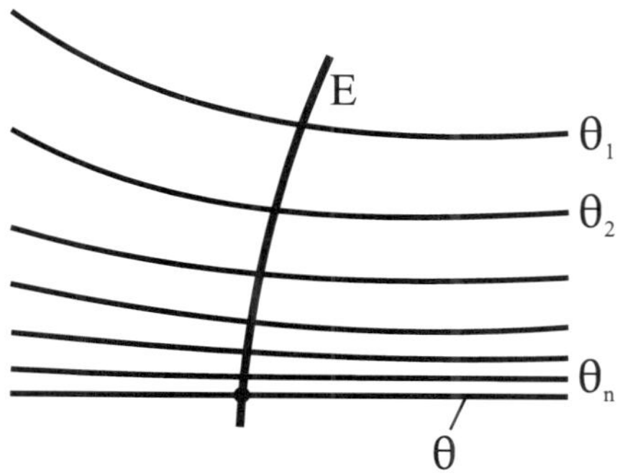

FIGURE 3.8. Intersections of the synchronization curve and the curves of sparkling saddle connections in the parameter plane.

The time intervals $\tau_{ij}^{\pm}(\varepsilon, \mu)$ are defined by (3.5), and identity (3.6) holds with ε replaced by ε, μ. The vector field $X_{\varepsilon,\mu}$ has two simultaneous saddle connections if and only if the values $T_{11}(\varepsilon, \mu)$, $T_{22}(\varepsilon, \mu)$ are both integers. By (3.6), this is equivalent to the following statement: the values $T_{11}(\varepsilon, \mu)$, $\tau_{12}^{+}(\varepsilon, \mu) - \tau_{12}^{-}(\varepsilon, \mu)$ are both integers. The violation of the Malta–Palis condition implies that

$$\tau_{12}^{+}(0,0) - \tau_{12}^{-}(0,0) = k$$

for some $k \in \mathbb{Z}$. Denote by E the "synchronization curve" in the parameter plane given by the equation

$$\tau_{12}^{+}(\varepsilon, \mu) - \tau_{12}^{-}(\varepsilon, \mu) = k$$

and considered near 0 (see Figure 3.8). The second genericity assumption in Theorem 3.3 is:

The curve E is smooth and passes through zero; the curves θ and E intersect transversally at 0.

Now consider the equation

$$T(\varepsilon, \mu) = n, \qquad T = T_{11} - T_{22}. \tag{3.7}$$

The following proposition claims that (3.7) defines a "curve of sparkling saddle connections" in the parameter plane.

PROPOSITION 3.2. *For any sufficiently large n, equation* (3.7) *determines a curve θ_n in the parameter plane. This curve is the graph of a C^1 function ε_n on μ. These functions are defined in the same neighborhood of zero and tend to zero together with their derivative as $n \to \infty$.*

The proposition is illustrated in Figure 3.8.

By identity (3.6), in the case when the synchronization curve intersects the curve of sparkling saddle connections, two simultaneous saddle connections occur. The second genericity assumption, together with the above proposition, imply that there is a countable set of such intersection points for a generic family.

PROOF OF PROPOSITION 3.2. The function T in equation (3.7) has the following properties:

1. $T(\,\cdot\,, \mu)$ tends to infinity uniformly with respect to μ as $\varepsilon \to 0$.

2. This function is monotonic in ε and its derivative with respect to ε tends to $-\infty$ as $\varepsilon \to 0^+$.
3. The derivative of T with respect to μ is bounded:

$$\frac{\partial T}{\partial \mu} = O(1).$$

Denote by ε_n the function given by equation (3.7), and by θ_n the corresponding curve:

$$\theta_n = \{(\varepsilon, \mu) \in (\mathbb{R}^2, 0) \mid \varepsilon = \varepsilon_n(\mu)\}$$

(see Figure 3.8). Statements 1 and 2 imply the existence of functions ε_n tending uniformly to 0 as $n \to \infty$. Statement 3, together with the implicit function theorem, implies that the derivatives of these functions have the same property.

Hence it remains to prove statements 1 to 3. The first two follow from Proposition 3.1. Let us prove the third. By (3.4),

$$T(\varepsilon, \mu) = F_{\varepsilon,\mu}(x^+(\varepsilon, \mu)) - F_{\varepsilon,\mu}(x^-(\varepsilon, \mu)), \qquad F_{\varepsilon,\mu} = F_1(\varepsilon, \mu, \cdot\,) + F_2(\varepsilon, \mu, \cdot\,),$$
$$F_1(\varepsilon, \mu, x) = \frac{1}{\sqrt{\varepsilon}} \arctan \frac{x}{\sqrt{\varepsilon}}, \qquad F_2(\varepsilon, \mu, x) = \frac{a(\varepsilon, \mu)}{2} \log(x^2 + \varepsilon).$$

The functions $x^{\pm}$ are smooth, as well as the function F_2, provided that $|x|$ is bounded away from zero. Moreover,

$$\frac{\partial F_1(\varepsilon, \mu, x^{\pm}(\varepsilon, \mu))}{\partial \mu} = \frac{1}{(x^{\pm}(\varepsilon, \mu))^2 + \varepsilon} \, \frac{\partial x^{\pm}(\varepsilon, \mu)}{\partial \mu}.$$

This function is bounded near zero. This proves the proposition.

Together with Proposition 3.2, Theorem 3.3 is proved.

CHAPTER 4

Homoclinic Orbits of Nonhyperbolic Singular Points

In this chapter we consider bifurcations of vector fields in $\mathbb{R}^n$ having homoclinic orbits of nonhyperbolic singular points. The main result generalizes the corresponding theorem from Chapter 3 to the multidimensional case. In §1 we study a homoclinic orbit of a saddlenode whose hyperbolic part is a node. In §2 we prove an important lemma on hyperbolicity of the product of linear maps which will be often used in Chapters 4, 5, 7. In §3 we analyze bifurcations of homoclinic orbits of a saddlenode whose hyperbolic part is a saddle. In §4 the case of several homoclinic orbits of a saddlenode is studied: we show that the horseshoe dynamics may be created. The same method allows us to prove the Birkhoff–Smale theorem in §5.

This chapter contains the main tools that allow the application of normal forms and hyperbolic theory to the study of nonlocal bifurcations.

§1. Homoclinic orbit of a saddlenode: the case of a nodal hyperbolic part

1.1. Creation of a stable (unstable) periodic orbit. The main result of §1 is the following

THEOREM 1.1. *Assume that in a generic one-parameter family $\{X_\varepsilon\}$ of smooth vector fields in $\mathbb{R}^n$ the zero value of the parameter corresponds to a vector field X_0 having a homoclinic orbit γ of a saddlenode singular point O, whose nonzero eigenvalues are all stable or all unstable. Then there exists a neighborhood $V \ni 0$ in the parameter space $\mathbb{R}^1$ and a neighborhood $U \supset \gamma \cup O$ in the phase space such that for all ε in one connected component of $V \setminus 0$ the nonwandering set of X_ε in U consists of two singular points converging to O as $\varepsilon \to 0$, while for ε in the second connected component, the nonwandering set of $X_\varepsilon|_U$ consists of a unique attracting or repelling periodic orbit which converges to $\gamma \cup O$ as $\varepsilon \to 0$.*

1.2. Genericity assumptions. We formulate explicitly three genericity-type conditions that must be imposed on the family X_ε for Theorem 1.1 to hold. The first two conditions are imposed on the vector field X_0, while the third concerns the derivative $(d/d\varepsilon)X_\varepsilon|_{\varepsilon=0}$.

1. *The singular point O of the vector field is a nonresonant saddlenode of multiplicity 2 which is a node with respect to the hyperbolic variables.*

This means that the eigenvalues of the singular point are 0 and a tuple λ_i with $\operatorname{Re}\lambda_i < 0$ for all $i = 1, \dots, n-1$ (resp., all $\operatorname{Re}\lambda_i$ positive), and the tuple $\operatorname{Re}\lambda = (\operatorname{Re}\lambda_1, \dots, \operatorname{Re}\lambda_{n-1})$ is nonresonant. The multiplicity assumption means that the restriction of the vector field to the (one-dimensional) center manifold has the form

$$\dot{x} = ax^2 + \cdots, \qquad a \neq 0,\ x \in W^c \simeq (\mathbb{R}^1, 0),$$

where the dots stand for terms of order $\geqslant 3$.

2. *The homoclinic orbit tends to the singularity as $t \to \pm\infty$, being tangent to the center manifold.*

This assumption rules out the exceptional case of an orbit converging to (resp., diverging from) the singular point along the hyperbolic stable (resp., unstable) manifold.

3. *The family X_ε transversally intersects at the point X_0 the surface of vector fields having degenerate singular points.*

This means that if we consider the center manifold of the local family X_ε and write the restriction of the family to this manifold as

$$\dot{x} = f(x, \varepsilon), \quad \dot{\varepsilon} = 0, \qquad (x, \varepsilon) \in (\mathbb{R}^2, 0),\ f(0,0) = 0,$$

then

$$\left. \frac{\partial f(0, \varepsilon)}{\partial \varepsilon} \right|_{\varepsilon=0} \neq 0.$$

Under this set of assumptions we prove Theorem 1.1 in the next three subsections for the case $\operatorname{Re} \lambda_i < 0$ (the second case can be reduced to the first by time reversal and the symmetry $x \mapsto -x$). Basically the proof follows the same lines as the proof for the 2-dimensional case; see Chapter 3.

1.3. The normal form for a perturbation of a saddlenode. The genericity assumptions 1 and 3 allow us to apply the theorem on the normal form for the perturbation X_ε; see §5 of Chapter 2. Using a finitely smooth change of coordinates and parameter, one may transform the original family to the form

$$\dot{x} = (x^2 + \varepsilon)(1 + a(\varepsilon)x)^{-1}, \qquad x \in \mathbb{R}^1, \tag{1.1}$$

$$\dot{y} = A(x, \varepsilon)y, \qquad y \in \mathbb{R}^{n-1},\ \varepsilon \in \mathbb{R}, \tag{1.2}$$

with finitely smooth a and A; all the eigenvalues of $A(0,0)$ have negative real parts. The system has a triangular structure: one may first integrate the separate equation (1.1) and then substitute it into equation (1.2), which then becomes a (nonautonomous) linear homogeneous system of equations. In the next subsection, we show that it has exponentially decreasing solutions as $t \to \infty$ (recall that we consider the case when all eigenvalues of $A(0,0)$ are stable).

Consider the subcase $\varepsilon < 0$. For such ε, system (1.1), (1.2) has two singular points $(\pm\sqrt{-\varepsilon}, 0)$; one of them is a stable node, the other is a saddle (see Figure 4.1a).

From this figure one can see that all orbits near γ that are not on the stable manifold of the saddle will converge to the stable node as $t \to +\infty$. Therefore the only nonwandering points are the two singularities, if the parameter ε is negative.

If the parameter ε is positive, then the singular point disappears. We construct a parameter-dependent family of Poincaré maps for a transversal to γ; these maps will be decomposed into the product of two mappings, the *singular* and the *regular* one. Roughly speaking, the singular part is a map of the cross-section to γ along the phase curves passing by the annihilated singular point (see Figure 4.1c). This map may be completely described in terms of the normal form (1.1), (1.2). The regular part is a map along the phase curves located near the homoclinic orbit γ

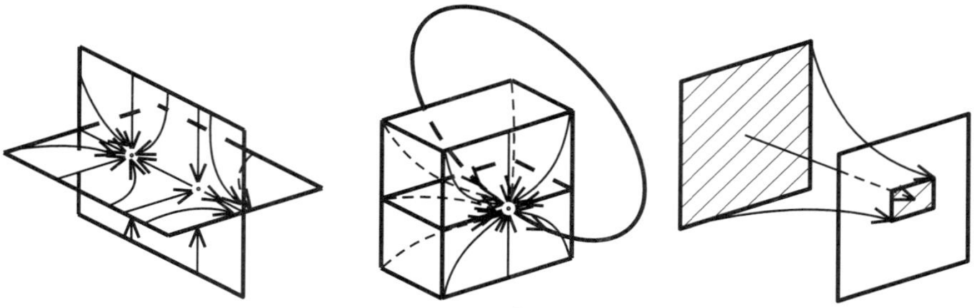

FIGURE 4.1. (a–c) The local bifurcation of a saddlenode with nodal hyperbolic part; (b) the homoclinic orbit of a saddlenode.

away from the singular point (see Figure 4.2). By virtue of general theorems, it is a diffeomorphism smoothly depending on the parameter.

The singular part is strongly contracting with the Lipschitz constant tending to zero as the parameter tends to zero from the right. The regular part has a uniformly bounded Lipschitz constant. Therefore the product is contracting for all small positive ε, hence it has a unique contracting fixed point.

1.4. The singular correspondence map $\Delta_\varepsilon^{\text{sing}}$. Let

$$\Gamma^+ = \{(x,y) \mid x = -\delta < 0,\ |y| \leqslant \delta\}, \qquad \Gamma^- = \{(x,y) \mid x = \delta > 0,\ |y| \leqslant \delta\}$$

be two cross-sections for system (1.1), (1.2). The first is the entrance gate: all trajectories crossing it enter the neighborhood of the fixed point. The other section is the exit in the same sense. Clearly, for $\varepsilon > 0$ the derivative $\dot{x}$ is positive; therefore each trajectory starting on Γ^+ will intersect Γ^- at a certain point; we denote this map by $\Delta_\varepsilon^{\text{sing}} \colon \Gamma^+ \to \Gamma^-$; it will be referred to as the *correspondence map* along the phase curves of the field (1.1), (1.2) for a given ε. Its properties will now be established.

LEMMA 1.2. *The correspondence map is defined for all positive ε and is linear*:

$$\Delta_\varepsilon^{\text{sing}} \colon y \mapsto \Lambda_\varepsilon y.$$

Moreover, $\|\Lambda_\varepsilon\| \to 0$ *as* $\varepsilon \to 0+$.

REMARK. In fact, the norm $\|\Lambda_\varepsilon\|$ decreases exponentially with $\varepsilon \to 0+$.

PROOF. System (1.1), (1.2) is of triangular form, so the time T_ε necessary for reaching Γ^- from Γ^+ can be explicitly computed:

$$T_\varepsilon = \int_{-\delta}^{\delta} \frac{1 + a(\varepsilon)x}{x^2 + \varepsilon}\, dx = \frac{\pi + o(1)}{\sqrt{\varepsilon}}.$$

The term $o(1)$ for any fixed δ converges to zero as $\varepsilon \to 0$.

Let $x_\varepsilon(t)$ be the solution of (1.1) with the initial condition $x_\varepsilon(0) = -\delta$. Denote $A_\varepsilon(t) = A(x_\varepsilon(t), \varepsilon)$, and let $Y_\varepsilon(t)$ be the fundamental matrix of solutions of the linear system

$$\dot{y} = A_\varepsilon(t)\, y, \qquad y \in \mathbb{R}^{n-1},$$

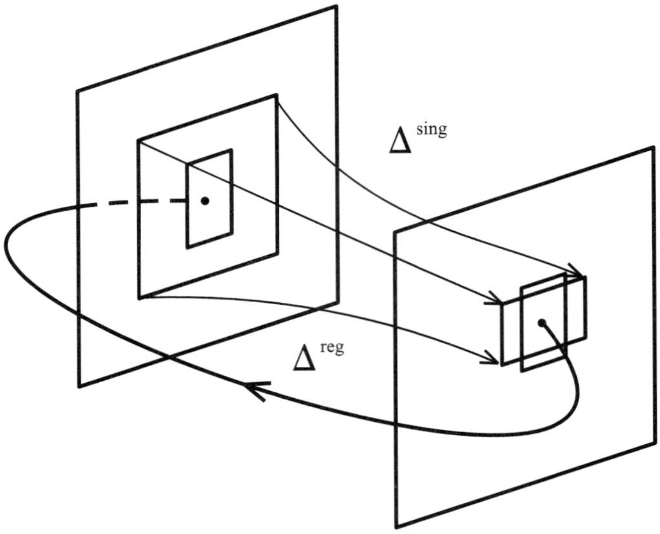

FIGURE 4.2. The regular and singular parts of the Poincaré map.

subject to the initial condition $Y_\varepsilon(0) = E$. Then $\Lambda_\varepsilon = Y_\varepsilon(T_\varepsilon)$. If the matrix function $A_\varepsilon(\,\cdot\,)$ were constant (and equal to A), then we would have $\Lambda_\varepsilon = \exp(T_\varepsilon \cdot A)$, and since all eigenvalues of A are stable, $\operatorname{Re}\lambda_i \leqslant -\varkappa < 0$, we would have

$$\|\Lambda_\varepsilon\| < \mathrm{const}\cdot\exp(-\varkappa/\sqrt{\varepsilon}) \to 0,$$

so the lemma would be proved.

In the general case, if we again denote $\varkappa = -\max_j \operatorname{Re}\lambda_j > 0$, where λ_j are eigenvalues of the matrix $A_0 = A(0,0)$, then by the Lyapunov lemma there exists a positive quadratic form W on $\mathbb{R}^{n-1}$ such that

$$L_{A_0 y} W = (\operatorname{grad} W, A_0 y) \leqslant -\frac{\varkappa}{2}\, W.$$

Choosing ε and δ small enough, one can guarantee the inequality

$$L_{A(x,\varepsilon)y} W = (\operatorname{grad} W, A(x,\varepsilon)\, y) \leqslant -\frac{\varkappa}{4}\, W,$$

which implies that

$$W(Y_\varepsilon(t)\, y) \leqslant \exp\Big(-\frac{\varkappa}{4}\, t\Big) W(y),$$

hence

$$W(Y_\varepsilon(T_\varepsilon)\, y) \leqslant \exp\Big(-\frac{\varkappa}{4}\, T_\varepsilon\Big) W(y).$$

1.5. The regular part and the Poincaré map. The genericity assumption 2 implies that for δ small enough the homoclinic orbit γ intersects the section Γ^+ at an inner point of the latter. Denote this point by p. The intersection of γ with Γ^- is the point $(\delta, 0)$, which we denote by q. The curve γ reaches p from q in finite time. So by the general theorem on the smooth dependence of solutions of ODE's on initial conditions and parameters, there exists a neighborhood $V \subseteq \Gamma^-$

of q on the cross-section Γ^- such that the map $\Delta_\varepsilon^{\text{reg}}: V \to \Gamma^+$ along the phase curves of the vector field X_ε is smooth and well defined for all sufficiently small ε (see Figure 4.2). Let $L > 0$ be a constant such that

$$\left\| \frac{\partial \Delta_\varepsilon^{\text{reg}}}{\partial y} \right\| < L.$$

For small ε, by virtue of Lemma 4.1.2, $\Delta_\varepsilon^{\text{sing}}(\Gamma^+) \subset V$, hence the Poincaré map

$$\Delta_\varepsilon = \Delta_\varepsilon^{\text{reg}} \circ \Delta_\varepsilon^{\text{sing}}: \Gamma^+ \to \Gamma^+$$

is well defined, and

$$\|D\Delta_\varepsilon\| = \|D\Delta_\varepsilon^{\text{reg}} \cdot D\Delta_\varepsilon^{\text{sing}}\| \leqslant L\|\Lambda_\varepsilon\| \to 0 \quad \text{as } \varepsilon \to 0+.$$

On the other hand, the region Γ^+ is convex, therefore the Lipschitz constant of Δ_ε tends to zero as $\varepsilon \to 0$. Thus we conclude that the Poincaré map Δ_ε has a unique stable fixed point. This proves Theorem 1.1.

§2. Lemma on the hyperbolicity of the product of linear maps

In this section we prove an important lemma on the hyperbolicity of the product of linear maps, which links the hyperbolic theory with normal form theory and allows us to prove numerous bifurcation results.

2.1. Linearization. The monodromy map for a homoclinic orbit admits a representation as the product of a regular smooth part associated with the "distant part" of the homoclinic loop, and a "singular part" associated with a small neighborhood of the singular point at the vertex of the loop. The singular part exhibits strong hyperbolicity: the rate of contraction along the stable manifold and the rate of expansion along the unstable one tend to infinity as the parameter approaches the bifurcation value. On the other hand, we have no definite information about the regular part except for the uniform boundedness of its derivative. It turns out that, under some transversality conditions, strong hyperbolicity overcomes any bounded term when composed with it, and the product will be hyperbolic again (and even strongly hyperbolic in the above sense). Below we prove this fact. Clearly, it is sufficient to prove it for linear maps, since the nonlinear case reduces to the linear one by computing derivatives using the chain rule.

2.2. The cone condition and statement of the model lemma. Let μ_h, μ_v be two positive constants satisfying the inequality $\mu_h \mu_v < 1$. Consider a decomposition $\mathbb{R}^n = \mathbb{R}^s \oplus \mathbb{R}^u$ and the corresponding decomposition of vectors in $\mathbb{R}^n$: $\xi = (\xi^-, \xi^+)$. Let

$$K^+ = \{|\xi^-| \leqslant \mu_v |\xi^+|\}, \qquad K^- = \{|\xi^+| \leqslant \mu_h |\xi^-|\}.$$

Consider an invertible linear operator $A: \mathbb{R}^n \to \mathbb{R}^n$. Denote: $\eta = A\xi$. We say that A satisfies (μ_h, μ_v) *cone condition*, if

1) $AK^+ \subset K^+$,
2) $A^{-1}K^- \subset K^-$,

and there exists a $\lambda > 1$ such that

3) $|\eta^+| \geqslant \lambda |\xi^+|$ for any $\xi \in K^+$,
4) $|\xi^-| \geqslant \lambda |\eta^-|$ for any $\eta \in K^-$.

LEMMA 2.1 (Model Lemma). *Consider two invertible linear operators from the space $\mathbb{R}^n$ to itself given by block matrices (corresponding to the above decomposition) of the form*

$$H = \begin{pmatrix} a & b \\ c & d \end{pmatrix}, \qquad \Sigma = \begin{pmatrix} \Lambda & 0 \\ 0 & M \end{pmatrix}.$$

Let $d\colon \mathbb{R}^u \to \mathbb{R}^u$ be invertible, and

$$\|H\| < L, \qquad \|H^{-1}\| < L, \tag{2.1}$$

$$\|d^{-1}\| < L. \tag{2.2}$$

Then for any $0 < \mu_h < \mu_v^{-1} \ll 1$ (provided that μ_v is greater than some constant depending on L only) there exists a positive δ such that under the assumptions

$$\|M^{-1}\| < \delta, \tag{2.3}$$

$$\|\Lambda\| < \delta \tag{2.4}$$

the linear map $A = H\Sigma$ satisfies the (μ_h, μ_v) cone condition with $\lambda = 3$.

Lemma 2.1 is proved in the next two subsections. Inequalities (2.3) and (2.4) mean that the operators Σ and Σ^{-1} are strong expansions in the spaces $\mathbb{R}^u$ and $\mathbb{R}^s$, respectively. The smaller is δ, the stronger are the expansions, and the choice of δ is arbitrary. Inequality (2.1) means that the operator H is of relatively weak distortion as compared to these expansions, so the strong hyperbolicity of Σ overcomes. Inequality (2.2) is the principal one; it means that the operator H does not merge the contracting and the expanding subspaces of Σ too much.

2.3. Action of the operator A. It is convenient to study the properties of the operators A and A^{-1} separately. Propositions 2.1 and 2.2 below provide not only the proof of Lemma 2.1, but of its generalization in §5.6 as well.

PROPOSITION 2.1. *Suppose the operators H and Σ are of the same form as in Lemma 2.1, satisfy conditions (2.1), (2.2), (2.3), and*

$$\|\Lambda\| \leqslant L. \tag{2.5}$$

Then for μ_v large enough and depending on L, for any $\mu_h < 1/\mu_v$ and for δ sufficiently small and depending on L, μ_v, and μ_h, the operator A satisfies assumptions 1 and 3 of the (μ_h, μ_v) cone condition.

PROOF. Before giving the formal proof, let us illustrate this statement (see Figure 4.3a).

Choosing δ, we control the expansion of the operator Σ in K^+. In particular, the smaller is δ, the narrower is the image ΣK^+, which is stretched along the plane $\mathbb{R}^u$. The operator H brings this narrow image into some cone that contains the plane $H\mathbb{R}^u$. The position of this plane is controlled by (2.2): no unit vector from $\mathbb{R}^u$ can be transformed by H to a vector with a plus-component smaller than L^{-1}. Choosing μ_v large enough, we guarantee the invariance of the cone K^+ under A.

COMMENTARY. The following remark is crucial for the proof of the lemma, as well as that of similar statements in the next chapters.

If $\|B\| \leqslant L$, and B is invertible, then for any ξ, we have

$$|B^{-1}\xi| \geqslant L^{-1}|\xi|.$$

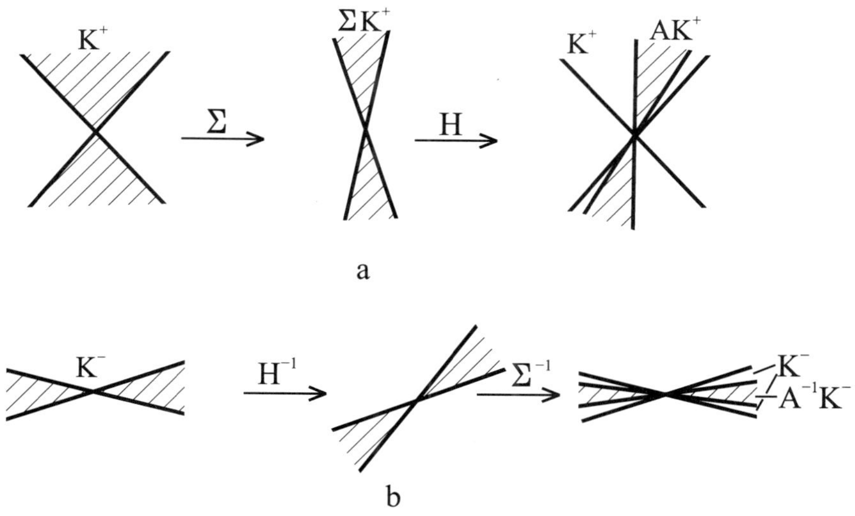

FIGURE 4.3. (a) Action of $A = H\Sigma$. (b) Action of $A^{-1} = \Sigma^{-1}H^{-1}$.

Let us now pass to the formal proof of Proposition 2.1. Let $\xi \in K^+$, $\eta = A\xi$. Then $|\xi^-| \leqslant \mu_v|\xi^+|$. Moreover, by (2.3) and (2.5),

$$|M\xi^+| \geqslant \delta^{-1}|\xi^+|, \qquad |\Lambda\xi^-| \leqslant L|\xi^-|.$$

Hence, taking the inequality $|\xi^+| = |M^{-1}M\xi^+| < \delta|M\xi^+|$ into account, we obtain:

$$|\eta^+| \geqslant |dM\xi^+| - |c\Lambda\xi^-| \geqslant L^{-1}|M\xi^+| - L^2\mu_v|\xi^+| \geqslant (L^{-1} - L^2\mu_v\delta)|M\xi^+|.$$

Consequently,

$$|\eta^+| \geqslant C|M\xi^+|, \quad \text{where } C = L^{-1} - \delta L^2\mu_v. \tag{2.6}$$

Similarly,

$$|\eta^-| \leqslant |a\Lambda\xi^-| + |bM\xi^+| \leqslant L^2\mu_v|\xi^+| + L|M\xi^+| \leqslant (L^2\mu_v\delta + L)|M\xi^+|.$$

Consequently,

$$|\eta^-| \leqslant D|M\xi^+|, \quad \text{where } D = L^2\mu_v\delta + L. \tag{2.7}$$

The invariance of K^+ under A is implied by the choice of $\mu_v \geqslant 2L^2$. After that a small $\delta = \delta(\mu_v, L)$ is chosen so that

$$D/C \leqslant 2L^2 \leqslant \mu_v.$$

Together with (2.6) and (2.7), this implies $\eta \in K^+$. Hence $AK^+ \subset K^+$.

Now assumption 3 of the (μ_h, μ_v) cone condition follows if we choose $\lambda = 3$ and δ so small that

$$\delta^{-1}C > \delta^{-1}L^{-1}/2 > 3.$$

This proves Proposition 2.1.

2.4. Action of the inverse operator. The action in the title is illustrated by Figure 4.3b. Namely, the cone K^- is moved off itself, but Σ^{-1}, being a strong expansion in the minus-direction and a moderate distortion in the plus-direction, returns the image $H^{-1}K^-$ inside K^-.

PROPOSITION 2.2. *Let H and Σ be the operators of the same form as in Lemma* 2.1. *Assume that they satisfy conditions* (2.1), (2.2), (2.4), *and*

$$\|M^{-1}\| \leqslant L. \tag{2.8}$$

Then for μ_v large enough and depending on L, for any $\mu_h < 1/\mu_v$ and for δ sufficiently small and depending on L, μ_v, and μ_h, the operator $A^{-1} = \Sigma^{-1}H^{-1}$ satisfies assumptions 2 *and* 4 *of the (μ_h, μ_v) cone condition.*

PROOF. Let $\eta \in K^-$, $\xi = A^{-1}\eta$. Then $|\eta^+| \leqslant \mu_h|\eta^{-1}|$. Moreover, by (2.4) and (2.8), we have

$$|\Lambda^{-1}\eta^-| \geqslant \delta^{-1}|\eta^-|, \qquad |M^{-1}\eta^+| \leqslant L|\eta^+|.$$

Let

$$H^{-1} = \begin{pmatrix} \bar{a} & \bar{b} \\ \bar{c} & \bar{d} \end{pmatrix}.$$

Below, at the beginning of 2.6, we will prove that

$$\bar{a}^{-1} = a - bd^{-1}c. \tag{2.9}$$

Hence $\|\bar{a}^{-1}\| \leqslant l = L + L^3$. Let us now estimate $|\xi^-|$ from below and $|\xi^+|$ from above. We have

$$|\xi^-| \geqslant |\Lambda^{-1}\bar{a}\eta^-| - |\Lambda^{-1}\bar{b}\eta^+| \geqslant \delta^{-1}(l - L\mu_h)|\eta^-|.$$

Consequently, $|\xi^-| \geqslant \delta^{-1}\bar{C}|\eta^-|$, where $\bar{C} = l - L\mu_h$. Similarly,

$$|\xi^+| \leqslant |M^{-1}\bar{c}\eta^-| + |M^{-1}\bar{d}\eta^+| \leqslant L^2(1 + \mu_h)|\eta^-|.$$

Consequently, $|\xi^+| \leqslant \bar{D}|\eta^-|$, where $\bar{D} = L^2(1 + \mu_h)$. The invariance of K^- under A^{-1} is implied by the choice of arbitrary $\mu_h < 1/\mu_v$ and sufficiently small δ, depending on μ_h and L. Proposition 2.2 is proved.

Propositions 2.1, 2.2 together imply Lemma 2.1.

REMARK. Denote by K_ε^+, K_ε^- the two cones defined just like K^+ and K^-, but with μ_v, μ_h replaced by $\mu_v(1-\varepsilon)$, $\mu_h(1-\varepsilon)$. In fact we have proved that by taking arbitrary $\varepsilon < 1$ and large μ_v, then small μ_h, then small δ, we can ensure that

$$AK^+ \subset K_\varepsilon^+, \qquad A^{-1}K^- \subset K_\varepsilon^-. \tag{2.10}$$

2.5. Generalization to the non-block-diagonal case.

LEMMA 2.2. *Consider two invertible linear operators from the space $\mathbb{R}^s \oplus \mathbb{R}^u$ to itself given by block matrices (corresponding to the above decomposition)*

$$H = \begin{pmatrix} a & b \\ c & d \end{pmatrix}, \qquad \Sigma = \begin{pmatrix} 1 & B \\ 0 & 1 \end{pmatrix}\begin{pmatrix} \Lambda & 0 \\ 0 & M \end{pmatrix}\begin{pmatrix} 1 & 0 \\ D & 1 \end{pmatrix}.$$

Here 1 *is the unit matrix having the same size as the block it is placed in. Suppose that $d\colon \mathbb{R}^u \to \mathbb{R}^u$ is invertible and assumptions* (2.1), (2.2) *hold. Then for any*

$0 < \mu_h < \mu_v^{-1} \ll 1$ (μ_v^{-1} is smaller than some constant depending on L only) there exists a positive δ such that under assumptions (2.3), (2.4), and

$$\|B\| < L, \quad \|D\| < \delta, \quad \|cB\| < \delta \tag{2.11}$$

the linear map $A = H\Sigma$ satisfies the (μ_h, μ_v) cone condition with $\lambda = 2$.

PROOF. Let us first prove the lemma for $D = 0$. Denote

$$H_1 = H\begin{pmatrix} 1 & B \\ 0 & 1 \end{pmatrix} = \begin{pmatrix} a' & b' \\ c' & d' \end{pmatrix}, \qquad \mathcal{B} = \begin{pmatrix} 1 & B \\ 0 & 1 \end{pmatrix}.$$

We will check that H_1 satisfies conditions (2.1) and (2.2) for H. First we check (2.1). For L large enough we have $\|H\| \leqslant L$, $\|\mathcal{B}\| \leqslant 2L$, hence $\|H_1\| \leqslant 2L^2$. Similarly, since

$$\mathcal{B}^{-1} = \begin{pmatrix} 1 & -B \\ 0 & 1 \end{pmatrix},$$

we have $\|H_1^{-1}\| \leqslant 2L^2$.

Now let us check (2.2) for d'. We have

$$d'^{-1} = (d + cB)^{-1} = (1 + d^{-1}cB)^{-1}d^{-1}.$$

By (2.11), for δ small enough,

$$\|d'^{-1}\| \leqslant L(1 - L\delta)^{-1} \leqslant 2L.$$

This verifies conditions (2.1) and (2.2) for H_1.

The matrices H_1 and $\Sigma_0 = \begin{pmatrix} \Lambda & 0 \\ 0 & M \end{pmatrix}$ satisfy the condition of Lemma 2.1 for H and Σ, respectively. Lemma 2.1 implies Lemma 2.2 for the case $D = 0$.

Let us now prove the lemma in the general situation. Denote

$$A_1 = H\begin{pmatrix} 1 & B \\ 0 & 1 \end{pmatrix}\begin{pmatrix} \Lambda & 0 \\ 0 & M \end{pmatrix}.$$

Then

$$A = A_1\mathcal{D}, \qquad \mathcal{D} = \begin{pmatrix} 1 & 0 \\ D & 1 \end{pmatrix}.$$

The matrix A_1 satisfies the cone condition with some constants μ_h, μ_v by the first part of the proof.

Let K^+, K^- be the corresponding cones. Let K_ε^+ and K_ε^- be the same cones as in the remark at the end of 2.4. If δ in (2.11) is small enough, then

$$\mathcal{D}K_\varepsilon^+ \subset K^+, \qquad \mathcal{D}^{-1}K^- \subset K_\varepsilon^+.$$

By (2.10),

$$A_1K^+ \subset K_\varepsilon^+, \qquad A_1^{-1}K^- \subset K_\varepsilon^-.$$

This implies assumptions 1 and 3 of the $(\mu_v(1-\varepsilon), \mu_h(1-\varepsilon))$ cone condition for A.

Suppose that assumptions 2 and 4 of the (μ_h, μ_v) cone condition hold for A_1 with $\lambda = 3$. Then for a sufficiently small δ in (2.11), assumptions 2 and 4 of the $(\mu_v(1-\varepsilon), \mu_h(1-\varepsilon))$ cone condition hold for A with $\lambda = 2$.

This proves Lemma 2.2.

2.6. Main lemma.

REMARK. A decomposition for a block matrix Σ required in Lemma 2.2 may be easily obtained. Indeed, let

$$\Sigma = \begin{pmatrix} \Lambda & B \\ D & M \end{pmatrix}.$$

Then

$$\Sigma = \begin{pmatrix} 1 & B' \\ 0 & 1 \end{pmatrix} \begin{pmatrix} \Lambda' & 0 \\ 0 & M \end{pmatrix} \begin{pmatrix} 1 & 0 \\ D' & 1 \end{pmatrix}, \tag{2.12}$$

where

$$B' = BM^{-1}, \quad D' = M^{-1}D, \quad \Lambda' = \Lambda - BM^{-1}D.$$

On the other hand,

$$\Sigma^{-1} = \begin{pmatrix} \bar{\Lambda} & \bar{B} \\ \bar{D} & \bar{M} \end{pmatrix}$$

with $\bar{\Lambda}^{-1} = \Lambda'$. The last formula proves (2.9).

Together with Lemma 2.2, this remark implies

LEMMA 2.3. *Consider two invertible linear operators in the space* $\mathbb{R}^s \oplus \mathbb{R}^u$ *given by the block matrices* (*corresponding to the above decomposition*)

$$H = \begin{pmatrix} a & b \\ c & d \end{pmatrix}, \qquad \Sigma = \begin{pmatrix} \Lambda & B \\ D & M \end{pmatrix}$$

with $d\colon \mathbb{R}^u \to \mathbb{R}^u$, $M\colon \mathbb{R}^u \to \mathbb{R}^u$ *such that* $|\det d| \cdot |\det M| \neq 0$. *Let* $L > 0$ *be a constant such that*

$$\|H\| < L, \qquad \|H^{-1}\| < L, \tag{2.1}$$

$$\|d^{-1}\| < L, \tag{2.2}$$

$$\|BM^{-1}\| < L. \tag{2.13}$$

Then for any $0 < \mu_h < \mu_v^{-1} \ll 1$ (μ_v^{-1} *is smaller than some constant depending on* L *only*) *there exists a positive* δ *depending on* L, μ_h, μ_v *such that under the assumptions*

$$\|M^{-1}\| < \delta, \tag{2.3}$$

$$\|\Lambda - BM^{-1}D\| < \delta, \tag{2.14}$$

$$\|M^{-1}D\| < \delta, \tag{2.15}$$

$$\|cBM^{-1}\| < \delta \tag{2.16}$$

the linear map $A = H\Sigma$ *satisfies the* (μ_h, μ_v) *cone condition with* $\lambda = 2$.

Indeed, (2.14) implies (2.4) in the notation of Lemma 2.2; relations (2.13), (2.15), and (2.16) together imply (2.11).

Lemma 2.3 will be the principal statement referred to when we check the cone condition for the Poincaré map. It will be applied to the case when the Poincaré map is equal to the product of regular and singular factors whose derivatives play the role of H and Σ, respectively.

The following proposition is necessary for the so-called *partially hyperbolic theory* sketched in §8.5 and used for the study of the bifurcation of several homoclinic surfaces of a saddlenode cycle.

PROPOSITION 2.3. *Let H and Σ have the same form as in Lemma 2.3. Let assumptions* (2.1), (2.2), (2.13), (2.15), *and* (2.16) *hold. Suppose that*

a) (2.3) *holds and we have*

$$\|\Lambda - BM^{-1}D\| \leqslant L, \tag{2.17}$$

or

b) (2.14) *holds and we have*

$$\|M^{-1}\| \leqslant L. \tag{2.18}$$

Then in case a) *the operator $A = H\Sigma$ satisfies assumptions* 1 *and* 3 *of the cone condition, and in case* b), *assumptions* 2 *and* 4. *In more detail, for any sufficiently large μ_v depending on L, any $\mu_h < 1/\mu_v$ and any sufficiently small δ depending on L, μ_v, and μ_h, the operator A satisfying the above hypothesis, at the same time satisfies assumptions* 1 *and* 3 *of the (μ_h, μ_v) cone condition with $\lambda = 2$ in case* a) *and* 2 *and* 4 *in case* b).

PROOF. This proposition reduces to Proposition 2.1 in case a) and to 2.2 in case b) in the very same way as the main lemma (Lemma 2.3) reduces to Lemma 2.2.

§3. Homoclinic orbit of a saddlenode: the case of a saddle hyperbolic part

A vector field with a saddlenode singular point possessing a saddle hyperbolic part may have any finite number r or even an infinite number of homoclinic orbits (see Figure 4.4 for $r = 1$ and Figure 4.11 for $r = 2$). These vector fields occur in generic one-parameter families. The cases $r = 1$ and $r > 1$ give rise to two different types of phenomena. In this section we consider the case $r = 1$.

The techniques developed in this section play a crucial role in the most of our subsequent investigations: §§4, 5 and Chapters 5, 7.

3.1. Creation of a periodic orbit of the saddle type.

THEOREM 3.1. *Assume that in a generic one-parameter family of vector fields X_ε in $\mathbb{R}^n$, the zero value of the parameter ε corresponds to a vector field X_0 possessing exactly one homoclinic orbit γ of the saddlenode O; we suppose that the point O has s-dimensional stable and u-dimensional unstable manifolds so that $s+u+1 = n$, and $s, u > 0$.*

Then there exists a neighborhood $U \supset \gamma \cup O$ and a neighborhood $V \ni 0$ on the parameter line such that for any $\varepsilon \in V$ on one side of the origin, the nonwandering set of X_ε in U consists of two saddles tending to O as $\varepsilon \to 0$. For $\varepsilon \in V$ on the other side of the origin, the corresponding nonwandering set consists of a unique hyperbolic periodic orbit which has $(s+1)$-dimensional stable and $(u+1)$-dimensional unstable manifolds. This orbit converges to $\gamma \cup O$ as $\varepsilon \to 0$.

3.2. Genericity assumptions. The family X_ε from Theorem 3.1 must satisfy four genericity assumptions. The first three are imposed on the field X_0, and the last one requires transversality of the family X_ε to a certain subset in the jet space. Assumptions 1, 2, and 4 coincide with assumptions 1, 2, 3 for Theorem 1.1, and we recall them briefly.

1. *The singular point O is a nonresonant saddlenode of multiplicity* 2.
2. *The homoclinic orbit γ tends to O along the center manifold as $t \to \pm\infty$.*

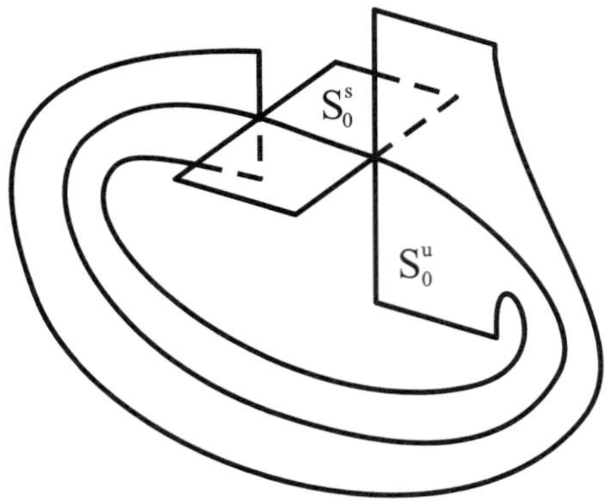

FIGURE 4.4. Saddlenode singular point with a transversal homoclinic orbit.

In order to formulate the third assumption, we need to recall that the stable and unstable *sets* S^s, S^u of a singular point O are sets of points on trajectories tending to O (respectively, tending to O in reverse time). They are $(s+1)$-dimensional (respectively, $(u+1)$-dimensional) manifolds with boundaries W^s (respectively, W^u), both diffeomorphic to the half-space.

3. *The sets S^u and S^s intersect transversally along the orbit γ* (see Figure 4.4).

4. *The family X_ε is transversal at $\varepsilon = 0$ to the subset of vector fields having a degenerate singular point.*

Theorem 3.1 will be proved in the next six subsections. The idea of the proof is roughly the same as in §1: first we use the normal form to study the local bifurcation near the saddlenode. When the singular point disappears, we consider the global bifurcation by studying the Poincaré map along the homoclinic orbit. Lemma 2.1 will be used to prove the hyperbolicity of the Poincaré map. Finally, by using the hyperbolic fixed point theorem, we prove the existence of a fixed point for the Poincaré map.

3.3. Normal form for the perturbation of a saddlenode: the case of a saddle hyperbolic part. The correspondence map. The genericity assumptions 1 and 4 allow us to apply the theorem on the local normal form for the perturbation of the saddlenode (see §5, Chapter 2). A finitely smooth change of coordinates and parameters brings the local family under consideration to the form

$$\begin{aligned} \dot{x} &= (x^2+\varepsilon)(1+a(\varepsilon)x)^{-1}, && x \in \mathbb{R}^1, \\ \dot{y} &= A(x,\varepsilon)y, && y \in \mathbb{R}^s, \\ \dot{z} &= B(x,\varepsilon)z, && z \in \mathbb{R}^u. \end{aligned} \tag{3.1}$$

All the eigenvalues of $A(0,0)$ ($B(0,0)$) have negative (respectively, positive) real parts. This family is studied in the same way as the system (1.1), (1.2). In the case $\varepsilon < 0$, it has two singular points of saddle type. They are the only nonwandering points in the neighborhood of the homoclinic orbit γ (see Figure 4.5a for the picture near the singular point).

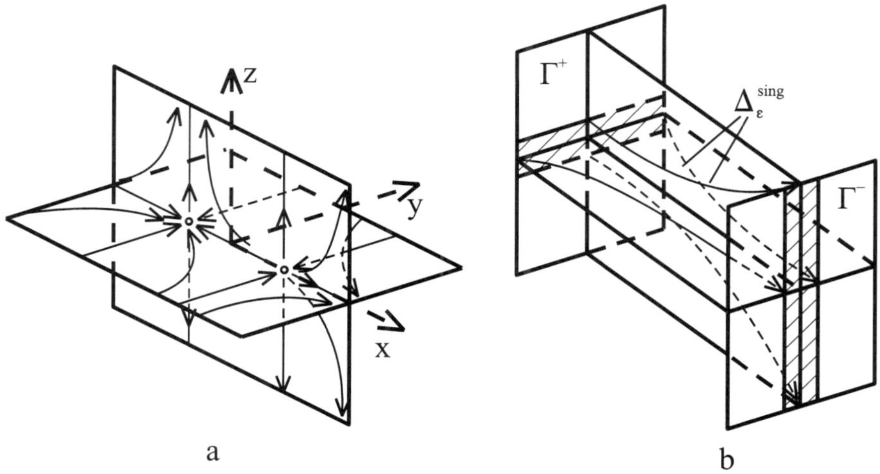

FIGURE 4.5. The local bifurcation of a saddlenode with saddle hyperbolic part.

In the case $\varepsilon > 0$, the family has no singular points at all. In this case the correspondence map $\Delta_\varepsilon^{\text{sing}}$ is well defined, as it is in the case of a nodal hyperbolic part. This map is also linear, but now it has hyperbolic character, being a strong contraction along the stable direction and a strong expansion in the unstable direction. Now we give a formal description related to Figure 4.5b.

In the same way as in 1.4, let

$$\Gamma^+ = \{(x, y, z) \in \mathbb{R}^s \mid x = -\delta,\ |y| \leqslant \delta,\ |z| \leqslant \delta\},$$
$$\Gamma^- = \{(x, y, z) \in \mathbb{R}^s \mid x = +\delta,\ |y| \leqslant \delta,\ |z| \leqslant \delta\},$$

Γ^- being the shift of Γ^+ by 2δ along the x-axis. The sets $\Gamma^\pm$ are the entrance and the exit cross-sections for the vector field (3.1) near the saddlenode.

LEMMA 3.1. *The singular correspondence map for system* (3.1) *is linear and splits into two blocks*:

$$\Delta_\varepsilon^{\text{sing}} \begin{pmatrix} y \\ z \end{pmatrix} = \begin{pmatrix} \Lambda_\varepsilon & 0 \\ 0 & M_\varepsilon \end{pmatrix} \cdot \begin{pmatrix} y \\ z \end{pmatrix}.$$

Moreover, $\|\Lambda_\varepsilon\| \to 0$, $\|M_\varepsilon^{-1}\| \to 0$ *as* $\varepsilon \to 0^+$.

PROOF. The first statement is an immediate consequence of the linearity of system (3.1) in y and z and the invariance of the planes $\mathbb{R}_x \times \mathbb{R}_y^s$ and $\mathbb{R}_x \times \mathbb{R}_z^u$. The second assertion is obtained by applying Lemma 1.2 to the restriction of system (3.1) to the plane (x, y) and (x, z) respectively (the second restriction must be considered in reverse time).

3.4. The regular map and a heuristic proof of the theorem. Now that we have an explicit description of the correspondence map, we can give a geometric "explanation" of Theorem 3.1. Let p and q be the intersection points

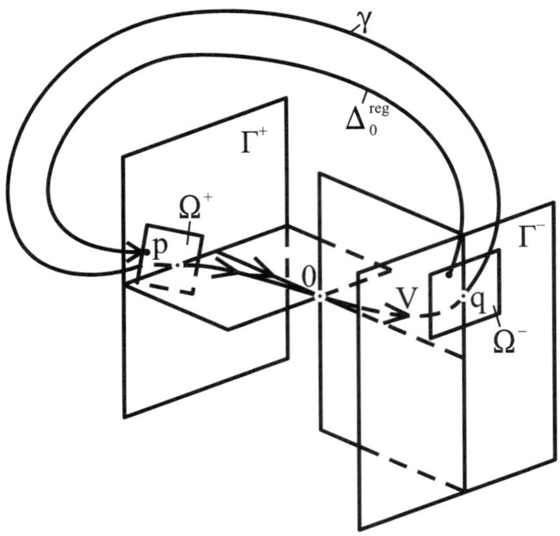

FIGURE 4.6. The intersection of the homoclinic orbit with the transversal planes.

of the homoclinic orbit γ with the interiors of Γ^+ and Γ^- respectively. By the genericity assumption 2 from 3.2, these points are well defined for small δ (see Figure 4.6). Moreover, the regular correspondence map along the phase curves of the field X_ε running near γ from Γ^- to Γ^+ is also well defined. Denote it by $\Delta_\varepsilon^{\text{reg}}$. It takes q to p (for $\varepsilon = 0$) and hence there exist two neighborhoods $\Omega^- = (\Gamma^-, q)$ and $\Omega^+ = (\Gamma^+, p)$ which are mapped one onto another by this map:

$$\Delta_\varepsilon^{\text{reg}} \colon \Omega^- \to \Omega^+$$

as shown in Figure 4.7 (Ω^+ depends on ε).

In order to show the geometric effects without struggling with technical difficulties, suppose for a moment that the map $\Delta_\varepsilon^{\text{reg}}$ is simply a translation in the normalizing charts on $\Gamma^\pm$.

Define a parameter-dependent domain of the Poincaré map by setting

$$D_\varepsilon = (\Delta_\varepsilon^{\text{sing}})^{-1}(\Omega^-),$$

so that the Poincaré map itself

$$\Delta_\varepsilon = \Delta_\varepsilon^{\text{reg}} \circ \Delta_\varepsilon^{\text{sing}} \colon D_\varepsilon \to \Gamma^+$$

is well defined (see Figure 4.8). In our particular case, this map is affine and has a hyperbolic fixed point O. This completes the heuristic investigation.

Now we proceed with the general case. The only difference with the previous model analysis is that the regular map is no longer linear. The picture will be qualitatively preserved, although slightly distorted and more detailed (see Figure 4.9). Our goal is to justify this observation by a formal proof.

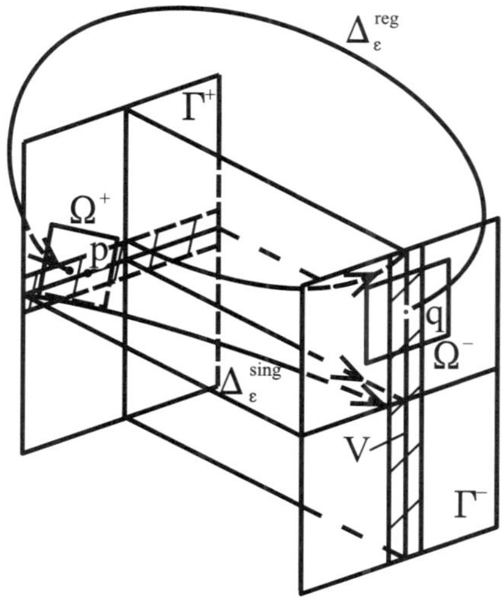

FIGURE 4.7. The regular and singular parts of the Poincaré map.

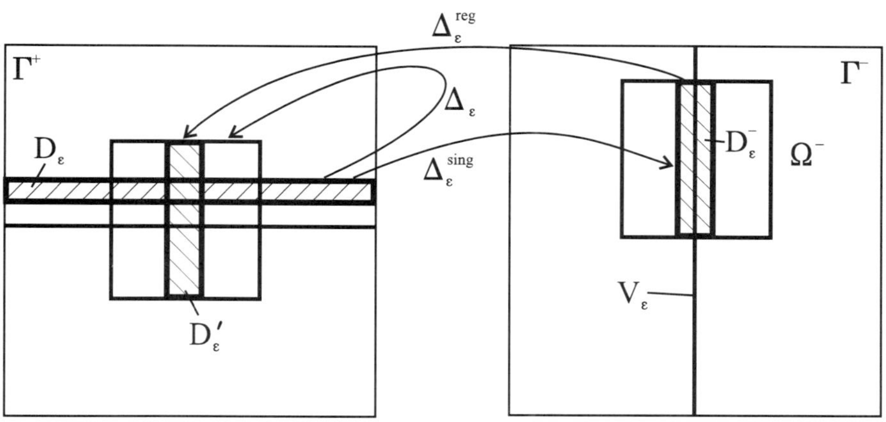

FIGURE 4.8. The Poincaré map in the heuristic proof.

3.5. Hyperbolic fixed point theorem and Poincaré map. Theorem 3.1 will be derived from the hyperbolic fixed point theorem (see Chapter 2, §4). Recall its statement together with the necessary definitions.

A region $B \subset \mathbb{R}^s \oplus \mathbb{R}^u$ is called a *standard rectangle* if $B = D_h \times D_v$, where D_h and D_v are compact connected manifolds with boundary homeomorphic to unit balls in $\mathbb{R}^s$ and $\mathbb{R}^u$ respectively. Denote

$$\partial_h B = D_h \times \partial D_v, \qquad \partial_v B = \partial D_h \times D_v$$

the horizontal and the vertical parts of the boundary of B.

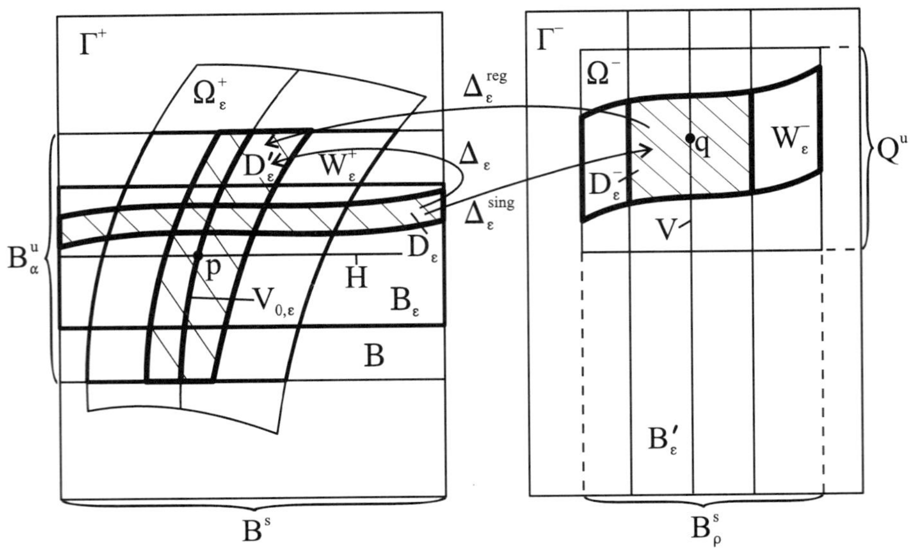

FIGURE 4.9. The Poincaré map in the general case. The same picture shows domains and maps in Theorem 3.3.

THEOREM 3.2 (Hyperbolic Fixed Point Theorem). *Let $B = D_h \times D_v$ be a standard rectangle. Let $D \subset B$ and $D' \subset B$ be v-cylindric and h-cylindric (μ_h, μ_v)-rectangles in B (with projections D_h and D_v, respectively). Let $f\colon D \to D'$ be a map satisfying the (μ_h, μ_v) cone condition. Then f has a unique fixed point O in D:*

$$O = \bigcap_{i=-\infty}^{\infty} f^i D.$$

We will apply Theorem 3.2 to obtain Theorem 3.1. This will be done by means of a general construction that will be used later in establishing the generation of a nontrivial hyperbolic set in §4 and the Birkhoff–Smale theorem in §5. To do this, we formulate a theorem that will be applied in all the three cases (see Figure 4.9).

THEOREM 3.3. *Let Γ^+ and Γ^- be two copies of the Cartesian product of unit balls $B^s \subset \mathbb{R}^s$ and $B^u \subset \mathbb{R}^u$. Let H be the horizontal fiber $B^s \times \{0\} \subset \Gamma^+$ and V the vertical fiber $\{0\} \times B^u \subset \Gamma^-$. Let $p \in H$, $q \in V$, and let $\Delta_\varepsilon^{\mathrm{reg}}$ be a parameter-dependent diffeomorphism of some neighborhood $\Omega^- \subset \Gamma^-$ of q to a neighborhood $\Omega_\varepsilon^+ \subset \Gamma^+$ of p with $\Delta_0^{\mathrm{reg}}(q) = p$ such that*

$$\Delta_0^{\mathrm{reg}}|V \text{ is transversal to } H \text{ at } q. \tag{3.2}$$

Suppose that for any small positive ε there exists:
a standard rectangle $B_\varepsilon \supset H$, vertically cylindric in Γ^+;
a standard rectangle $B'_\varepsilon \supset V$, horizontally cylindric in Γ^-;
a linear map $\Delta_\varepsilon^{\mathrm{sing}}\colon B_\varepsilon \to B'_\varepsilon$ with the following property:

$$\Delta_\varepsilon^{\mathrm{sing}} = \begin{pmatrix} \Lambda_\varepsilon & 0 \\ 0 & M_\varepsilon \end{pmatrix}, \quad \text{where } \|\Lambda_\varepsilon\| \to 0 \text{ and } \|M_\varepsilon^{-1}\| \to 0 \text{ as } \varepsilon \to 0^+. \tag{3.3}$$

Then there exists an $\alpha > 0$ such that for any standard rectangle B of the form

$$B = B^s \times B^u_\alpha, \qquad B^u_\alpha = \{|z| \leqslant \beta < \alpha\},$$

there exist domains D_ε, D'_ε such that the map

$$f = \Delta_\varepsilon = \Delta_\varepsilon^{\text{reg}} \circ \Delta_\varepsilon^{\text{sing}} \colon D_\varepsilon \to D'_\varepsilon$$

and the domains $D_\varepsilon = D$, $D'_\varepsilon = D'$, B for sufficiently small ε satisfy the assumptions of Theorem 3.2. In particular, D is horizontally cylindric and D' is vertically cylindric in B.

Theorems 3.3 and 3.2 imply Theorem 3.1. Indeed, the normalizing coordinates for the Poincaré map allow us to introduce charts on the cross-sections Γ^+, Γ^- so that the map $\Delta_\varepsilon^{\text{sing}}$ from (3.3) satisfies all the related assumptions of Theorem 3.3. The genericity assumptions 2 and 3 imply the hypotheses of Theorem 3.3 for $\Delta_\varepsilon^{\text{reg}}$. Hence, all the conditions of Theorem 3.3 hold for the regular and singular correspondence maps. Then their product Δ_ε satisfies the assumptions of Theorem 3.2. This theorem claims the existence and uniqueness of a hyperbolic fixed point for the map Δ_ε. This proves Theorem 3.1.

Let us now prove Theorem 3.3. This will be done in the next three subsections.

3.6. Constructions of domains and maps in Theorem 3.3. Without loss of generality, the neighborhood Ω^- of q may be chosen as the standard rectangle

$$\Omega^- = B^s_\rho \times Q^u, \quad \text{where } B^s_\rho = \{|y| \leqslant \rho\},\ Q^u = \{|z - z(q)| \leqslant \rho\}.$$

By the transversality assumption (3.2), for all sufficiently small ρ there exists an $\alpha > 0$ such that the image $\Delta_\varepsilon^{\text{reg}}(\{y\} \times Q^u) = V^+_{y,\varepsilon}$ contains the graph $V_{y,\varepsilon}$ of a smooth map

$$v_{y,\varepsilon} \colon B^u_\alpha \to B^s,$$

where $B^u_\alpha = \{|z| \leqslant \alpha\}$ (see Figure 4.9 for $V_{0,\varepsilon}$ and Figure 4.10 for $V_{y',\varepsilon}$). Let $B = B^s \times B^u_\alpha$. Then $V_{y,\varepsilon}$ is a μ_v-vertical surface for a sufficiently large μ_v and any $y \in B^s_\rho$. By the definition of this surface,

$$\partial V_{y,\varepsilon} \subset \partial_h B. \tag{3.4}$$

Let us take $\Omega^+_\varepsilon = \Delta_\varepsilon^{\text{reg}} \Omega^-_\varepsilon$, $W^+_\varepsilon = \Omega^+_\varepsilon \cap B$ (see Figure 4.9). Let $W^-_\varepsilon = (\Delta_\varepsilon^{\text{reg}})^{-1} W^+_\varepsilon$. Let B_ε and B'_ε be the same as in Theorem 3.3. Let

$$D^-_\varepsilon = B'_\varepsilon \cap W^-_\varepsilon, \quad D_\varepsilon = (\Delta_\varepsilon^{\text{sing}})^{-1} D^-_\varepsilon, \quad D'_\varepsilon = \Delta_\varepsilon^{\text{reg}} D^-_\varepsilon,$$

as shown in Figure 4.9. Let $\Delta_\varepsilon = \Delta_\varepsilon^{\text{reg}} \circ \Delta_\varepsilon^{\text{sing}} \colon D_\varepsilon \to D'_\varepsilon$. The desired domains and map are constructed.

3.7. Rectangular structure. Recall the necessary definition.

DEFINITION 2.4.4. A (μ_h, μ_v)-*rectangle* D in $\mathbb{R}^s \oplus \mathbb{R}^u$ is a domain of the form

$$D = F(B^s \times B^u),$$

where B^s and B^u are the unit balls in $\mathbb{R}^s$ and $\mathbb{R}^u$ respectively, F is a homeomorphism $B^s \times B^u \to \mathbb{R}^s \oplus \mathbb{R}^u$ and the surfaces $F(B^s \times \{y\})$ and $F(\{x\} \times B^u)$ are μ_h-horizontal and μ_v-vertical respectively for any $y \in B^u$, $x \in B^s$. The horizontal and vertical parts of the boundary of D are defined as the images under F of the horizontal and vertical parts of the boundary of $B^s \times B^u$ respectively:

$$\partial_h D = F(B^s \times \partial B^u), \qquad \partial_v D = F(\partial B^s \times B^u).$$

LEMMA 3.2. *There exist positive constants μ_h, μ_v, $\mu_h \mu_v < 1$, such that D_ε, D'_ε are (μ_h, μ_v)-rectangles horizontally and vertically cylindric in B respectively.*

PROOF. The proof is illustrated in Figure 4.10.

The constant μ_v will be the same as in 3.6. Choose any μ_h satisfying the inequalities $0 < \mu_h < 1/\mu_v$.

Let us construct the auxiliary horizontal surfaces. The transversality assumption (3.2) implies that $H^- = (\Delta_\varepsilon^{\mathrm{reg}})^{-1} H \cap \Omega^-$ is the graph of a smooth map $h_\varepsilon^- \colon B_\rho^s \to Q^u$ for a small ρ. We define $H_{z,\varepsilon}^- \subset \Gamma^-$ as

$$H_{z,\varepsilon}^- = (\Delta_\varepsilon^{\mathrm{reg}})^{-1}(B^s \times \{z\}), \qquad z \in B_\alpha^u.$$

By the previous statement, this surface is the graph of a smooth map

$$h_{z,\varepsilon}^- \colon B_\rho^s \to Q^u, \qquad h_{0,\varepsilon}^- = h_\varepsilon^-$$

for sufficiently small ρ and α. Let $H'_{z,\varepsilon} = H_{z,\varepsilon}^- \cap B'_\varepsilon$. The surface $H'_{z,\varepsilon}$ is a manifold with boundary $\partial H'_{z,\varepsilon}$, and

$$\partial H'_{z,\varepsilon} \subset \partial_v B'_\varepsilon \tag{3.5}$$

as shown in Figure 4.10.

The desired auxiliary surfaces are defined as $H_{z,\varepsilon} = (\Delta_\varepsilon^{\mathrm{sing}})^{-1} H'_{z,\varepsilon}$. By (3.5),

$$\partial H_{z,\varepsilon} \subset \partial_v B_\varepsilon \subset \partial_v B. \tag{3.6}$$

The surface $H_{z,\varepsilon}$ is the graph of the map

$$h_{z,\varepsilon} \colon B^s \to B_\alpha^u, \qquad h_{z,\varepsilon} = M_\varepsilon^{-1} \circ h_{z,\varepsilon}^- \circ \Lambda_\varepsilon.$$

The Lipschitz constant of this map tends to zero as $\varepsilon \to 0$ by (3.3). Hence, $H_{z,\varepsilon}$ is a μ_h-horizontal surface for arbitrary $\mu_h > 0$ and sufficiently small ε depending on μ_h.

Now we can introduce the rectangular structure on D_ε and D'_ε. The corresponding maps are

$$\begin{aligned} F_\varepsilon \colon B^s \times B_\alpha^u \to D_\varepsilon, &\qquad (y, z) \mapsto (y, h_{z,\varepsilon}(y)), \\ F'_\varepsilon \colon B^s \times B_\alpha^u \to D'_\varepsilon, &\qquad (y, z) \mapsto (v_{y',\varepsilon}(z), z), \\ &y' = \Lambda_\varepsilon y. \end{aligned}$$

The map F_ε satisfies the assumptions of Definition 2.4.4 for $F = F_\varepsilon$ and ε small enough. Indeed, $F_\varepsilon(\{y\} \times B_\alpha^u) \subset \{y\} \times B^u$. This surface is μ_v-vertical for any μ_v. Further, $F_\varepsilon(B^s \times \{z\}) \subset H_{z,\varepsilon}$. This surface is μ_h-horizontal by the above reason. Finally, $\partial_v D_\varepsilon \subset \partial_v B$ by (3.6). Hence, D_ε is (μ_h, μ_v)-rectangle, vertically cylindric in B.

In the same way

$$F'_\varepsilon(\{y\} \times B_\alpha^u) \subset V_{y,\varepsilon}, \qquad F'_\varepsilon(B^s \times \{z\}) \subset B^s \times \{z\}.$$

The first surface is μ_v-vertical by the choice of μ_v, and the second is μ_h-horizontal for any $\mu_h > 0$. By (3.4), D'_ε is horizontally cylindric in B. This proves Lemma 3.2.

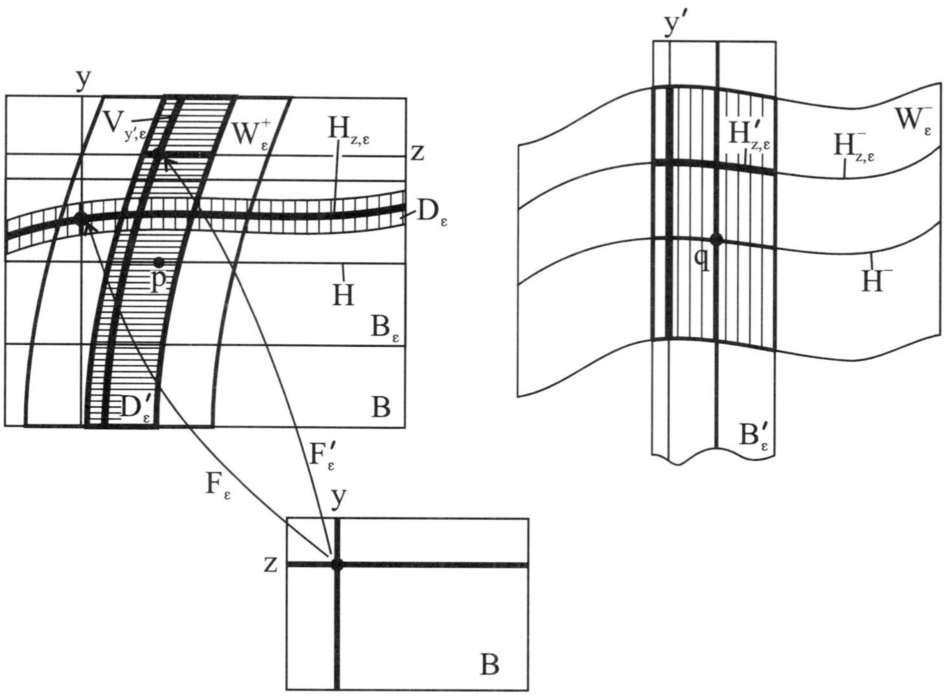

FIGURE 4.10. The rectangular structure on D_ε and D'_ε.

3.8. Cone condition for the principal correspondence (Poincaré) map.

LEMMA 3.3. *The map Δ_ε from Theorem 3.3 satisfies the (μ_h, μ_v) cone condition for appropriate constants μ_h, μ_v, $\mu_h\mu_v < 1$.*

PROOF. The cone condition for the principal correspondence map is implied by Lemma 2.1 and the transversality condition (3.2). Applying the chain rule to the principal correspondence map, we obtain

$$D\Delta_\varepsilon = D\Delta_\varepsilon^{\text{reg}}(\Delta_\varepsilon^{\text{sing}}) \cdot D\Delta_\varepsilon^{\text{sing}}. \tag{3.7}$$

Denote $D\Delta_\varepsilon^{\text{sing}}(u)$ by Σ_u, $\Delta_\varepsilon^{\text{sing}}(u)$ by v and $D\Delta_\varepsilon^{\text{reg}}(v)$ by H_v. Then for all small ε and for any choice of $u \in D_\varepsilon$, the matrices $\Sigma = \Sigma_u$ and $H = H_v$ satisfy the assumptions of Lemma 2.1. Indeed, the splitting $T_uD_\varepsilon = \mathbb{R}^s \oplus \mathbb{R}^u$ determines the block structure of the matrices Σ and H:

$$H = \begin{pmatrix} a & b \\ c & d \end{pmatrix}, \qquad \Sigma = \begin{pmatrix} \Lambda & 0 \\ 0 & M \end{pmatrix}. \tag{3.8}$$

By (3.3), $\Lambda = \Lambda_\varepsilon$ and $M = M_\varepsilon$ satisfy $\|\Lambda_\varepsilon\| \to 0$, $\|M_\varepsilon^{-1}\| \to 0$ as $\varepsilon \to 0$. By the regularity of the map $\Delta_\varepsilon^{\text{reg}}$, there exists a constant L such that inequality (2.1) holds uniformly in ε.

The most interesting condition is (2.2), which reduces to $\|d^{-1}\| \leqslant L$. It is a consequence of the transversality condition for sufficiently small ρ. Indeed, let Ω^-

be the same as at the beginning of 3.6. Define a map $\pi_z\colon \Gamma^+ \to \mathbb{R}^u$, $(y, z) \mapsto z$. Let

$$g_\varepsilon = \pi_z \Delta_\varepsilon^{\mathrm{reg}}|_V \colon V \to \mathbb{R}^u. \tag{3.9}$$

The transversality condition (3.2) means that $\det Dg_0(z(q)) \neq 0$. But

$$Dg_0(z(q)) = d(q)|_{\varepsilon=0},$$

where d is the block from (3.8). Thus $d(q)^{-1}$ exists and, by the regularity assumption, there is an L such that $\|d^{-1}\| \leqslant L$. This inequality holds for all sufficiently small ε everywhere in Ω^- for small ρ.

We have verified all the assumptions of Lemma 2.1. This means that the cone condition holds for the map Δ_ε, and proves Lemma 3.3.

All the assumptions of Theorem 3.2 for Δ_ε follow from Lemmas 3.2 and 3.3. This implies Theorem 3.3.

Theorem 5.2 is proved below in the same way. In this theorem $\varepsilon \to 0$ is replaced by $k \to \infty$, $k \in \mathbb{N}$; the domain $B = B^s \times B^u_\alpha$ is replaced by $B = B_y \times B_\alpha$, where B_y is an arbitrary ball in B^s centered at p.

§4. Several homoclinic orbits of a saddlenode

In this section we consider bifurcations of a saddlenode with $r > 1$ homoclinic orbits. A vector field exhibiting such a degeneracy occurs in typical one-parameter families and belongs to the boundary of the MS-set. Bifurcations of such saddlenodes display the simplest example of an Ω-explosion, a sudden increase of the nonwandering set after a small change of the parameters.

4.1. Generation of a nontrivial invariant hyperbolic set: Ω-explosion.

THEOREM 4.1. *Assume that in a generic one-parameter family of vector fields X_ε in $\mathbb{R}^n$ for $\varepsilon = 0$ the vector field X_0 has $r \geqslant 2$ homoclinic orbits γ_i, $i = 1, \dots, r$, of a saddlenode O at the origin (see Figure 4.11). Denote as usual by s and u the dimensions of the stable and unstable invariant manifolds of O and assume that $s, u > 0$, so that $s + u + 1 = n$.*

Then there exist

- *a neighborhood U of the union of the homoclinic orbits and the singular point,*
- *a neighborhood V of the critical parameter value $\varepsilon = 0$ in the parameter line,*

such that:

1. *For all $\varepsilon \in V$ on one side of zero, the nonwandering set of the field in U consists of two saddles converging to the origin as $\varepsilon \to 0$.*
2. *For all $\varepsilon \in V$ on the other side of zero, the nonwandering set Ω of X_ε in U is hyperbolic, and the restriction $X_\varepsilon|_\Omega$ is topologically equivalent to the suspension over the shift map of the set of bi-infinite sequences with r symbols.*

The genericity assumptions for the family X_ε in Theorem 4.1 are the same as in 3.2. More precisely, the vector field must satisfy conditions 2 and 3 for any homoclinic orbit γ_i. Assumptions 1 and 4 are preserved verbatim.

In the proof we use the following

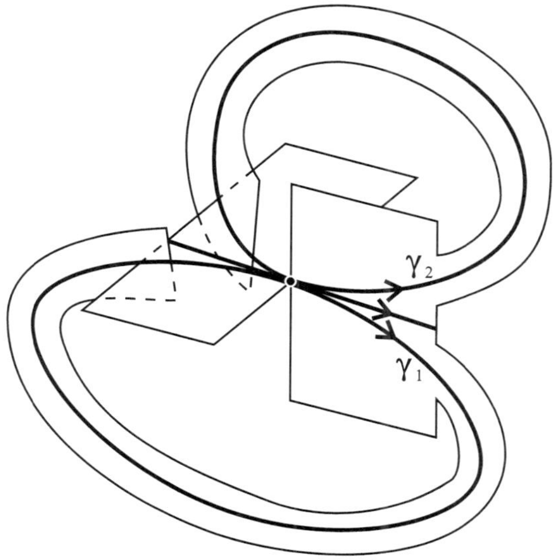

FIGURE 4.11. Several homoclinic orbits of a saddlenode.

THEOREM 4.2 (Smale Horseshoe Existence Theorem). *Let μ_h, μ_v be two positive constants with $\mu_h\mu_v < 1$, and B a standard rectangle. Let $D_i \subset \mathbb{R}^s \oplus \mathbb{R}^u$, $i = 1, \dots, N$, $N > 1$, be N pairwise disjoint (μ_h, μ_v)-rectangles. Let*

$$f \colon D = \bigcup_{i=1}^{N} D_i \to f(D) \subset B$$

be a diffeomorphism of D onto its image satisfying the following assumptions for $i, j = 1, \dots, N$:

1) *the map f satisfies the (μ_h, μ_v) cone condition in D;*
2) *the domains D_i are vertically cylindric and $D_i' = f(D_i)$ are horizontally cylindric in B.*

Then the set

$$\Lambda = \bigcap_{-\infty}^{\infty} f^k D$$

is a hyperbolic invariant Cantor set. The restriction of f to Λ is topologically conjugated to the Bernoulli shift σ on the space Σ^N of sequences of elements taking N different values $1, \dots, N$. This means that there exists a homeomorphism ϕ such that the diagram

$$\begin{array}{ccc} \Lambda & \xrightarrow{\ f\ } & \Lambda \\ {\scriptstyle\phi}\downarrow & & \downarrow{\scriptstyle\phi} \\ \Sigma^N & \xrightarrow{\ \sigma\ } & \Sigma^N \end{array}$$

is commutative.

4.2. Proof of Theorem 4.1. The proof of Theorem 4.1 is mostly contained in the proof of Theorem 3.1; only a few additions are needed. Suppose that $\Gamma^{\pm}$,

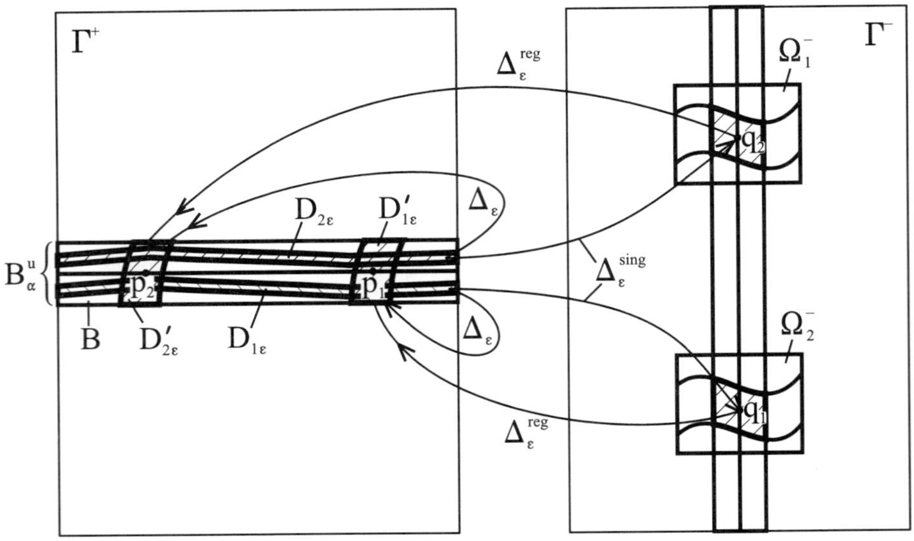

FIGURE 4.12. The Poincaré map of several homoclinic orbits of a saddlenode as a Smale horseshoe map.

y, and z are the same as in 3.3 and $V = \{0\} \times B^u$. Let p_i, q_i be the intersection points of γ_i with Γ^+ and Γ^-, respectively, $q_i \in V$.

For any i we will repeat the constructions of §3 and obtain domains $D_{i\varepsilon}$, $D'_{i\varepsilon}$ and a map Δ_ε that, playing the role of D_i, D'_i, and f, satisfy the assumptions of Theorem 4.1 for small $\varepsilon > 0$ (see Figure 4.12). The only thing that remains to be checked is that the domain B in this construction may be taken independently of i.

In more detail, let Ω_i^- be the pairwise disjoint domains of $\Delta_\varepsilon^{\text{reg}}$ in Γ^- near q_i of the form:

$$\Omega_i^- = \{(y,z) : |y| \leqslant \rho,\ |z - z(q_i)| \leqslant \rho\}.$$

Let $\Delta_\varepsilon^{\text{reg}}$ be a diffeomorphism in Ω_i^-, and $\Omega_{i\varepsilon}^+ = \Delta_\varepsilon^{\text{reg}}\Omega_i^-$. Put

$$V_{i0} = \Delta_\varepsilon^{\text{reg}}(\Omega_i^- \cap V).$$

Suppose ρ in the definition of Ω_i^- is so small that $V_{i0} \ni p_i$ contains the graph of a map v_i of the same ball B_α^u: $|z| \leqslant \alpha$, not depending on i, in B^u. Let $B = B^s \times B_\alpha^u \subset \Gamma^+$. Then the construction of §3 produces domains $D_{i\varepsilon}$ and $D'_{i\varepsilon}$ in B, for each i, which satisfy all the assumptions of Theorem 4.2. This proves Theorem 4.1.

§5. Birkhoff–Smale theorem

In this section we prove the Birkhoff–Smale theorem stated in §4 of Chapter 1, with some additional assumptions allowing us to linearize the Poincaré map of the periodic orbit with a transversal homoclinic curve.

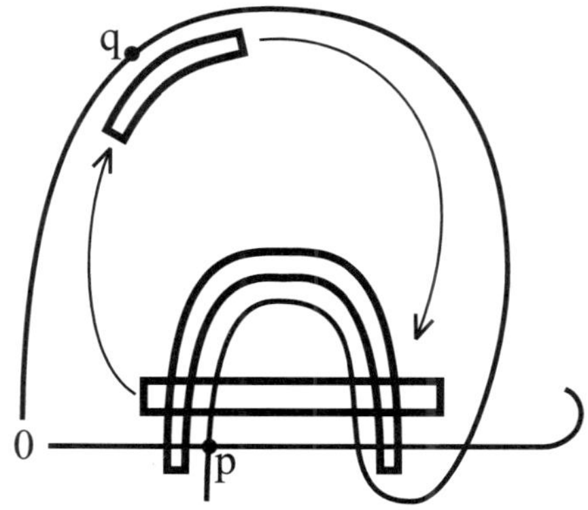

FIGURE 4.13. Heuristic proof of the Birkhoff–Smale theorem.

5.1. Statement of the theorem.

THEOREM 5.1 (Birkhoff–Smale Theorem). *Let X be a smooth vector field in $\mathbb{R}^s$ with a hyperbolic periodic orbit σ. If σ has a transversal homoclinic orbit γ, then X has a nontrivial hyperbolic invariant set in any neighborhood of $\gamma \cup \sigma$.*

The proof is based on the fact that for a cross-section to the periodic orbit σ, one can construct a Poincaré map for the polycycle formed by the union $\sigma \cap \gamma$ whose restriction to some subset is a Smale horseshoe. Heuristically, the construction is shown in Figure 4.13.

The rigorous proof involves some specific techniques. Below we show that we can apply the one already developed in §3.

We will prove the above theorem under the additional genericity assumption that the Poincaré map of σ can be C^1 linearized.

REMARK. A sufficient condition for such linearizability is that the fixed point of the Poincaré map is nonresonant. This is a consequence of the Sternberg theorem proved in Chapter 9. In fact, a stronger result is true. The germ of a smooth map at the hyperbolic fixed point can be C^1-linearized if its multipliers λ_i, $i = 1, \dots, n$, satisfy the inequalities

$$|\lambda_i| \neq |\lambda_j| \cdot |\lambda_k| \quad \text{for } |\lambda_j| < 1 < |\lambda_k|, \ i, j, k \in \{1, \dots, n\}.$$

See [**Go**]. The last result is beyond the scope of this book.

Theorem 5.1 is derived below from the Smale horseshoe existence theorem (Theorem 4.2). To check the conditions of this theorem, we make use of the following analog of Theorem 3.3 shown in Figure 4.14.

THEOREM 5.2. *Let Γ be the Cartesian product of the unit balls $B^s \subset \mathbb{R}^s$ and $B^u \subset \mathbb{R}^u$. Let H be the horizontal fiber $B^s \times \{0\} \subset \Gamma$ and V the vertical fiber $\{0\} \times B^u$. Let $p \in H$, $q \in V$, and let Δ^{reg} be a diffeomorphism of some neighborhood $\Omega^- \subset \Gamma$ of q onto a neighborhood $\Omega^+ \subset \Gamma$ of p, with $\Delta^{\text{reg}}(q) = p$ and*

$$\Delta^{\text{reg}}|_V \ \textit{transversal to } H \textit{ at } q. \tag{5.1}$$

Suppose that for any large $k \in \mathbb{N}$ there exist:

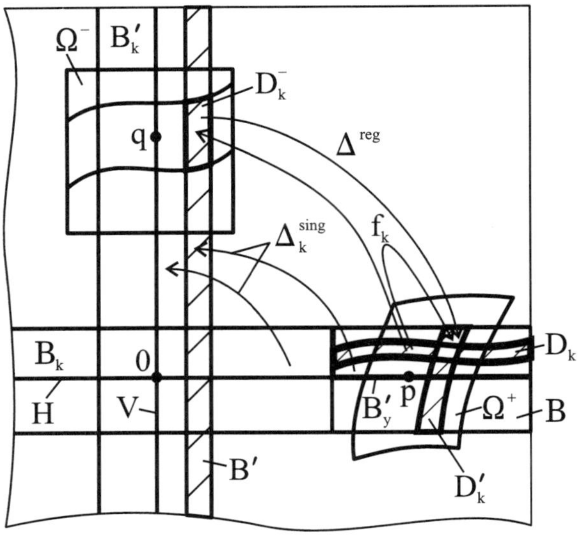

FIGURE 4.14. Construction of domains for the horseshoe map.

a standard rectangle $B_k \supset H$ *vertically cylindric in* Γ;
a standard rectangle $B'_k \supset V$ *horizontally cylindric in* Γ;
a linear map $\Delta_k^{\text{sing}}\colon B_k \to B'_k$ *given by a block matrix*

$$\Delta_k^{\text{sing}} = \begin{pmatrix} \Lambda_k & 0 \\ 0 & M_k \end{pmatrix} \tag{5.2}$$

such that $\|\Lambda_k\| \to 0$, $\|M_k^{-1}\| \to 0$ *as* $k \to \infty$.

Then there exists an $\alpha > 0$ *such that for any ball* $B_y \subset H$ *containing* p *in its interior, any* $\beta < \alpha$ *and* $B = B_y \times B^u_\beta$, *where* $B^u_\beta = \{z : |z| \leqslant \beta\}$, *there exist domains* D_k, D'_k *for which the map*

$$f_k = \Delta^{\text{reg}} \circ \Delta_k^{\text{sing}}\colon D_k \to D'_k$$

and the domains $D = D_k$, $D' = D'_k$, B *satisfy the assumptions of Theorem* 3.2 *for any sufficiently large* k.

The proof is practically the same as for Theorem 3.3. It is illustrated in Figure 4.14.

The double application of Theorem 5.2 yields the situation described in the Smale horseshoe existence theorem (see Figure 4.15).

5.2. Checking the assumptions of Theorem 5.2. By assumption, the Poincaré map of the periodic orbit σ is C^1-linearizable. This means that there exists a cross-section Γ of the phase curves together with the chart (y, z) such that

1) the intersection point of the cross-section with the periodic orbit σ corresponds to the origin;
2) the Poincaré map P is a linear one:

$$P(y, z) = (Ay, Bz), \tag{5.3}$$

where $\|A\|, \|(B)^{-1}\| < \lambda < 1$.

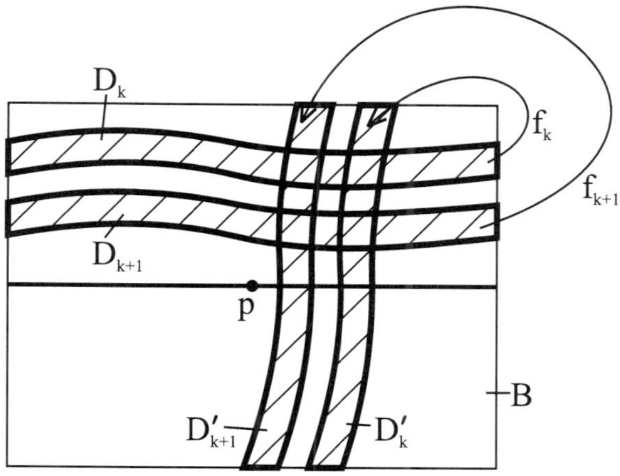

FIGURE 4.15. Smale horseshoe map generated by a homoclinic orbit. Only the domains inside B are shown.

Take points $p \in H$ and $q \in V$ on the homoclinic orbit of σ. Then a germ of a diffeomorphism, the correspondence map along the orbits running outside some neighborhood of σ, is well defined:

$$\Delta^{\mathrm{reg}} : (\Gamma, q) \to (\Gamma, p).$$

By the transversality assumption in Theorem 5.1, the hypothesis (5.1) of Theorem 5.2 holds. Let $\Delta_k^{\mathrm{sing}} = P^k$. Then the assumption on B_k, B'_k, and Δ_k^{sing} follows from (5.3). Therefore, all the assumptions of Theorem 5.2 hold. Hence, for any sufficiently large k there exist domains B, D_k, D'_k and a map $f_k : D_k \to D'_k$ that satisfies the conditions of Theorem 3.3. Namely, D_k and D'_k are vertically cylindric and horizontally cylindric in B respectively; the map $f_k : D_k \to D'_k$ satisfies the cone condition. The domain B does not depend on k. Theorem 5.2 gives some freedom in the choice of B. This freedom will be used in 5.3 below.

Now let $D = D_k \cup D_{k+1}$, $D' = D'_k \cup D'_{k+1}$, and take f equal to f_k on D_k and f_{k+1} on D_{k+1}. The properties of these domains and maps constitute almost all the assumptions of the Smale horseshoe existence theorem (see Figure 4.15). The only assumption remaining to be checked is

$$D_k \cap D_{k+1} = \varnothing, \qquad D'_k \cap D'_{k+1} = \varnothing. \tag{5.4}$$

This assumption is satisfied by a suitable choice of B.

5.3. Checking the assumptions of the Smale horseshoe existence theorem. Choose B in such a way that

$$PB \cap B = 0. \tag{5.5}$$

Let Ω^- be the domain of Δ^{reg}. Choose this domain so small that $P\Omega^- \cap \Omega^- = \varnothing$. The construction that proves Theorem 5.2 yields domains D_k, D_k^-, D'_k with

$$D_k \subset B, \quad D_k^- = P^k D_k \subset \Omega^-, \quad D'_k \subset (\Delta^{\mathrm{reg}})^{-1} D_k^-,$$

as shown in Figure 4.14. By (5.5),

$$P^k B \cap P^{k+1} B = \varnothing.$$

Hence, $D_k^- \cap D_{k+1}^- = \varnothing$. Consequently, the second equality of formula (5.4) holds.

On the other hand, suppose that $\varnothing \neq D_k \cap D_{k+1} \ni a$. Then

$$P^k a \in D_k^- \in \Omega^-, \qquad P^{k+1} a \in D_{k+1}^- \in \Omega^-.$$

Hence $P\Omega^- \cap \Omega^- \ni P^{k+1} a \neq \varnothing$, a contradiction.

This verifies (5.4) and hence all the assumptions of the Smale horseshoe existence theorem for the map f. The Birkhoff–Smale theorem is proved.

CHAPTER 5

Homoclinic Tori and Klein Bottles of Nonhyperbolic Periodic Orbits: Noncritical Case

In this and the next chapters we consider bifurcations of compact homoclinic surfaces of a nonhyperbolic cycle with multiplier one. We begin in §1 by classifying the union of homoclinic orbits and describe the concept of criticality. This chapter studies the noncritical case, and the next one deals with the critical case.

In §2 we use Fenichel's invariant manifold theorem to prove the persistence of noncritical homoclinic tori and Klein bottles. This reduces the study of the simplest bifurcations to two-dimensional problems. In §3, we prove that the rotation number as a function of the parameter of a generic one-parameter family of diffeomorphisms of a circle is a Cantor function. In §4 we use the results of §3 to study global bifurcation of homoclinic tori. In §5 we consider the blue sky catastrophe created by bifurcation of a semistable cycle on a Klein bottle filled with homoclinic orbits of this cycle. In §6 we generalize the Smale horseshoe existence theorem to the *partially hyperbolic* case. In §7 we use the result of §6 to describe the complicated nonhyperbolic invariant set created as a result of bifurcation of several noncritical homoclinic tori or Klein bottles. In §8 we demonstrate the generation of a strange attractor through the bifurcation of a "twisted homoclinic surface".

§1. The topological and smooth structure of the union of homoclinic orbits

1.1. Fundamental assumptions. In this subsection we specify the assumptions that the vector fields studied in this and the next chapter must satisfy. The first four assumptions concern the critical vector field. The last one is related to a transversality property in the space of vector fields.

FUNDAMENTAL ASSUMPTIONS. Let X_ε be a one-parameter family of C^∞ vector fields in $\mathbb{R}^n$, $n \geqslant 3$, such that

1. X_0 has a nonhyperbolic cycle L with multipliers $1, \lambda_1, \dots, \lambda_s, \mu_1, \dots, \mu_u$, $u, s \geqslant 0$, $s + u = n - 2$, satisfying

$$|\lambda_i| < 1 < |\mu_j|, \qquad i \in \{1, \dots, s\},\ j \in \{1, \dots, u\},$$

where the $n-2$ tuple $\{|\lambda_i|, |\mu_j|\}$ is multiplicatively nonresonant in the sense of Definition 2.5.2. Moreover, the cycle L is of multiplicity two, i.e., the Poincaré map of L restricted to its center manifold has the form

$$x \mapsto x + ax^2 + \cdots, \qquad a \neq 0,\ x \in (\mathbb{R}, 0),$$

where the dots denote the terms of higher order. Under this assumption, we call L a *saddlenode cycle*.

2. The union H of the homoclinic orbits of L is nonempty and does not intersect the stable and unstable manifolds of L.
3. The union $H \cup L$ is compact.

To state the last assumption on the unperturbed system, let us recall the definition of stable and unstable set introduced in Chapter 1.

DEFINITION 1.1. The *stable* (*unstable*) *set* of the cycle L is defined as the union of the orbits which tend to L as $t \to \infty$ (resp., $-\infty$).

By assumption 1, the stable (unstable) set of L is an $s+2$ (resp., $u+2$)-dimensional manifold with boundary. The boundary is the stable (resp., unstable) manifold of L. This was mentioned in Chapter 1 and follows from the study of the Poincaré map in 1.2 below.

4. The stable and unstable sets of L intersect along the set H transversally.
5. The family X_ε at $\varepsilon = 0$ is transversal to the hypersurface located in the space of vector fields and consisting of fields with nonhyperbolic cycle close to L. More precisely, if we denote by $f(x,\varepsilon)$ the restriction of the local family of ε-depending Poincaré maps of L to its center manifold, then

$$\left.\frac{\partial f}{\partial \varepsilon}\right|_{(0,0)} \neq 0.$$

In the next part of this section, we will study the topological and smooth structure of the closure of the union of homoclinic orbits of the unperturbed system under assumptions 1–4 above.

1.2. Normal form for the Poincaré map and structure of the union of homoclinic orbits. The union in the subhead is the intersection of the stable and unstable sets S_L^s and S_L^u of the saddlenode cycle. For topological reasons, they are continuous manifolds with boundary of dimensions $s+2$ and $u+2$ respectively. We shall now prove that they are infinitely smooth.

Let Γ be a section transversal to L at a point of L; denote by P_ε the Poincaré map of L defined on Γ. Assumption 1 allows us to apply the theorem on the local normal form of a saddlenode family of maps (see §5 of Chapter 2). We use this theorem for the zero parameter value only. A finitely smooth change of coordinates brings the Poincaré map P_0 to the form

$$x' = f(x), \quad y' = A(x)y, \quad z' = B(x)z, \qquad (x,y,z) \in (\mathbb{R} \times \mathbb{R}^s \times \mathbb{R}^u, 0), \tag{1.1}$$

where $\|A(x)\|$, $\|B(x)^{-1}\| < \lambda < 1$, and $f(x)$ is the time one phase flow transformation of the equation

$$\dot{x} = w(x), \qquad w(x) = x^2(1+ax)^{-1}.$$

The stable and unstable sets in the local chart above are

$$S^s = \{(x,y,z) \mid x \leqslant 0,\ z = 0\} \quad \text{and} \quad S^u = \{(x,y,z) \mid x \geqslant 0,\ y = 0\}$$

respectively. The set S^s is uniquely determined. The iterates of P applied to the intersection of S^s with an arbitrary small neighborhood V of zero generate the whole set S^s. The smoother this intersection, the smaller the neighborhood V. Hence, S^s is infinitely smooth. The same arguments prove the infinite smoothness of S^u.

This implies the infinite smoothness of S_L^s and S_L^u as manifolds with boundaries. The transversal intersection of these manifolds that does not touch the boundary is a smooth 2-manifold. Denote it by H.

THEOREM 1.1. *Let H be the union of homoclinic orbits of a saddlenode cycle of a vector field that satisfies the fundamental assumptions* 1–4. *Then each connected component of H is an embedded C^∞ cylinder.*

PROOF. It was already proved at the beginning of the subsection that H is a smooth 2-manifold. Suppose that h is an arbitrary connected component of H. Let $\gamma = \Gamma \cap h$. The orbits of X are transversal to Γ, hence h has the same property, and γ is smooth.

Suppose first that γ has a compact component γ_0. Then γ_0 is a circle, and the saturation of γ_0 by the orbits of X is diffeomorphic to a cylinder. This proves the theorem in the first case.

Suppose now that no component of γ is compact. Then any connected component of $\gamma^- = \gamma \cap \{x < 0\}$ sufficiently close to zero has zero as a limit point. In the converse case, assumption 2 would be violated. On the other hand, h is a homoclinic surface, so we can neglect the components of γ distant from zero. Compactness of the union $h \cup L$ implies that the number of the components γ_j of γ^- is finite. Hence, the Poincaré map P produces a cyclic permutation of the curves γ_j. Consequently, for some k, any curve γ_j is invariant under the action of P^k.

Take a tubular neighborhood U of L and extend the normalizing coordinates from Γ to U in such a way that the hypersurface $x = 0$ in U will be X-invariant, and the x-coordinate will grow along the orbits of X. The latter is possible, because x grows along the orbits of P. Take the hypersurface $\Gamma^+ = \{x = -\delta\}$ in U for small positive δ. The superscript $+$ reminds us that the orbit of X enters the neighborhood of the circle through Γ^+. Let π be the projection of γ into Γ^+ along the orbits of X located in U. This map is well defined, because the orbits of X located in $U \cap S_L^s$ intersect Γ^+ exactly once, for δ small enough. The image $C^+ = \pi\gamma$ is diffeomorphic to the circle γ_j/P^k. The saturation of C^+ by the orbits of X is diffeomorphic to a cylinder. This proves the theorem.

1.3. Topological classification of the closure of the union of homoclinic orbits. Let $\Gamma^- = \{x = \delta\}$ in U for small positive δ. Let $C^- = h \cap \Gamma^-$ be a circle similar to C^+ and located in the other half-space $x > 0$.

The natural projections of the circles C^+ and C^- onto L have degrees that we denote by p and q. They are integers, not necessary positive.

These integers form a complete set of topological invariants of the union $h \cup L$. This will be illustrated, but not proved, below. Any pair (p, q) may be realized in a phase space of sufficiently large dimension. The least possible dimension n is indicated in the examples that follow.

1. $(1, 1)$, a homoclinic torus, $n = 3$.
2. $(1, -1)$, a homoclinic Klein bottle, $n = 4$.
3. $(1, 0)$, a half-cylinder with boundary L, whose "infinite part" winds around L, $n = 3$.
4. $(0, 0)$, a "snake" with both head and tail winding around L, $n = 4$.
5. $(p, 1)$, $p > 1$, a non-Hausdorff torus with p leaves glued to one leaf along L, $n = 4$.

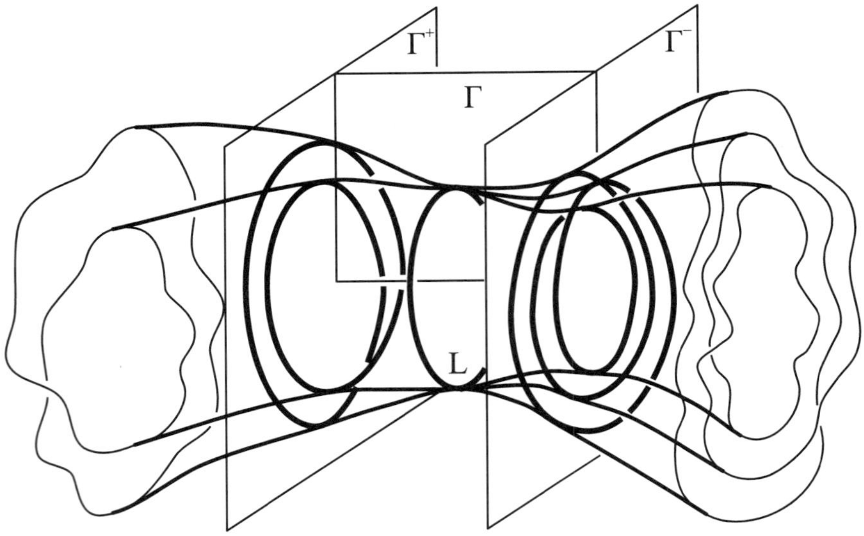

FIGURE 5.1. The intersection of a homoclinic surface of type $(2, 3)$ or $(2, -3)$ with a neighborhood of L.

6. $(p, 1)$, $p < -1$, a non-Hausdorff Klein bottle with p leaves glued to one leaf along L, $n = 4$.

7. $(p, 0)$, a snake with p heads, the tail winding around the heads, glued together along L, $n = 5$.

8. (p, q) arbitrary, $|p| > 1$, $|q| > 1$, a non-Hausdorff torus for $pq > 0$, a Klein bottle for $pq < 0$, with p leaves glued to q leaves along L, $n = 6$ (see Figure 5.1).

In this and the next chapter we mostly consider the bifurcation of components of type $(1, 1)$ and $(1, -1)$, which will be called homoclinic torus bifurcation and Klein bottle bifurcation, respectively. In §8 the transition "from Morse–Smale to Smale–Williams" in the case $(p, \pm 1)$ is presented.

1.4. Criticality. By (1.1), the Poincaré map P_0 has a family of invariant leaves of codimension one: $x = \text{const}$. Therefore there exists a strongly stable foliation (denoted by $\mathcal{F}^{ss}$) on the stable set S^s:

$$S^s \cap \{x = \text{const}\}.$$

Similarly, on the unstable set S^u there exists a strongly unstable foliation denoted by $\mathcal{F}^{uu}$.

DEFINITION 1.2. A connected component h of the set H is said to be s- or *u-critical* if the manifold h has a nontransversal intersection with some leaf of $\mathcal{F}^{ss}$ or $\mathcal{F}^{uu}$ respectively. The cylinder h is said to be *bicritical* if it is both s- and u-critical; h is said to be *semicritical* if it is s- or u-critical but not bicritical. It is said to be *noncritical* if it is neither s- nor u-critical.

Obviously, a component of H of type $(0, 0)$ must be bicritical. If L is a node in its hyperbolic variables, then H is connected and cannot be bicritical.

In this chapter we will consider the noncritical homoclinic torus or Klein bottle. The bifurcation of a semicritical torus is the subject of the next chapter.

1.5. Smoothness of the noncritical homoclinic torus or Klein bottle.

THEOREM 1.2. *The noncritical homoclinic torus or Klein bottle is of C^∞ class.*

Let h be a noncritical component of H of type $(1,1)$ or $(1,-1)$. In order to prove Theorem 1.2, by Theorem 1.1, it is sufficient to show that $h \cup L$ is smooth at the points of L. Let Γ and γ be the same as in 1.2. The curve γ is smooth. In what follows we will show that γ is smooth at the point O also. This is equivalent to showing that $\gamma \cup \{O\}$ is of class C^k at O for any $k \in \mathbb{N}$.

Let k be a given natural number. By assumption 1 and Theorem 2.5.4 on the normal form for a local family of maps, there exists a C^k chart on Γ such that the Poincaré map has the form (1.1). Let

$$\gamma^- = \gamma \cap \{x < 0\}, \qquad \gamma^+ = \gamma \cap \{x > 0\}.$$

We claim that the curves $\gamma^\pm$ and the x-axis have contact of order k at the point O. We will prove our statement only for γ^-. The claim for γ^+ can be proved similarly. Note that γ^- lies in the (x, y)-plane, which is invariant under the action of the Poincaré map. Since h is noncritical, γ^- is the graph of a C^k map:

$$y = \varphi(x), \qquad x \in (\mathbb{R}^-, 0).$$

Now Theorem 1.2 is a corollary of the following

LEMMA 1.1. *Let φ be a C^k map $(\mathbb{R}^-, 0) \to \mathbb{R}^s$ whose graph is invariant under the C^k transformation $(x, y) \mapsto (f(x), A(x)y)$ with $f(0) = 0$, $f'(0) = 1$, $f(x) - x > 0$ for $x < 0$, $\|A(x)\| < \lambda < 1$. Then φ may be extended to $(\mathbb{R}^+, 0)$ to a C^k map by setting it identically equal to zero.*

PROOF. Let us prove by induction that all the derivatives of φ up to order k tend to zero as $x \to -0$.

Base. Let $s_0 = [-\delta, 0)$. Then $f(s_0) \subset s_0$. Let $s_{n+1} = f(s_n)$. Denote

$$M_n = \sup_{x \in s_n} |\varphi(x)|.$$

The following equation is equivalent to the invariance of the graph of φ under P_0:

$$A \circ g(x)(\varphi \circ g(x)) = \varphi(x), \tag{1.2}$$

where $g = f^{-1}$. By equation (1.2), we have $M_{n+1} \leqslant \lambda M_n$. The base is established.

Induction step (from $i-1$ to i). Equation (1.2) implies:

$$\varphi^{(i)}(x) = A \circ g(x) \varphi^{(i)} \circ g(x)(g'(x))^i + P_i,$$

where P_i is a polynomial in $\varphi, \varphi', \dots, \varphi^{i-1}$ with bounded x-dependent coefficients and zero constant term. By the induction assumption, it is small on a segment s_0 small enough. More precisely, for any ε there exists an s_n such that $\max_{s_n} |P_i| < \varepsilon$. Denote

$$m_n = \sup_{x \in s_n} |\varphi^{(i)}(x)|.$$

For what follows, it is important that $g'(x) \to 1$ as $x \to 0$. Hence, for any $q \in (\lambda, 1)$ there exists an $N = N(i)$ such that if $n > N$, then $m_n \leqslant q m_{n-1} + \varepsilon$. Hence,

$$m_{N+k} \leqslant q^k m_n + \frac{\varepsilon}{1-q}.$$

Since ε is arbitrarily small, this implies that $m_n \to 0$ as $n \to \infty$.

Hence, the union $h \cup L$ is indeed C^∞. The induction step and the proof are completed.

§2. Persistence of noncritical homoclinic tori and Klein bottles

Suppose that in a generic one-parameter family the critical value of the parameter corresponds to a saddlenode cycle of multiplicity two. Suppose also that the union of the homoclinic curves of this cycle and the cycle itself is a torus or a Klein bottle. In terms of 1.3 this means that this union is a homoclinic surface of type $(1, 1)$ or $(1, -1)$. Suppose also that it is noncritical. Then by Theorem 1.2, it is smooth. In this section we prove that this smooth homoclinic surface persists in the generic family under small changes of the parameter value. The smoothness of the persisting manifolds of the family tends to infinity as the parameter value tends to the critical one.

2.1. Statement of the persistence theorem.

THEOREM 2.1. *In a generic one-parameter family of vector fields let a vector field with the following properties occur.*

1. *This vector field has a saddlenode cycle of multiplicity two.*
2. *The union of the cycle and its homoclinic orbits has type* $(1, 1)$ *or* $(1, -1)$*, and is noncritical.*

Then all the nearby vector fields of the family have an invariant manifold diffeomorphic to a torus or to a Klein bottle (corresponding to types $(1, 1)$ *and* $(1, -1)$ *respectively). This manifold is finitely smooth. The rate of smoothness tends to infinity as the parameter tends to the critical value.*

The rest of the section contains the proof of this theorem.

The extra genericity assumptions are just the fundamental assumptions from 1.1 (the transversality assumption 5 is of no need here). Denote the saddlenode cycle in Theorem 2.1 by L. By the fundamental assumptions, the Poincaré map for L is finitely smoothly equivalent to

$$(x, y, z) \mapsto (f(x), A(x)y, B(x)z), \qquad \|A\| < \lambda < 1, \ \|B^{-1}\| < \lambda < 1 \tag{2.1}$$

(see 1.1 for more details).

The assumptions of Theorem 2.1 allow us to apply the results of §1. They imply that the critical vector field has an infinitely smooth invariant 2-torus or Klein bottle. The persistence of this manifold under the perturbation will be proved by using the Fenichel theorem stated below.

If the transversality assumption mentioned above is fulfilled, then the bifurcations on the persistent invariant manifolds admit a detailed description. This is done in §§3, 4 for the torus, and in §5 for the Klein bottle.

2.2. Persistence of invariant manifolds by the Fenichel theorem. The theorem below claims that the attracting invariant manifold persists under a small perturbation, provided that the rate of approximation of the orbits to the manifold from the outside is greater than the rate at which the orbits approach each other on the manifold itself. The parameters characterizing these rates are called "Lyapunov type exponents". They are defined in the following way.

DEFINITION 2.1. A manifold with boundary is called *overflowing* or *negatively invariant* for a vector field if the field is tangent to the manifold at the interior points, and on the boundary also tends to the manifold and points outward.

DEFINITION 2.2. A negatively invariant manifold M with boundary of the field v is called *attracting* provided that there exist a neighborhood of this manifold, a nonnegative function ρ in this neighborhood, and a positive constant t such that

$$\rho(x) = 0 \text{ if and only if } x \in g_v^t M, \quad L_v\rho < 0 \text{ outside } g_v^t M.$$

Denote by TM and N the tangent and normal bundles to M respectively. Let T be the restriction to M of the tangent bundle to the phase space. Denote by $p\colon T \to N$ the projection along TM.

DEFINITION 2.3. The *rate of attraction* of the negatively invariant manifold M of the vector field v is

$$\lambda_N = \inf_{a\in M} \varliminf_{t\to+\infty} \frac{-\log \|pdg^t(g^{-t}a)\|}{t}. \tag{2.2}$$

DEFINITION 2.4. The *rate of contraction* onto the negatively invariant manifold M of the vector field v is

$$\lambda_T = \sup_{a\in M,\, \xi\in T_aM} \varlimsup_{t\to+\infty} \frac{\log |dg^{-t}(a)\xi|}{t}, \qquad \xi \neq 0. \tag{2.3}$$

The notation reminds us that the first exponent characterizes the contraction in the normal direction to M and the second in the tangent direction.

THEOREM 2.2 (Fenichel). *Suppose that v is a smooth vector field, M is a smooth manifold with boundary negatively invariant for the field v, where λ_N and λ_T are the corresponding exponents, $\lambda_N > 0$, and the positive integer r satisfies the inequality $r\lambda_T < \lambda_N$. Then any vector field C^r-close to v has a C^r smooth negatively invariant manifold close to M.*

2.3. Heuristic proof of the persistence theorem. Suppose first that in Theorem 2.1 the saddlenode cycle is a node in its hyperbolic variables. Then the Fenichel theorem may be applied to the homoclinic surface $\bar{H}$ itself. Indeed, the rate of contraction for this surface equals zero, while the rate of approximation is strictly positive. To explain this, consider first the rate of contraction λ_T on $\bar{H}$. In past time, all the orbits on this surface wind around the saddlenode cycle L. Hence, the exponential expansion of the orbits on M in past time is impossible. Therefore, the limit in (2.3) is zero.

The same reasoning shows that

$$\lambda_N > 0. \tag{2.4}$$

Indeed, the negative semiorbit of any point a on the homoclinic surface spends most of the time near the saddlenode cycle L. Let $Q \in (\|A(0)\|, 1)$. Let $t_n(a) < 0$ be the time necessary for the point $a \in \bar{H}$ to come to the Poincaré section Γ and to perform n revolutions around L in the opposite direction. The time $-t_n$ grows no faster than a linear function in n. Hence, $\|pdg^{-t_n}(g^{t_n}a)\| \leq CQ^{\alpha n}$ for some $\alpha > 0$. This implies (2.4).

By the Fenichel theorem, the homoclinic surface persists under small perturbations in a generic one-parameter family. For any r there exists a neighborhood of the critical parameter value such that the corresponding vector field has an r-smooth invariant manifold close to the homoclinic surface. This proves the theorem in the case of a node in hyperbolic variables.

For a saddlenode cycle which is a saddle in its hyperbolic variables, the same arguments are carried out in three steps.

The homoclinic surface in the previous argument must be replaced by the unstable set of the saddlenode cycle, or more precisely, by the intersection of this set with an appropriate neighborhood of the homoclinic surface. The first step is to prove that this intersection U is smooth and satisfies the assumptions of the Fenichel theorem. Smoothness is proved in 2.4 below. The exponents λ_T, λ_N are estimated in 2.5. It turns out that they behave in the same way as in the case of a node in hyperbolic variables. The manifold U turns out to be negatively invariant. By the Fenichel theorem, it is persistent, and has an arbitrary high rate of smoothness for sufficiently small parameter values.

The second step is to prove the same thing for the stable set of a saddlenode cycle. This is done by time reversal and a mere reference to Step 1. The manifold generated by the stable set is persistent too.

The third step is to take the intersection of these invariant manifolds. It is invariant itself. For the critical parameter value, this intersection is transversal and coincides with the homoclinic surface $\bar{H}$. For all the nearby parameter values, it is a finitely smooth manifold close to $\bar{H}$. This completes the heuristic proof of the theorem. Now we pass to the rigorous proof.

2.4. Global unstable set of the saddlenode cycle. Let U_{loc} be the local stable set of the saddlenode cycle L. Denote by U_g, g for *global*, the union of all the phase curves through U_{loc}. There exists a neighborhood W of the open homoclinic surface H such that the connected component $U \supset H$ of the intersection $W \cap U_g$ is smooth. We will prove that this intersection may be smoothly extended to a manifold containing the whole homoclinic surface $\bar{H} = H \cup L$.

LEMMA 2.1. *Under the assumptions of Theorem 2.1 and the fundamental assumptions, the critical vector field v has a smooth negatively invariant manifold U that contains the homoclinic surface and is the closure of the intersection of the global unstable set with an appropriate neighborhood of the homoclinic surface $\bar{H}$.*

PROOF. It is sufficient to prove that the intersection $U_\Gamma = U \cap \Gamma$ is a smooth manifold (see Figure 5.2).

Let x, y, z be the normalizing coordinates for the Poincaré map P of the cycle L that bring P to the form (2.1) on Γ.

Let $\gamma = \Gamma \cap H$, $\gamma^+ = \gamma \cap \{x < 0\}$ be the same as at the beginning of 1.5. The intersection U_Γ contains the sets $U^- = \{y = 0, x > 0\}$ and $U^+ = U_\Gamma \cap \{x < 0\}$. The surface U^+ is transversal to the y-planes at the points of γ^+. Then there exists

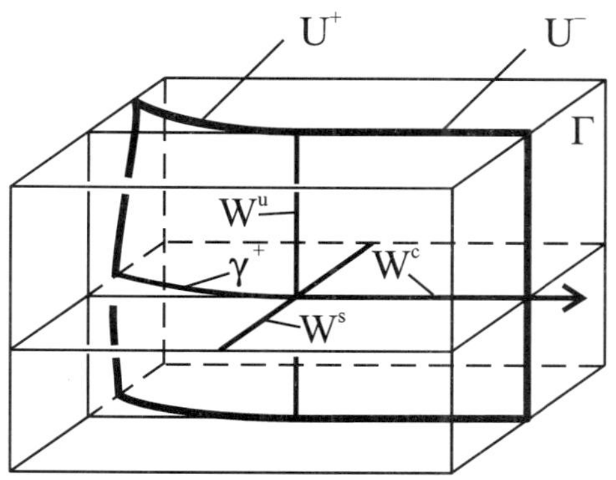

FIGURE 5.2. The intersection of a negatively invariant manifold containing the homoclinic surface, with the Poincaré section of a saddlenode cycle in dimension 4.

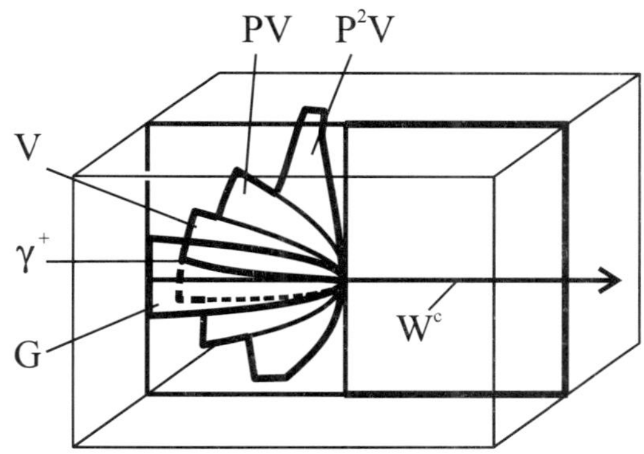

FIGURE 5.3. Construction of the negatively invariant manifold in the persistence theorem.

a domain G in the half-space $y = 0$, $x < 0$, and a subset $V \subset U^+$, which contain γ^+, such that V is the graph of a map $S \to \mathbb{R}^s$, $y = \varphi(x, z)$ defined on G.

The set V is invariant under P in the sense that PV is the graph of a map $PG \to \mathbb{R}^s$ that coincides with φ on the intersection $G \cap PG$ (see Figure 5.3).

Let $G_0 = G$, $G_{n+1} = PG_n \cap \Gamma$. By (2.1), $P(x, 0, z) = (f(x), 0, B(x)z)$, and $B(x)$ is a linear expansion. Hence, the union $G_\infty = \bigcup_0^\infty G_n$ contains a half-neighborhood of 0 in $\{y = 0\} \cap \Gamma$ (see Figure 5.3). The map φ in G_{n+1} is defined by induction:

$$\text{graph of } \varphi | G_{n+1} = P(\text{graph of } \varphi | G_n). \tag{2.5}$$

PROPOSITION 2.1. *The map* $\varphi\colon G_\infty \to W^s$ *having the property* (2.5) *may be smoothly extended to* $x \geqslant 0$, $y = 0$, $(x, y) \in \Gamma$, *by setting it identically equal to zero.*

This proposition immediately implies Lemma 2.1, because the intersection U^+ is given by $y \equiv 0$.

PROOF OF PROPOSITION 2.1. Denote by φ_n the restriction $\varphi|S_n$ and by P_0 the restriction of P to $y=0$: $P_0(x,z)=(f(x),B(x)z)$. The graph invariance equation (2.5), after substituting P from (2.1), takes the form:

$$A(x)\varphi_n(x,z)=\varphi_{n+1}\circ P_0(x,z).$$

Let g be as in (1.2). Then

$$\varphi_{n+1}(x,z)=A(g(x))\varphi_n\circ P_0^{-1}(x,z). \tag{2.6}$$

We will prove that all the derivatives of φ_n uniformly tend to 0 as $n\to\infty$. More precisely, let α be a multi-index and $M_{\alpha,n}=\max_{S_n}|D^\alpha\varphi_n|$. We will prove by induction on α that

$$M_{\alpha,n}\to 0 \text{ as } n\to\infty. \tag{2.7}$$

This will imply Proposition 2.1.

To carry the induction, let us split the multi-index α in the following way: (m,β), $m\in\mathbb{Z}_+$, $\beta\in\mathbb{Z}_+^u$ and order these multi-indices as follows:

$$(m,\beta)\geqslant(m',\beta') \quad \text{if } m\geqslant m' \text{ or } m=m',\ \beta-\beta'\in\mathbb{Z}_+^u.$$

Base of induction: $\alpha=0$, $\max_{S_n}|\varphi_n|\to 0$ as $n\to\infty$. This is an evident consequence of (2.6) and (2.1): $\|A\|\leqslant\lambda<1$.

Induction step. By the chain rule applied to (2.6) for $\alpha=(m,\beta)$, we have:

$$|D^\alpha\varphi_{n+1}|\leqslant\|A\|\|(D^\alpha\varphi_n)\circ P_0^{-1}\|\|g'|^m\|B^{-1}\|^{|\beta|}+|\mathcal{P}_\alpha|,$$

where $\mathcal{P}_\alpha$ is a polynomial on S_n in the derivatives of φ_n having lower order than α in the sense of the ordering of multi-indices defined above. The coefficients of $\mathcal{P}_\alpha$ depend on x and, by the induction assumption, this polynomial is small on S_n for large n. On the other hand, $g'(x)\to 1$ as $x\to 0$. Hence, for some $q<1$ and sufficiently large n depending on m, we have

$$\|A\|(g')^m|S_n\leqslant q.$$

Then for any $\varepsilon>0$ and n sufficiently large, $M_{\alpha,n+1}\leqslant qM_{\alpha,n}+\varepsilon$. This proves (2.7), hence Proposition 2.1 and Lemma 2.1.

2.5. Lyapunov type exponents for the global unstable set. Consider a generic one-parameter family from Theorem 2.1. If we multiply all the vector fields of the family by a nonzero function, the new family will have same orbits and invariant surfaces as the previous one. The function mentioned above may be taken in such a way that the saddlenode cycle L of the vector field v will have period 1. Moreover, the return time from the cross-section Γ to itself may be assumed equal to 1. More precisely, the germ of the Poincaré map of the vector field v on Γ at $0=\Gamma\cap L$ satisfies the equality

$$P(a)=g_v^1(a), \qquad a\in(\Gamma,0). \tag{2.8}$$

Without loss of generality, we may assume that relation (2.8) holds in Theorem 2.1.

LEMMA 2.2. *Let the vector field v be the same as in Theorem 2.1 with the additional property (2.8). Then the rate of contraction λ_T and the rate of approximation λ_N for the global unstable manifold U_g of v are zero and positive respectively.*

PROOF. The proof is carried out by replacing the phase flow transformations by iterates of the Poincaré map and making use of the normal form (2.1).

Let $a \in U_g$ be an arbitrary point. Then after some negative time the orbit starting at a will come to the point

$$b \in U^- = U_g \cap \Gamma \cap \{x > 0\}.$$

Replacing a by b will not change the limits in (2.2) and (2.3). Moreover, we have

$$g_v^{-n} b = P^{-n} b$$

by (2.8). Replacing the continuous time $t \to +\infty$ in (2.2), (2.3) by the discrete time $n \to +\infty$, we will not change the result. Hence,

$$\lambda_T = \sup_{b \in U^-, \, \xi \in T_b U^-} \overline{\lim_{n \to +\infty}} \frac{\log |dP^{-n}(b)\xi|}{n}, \qquad \xi \neq 0,$$

$$\lambda_N = \inf_{b \in U^-} \underline{\lim_{t \to +\infty}} \frac{-\log \|pdP^n(P^{-n}b)\|}{n}.$$

In the normalizing chart (2.1)

$$U^- = \{y = 0, \, x > 0\}, \quad \xi = (dx, dy, dz), \quad p\xi = dy.$$

On the other hand, for $b = (x_0, 0, z_0) \in U^-$, let $x(P^{-n}b) = x_n$. Then $x_n \to 0$.

Let $\xi = (dx, 0, dz) \in T_b U^-$. Then

$$\eta_n = dP^{-n}(b)\xi = \left((f^{-n})'(x_0)\, dx, 0, \left(\prod_0^{n-1} B^{-1}(x_j) \right) dz \right).$$

By (2.1), $x_j \to 0$, $(f^{-1})'(x_j) \to 1$ as $j \to \infty$, $\|B^{-1}\| < 1$. Hence,

$$\lim_{n \to \infty} \frac{\log |\eta_n|}{n} \leqslant 0.$$

The equality holds for $dx \neq 0$. This proves the first statement of Lemma 2.2: $\lambda_T = 0$.

Let us prove the second statement. Let $\xi = (dx, dy, dz)$, $x(P^{-n}b) = x_n$. Then

$$\mathbf{A}_n = pdP^n(P^{-n}b) = \prod_0^{n-1} A(x_j).$$

Since $\|A\| \leqslant \lambda < 1$, we have $\|\mathbf{A}_n\| \leqslant \lambda^n$. Hence, $\lambda_N \geqslant -\log \lambda > 0$. This proves Lemma 2.2.

As explained in 2.3, Lemmas 2.1 and 2.2 together prove the persistence theorem (Theorem 2.1).

§3. The rotation number as a function of the parameter in the family of diffeomorphisms of the circle

In this section we present some classical material on the rotation number of degree one monotonic (possibly nonbijective) maps of the circle. In the second half of the section, we discuss properties of the rotation number as a function of a parameter in generic one-parameter families. The main result of the section is Theorem 3.3. It asserts that the rotation number of an orientation-preserving diffeomorphism of the circle that depends monotonically on the parameter generically looks like a Cantor function of the parameter. Namely, it takes any rational value on some segment of nonzero length, and an arbitrary irrational value at a single point.

3.1. Definition and existence of the rotation number. Suppose that $f\colon S^1 \to S^1$ is a map and $\Pi\colon \mathbb{R} \to S^1$ is the covering map: $\Pi(t) = \exp(2\pi i t)$. Then there is a map $\bar{f}\colon \mathbb{R} \to \mathbb{R}$ such that $\Pi \circ \bar{f} = f \circ \Pi$; $\bar{f}$ is called a *lift* of f and is unique up to the addition of an integer. Any such $\bar{f}$ satisfies $\bar{f}(x+1) = \bar{f}(x) + n$, where $n \in \mathbb{Z}$ is called the *degree* of f. We say that f is *monotonic* if its lift $\bar{f}$ is a monotonic function.

LEMMA 3.1. *Let $f\colon \mathbb{R} \to \mathbb{R}$ be a monotonic continuous function such that*

$$f(x+1) = f(x) + 1 \tag{3.1}$$

for all $x \in \mathbb{R}$. Then

1. *The limit*

$$r(f) = \lim_{n\to\infty} \frac{f^n(x)}{n}$$

exists and is independent of the choice of the point $x \in \mathbb{R}$.

2. *$r(f+n) = r(f) + n$ for any $n \in \mathbb{Z}$.*

PROOF. First we prove assertion 1.

For any monotonic function f with property (3.1), the oscillation of its difference with the identity is no greater than 1. Indeed, the difference $\varphi(x) = f(x) - x$ is a one-periodic function. We shall prove that for arbitrary $x, y \in \mathbb{R}$

$$|\varphi(x) - \varphi(y)| < 1. \tag{3.2}$$

By periodicity, we may assume, without loss of generality, that $0 \leqslant x < y < 1$. By the monotonicity of f and by (3.1), we have:

$$f(x) \leqslant f(y) \leqslant f(x+1) = f(x) + 1.$$

Therefore,

$$x - y \leqslant \varphi(y) - \varphi(x) \leqslant x - y + 1.$$

This implies (3.2).

Now we apply (3.2) to f^k instead of f, $x = 0$, and $y = f^m(0)$ for arbitrarily chosen k and m. We obtain

$$|f^{m+k}(0) - f^m(0) - f^k(0)| < 1.$$

Hence,

$$|f^{nk}(0) - nf^k(0)| = \left|\sum_{j=1}^{n}(f^{kj}(0) - f^{k(j-1)}(0)) - nf^k(0)\right|$$
$$\leqslant \sum_{j=1}^{n} |f^{kj}(0) - f^{k(j-1)}(0) - f^k(0)| < n.$$

Therefore,

$$\left|\frac{f^{nk}(0)}{nk} - \frac{f^k(0)}{k}\right| < \frac{1}{k}.$$

Similarly,

$$\left|\frac{f^{nk}(0)}{nk} - \frac{f^n(0)}{n}\right| < \frac{1}{n}.$$

Hence, $\{f^n(0)/n\}$ forms a Cauchy sequence and therefore converges. Denote the corresponding limit by $r(f)$. On the other hand, for any $x \in [0, 1)$,

$$f^n(0) \leqslant f^n(x) < f^n(1) = f^n(0) + 1.$$

Consequently, $f^n(x)/n \to r(f)$. This proves assertion (1).

Now we prove (2). By induction we have

$$(f + n)^k(x) = f^k(x) + kn.$$

Thus

$$r(f + n) = \lim_{k\to\infty} \frac{(f + n)^k(0)}{k} = \lim_{k\to\infty} \frac{f^k(0) + kn}{k} = r(f) + n.$$

The proof of Lemma 3.1 is complete.

The lemma above allows us to define the rotation number for any monotonic degree one map of the circle. Let f be such a map and $\bar{f}$ its lift. We then define the rotation number of f as $r(\bar{f})$, which, by 2 of Lemma 3.1, is unique up to the addition of an integer. The rotation number is an important characteristic of degree one homeomorphisms of the circle. It determines the topological and dynamical properties of the map. Let us first consider two simple examples.

EXAMPLE 1. Let f be a rigid rotation of the circle by the angle $2\alpha\pi$; then $\bar{f} = x + \alpha$ is a lift of f. Thus

$$r(\bar{f}) = \lim_{n\to\infty} \frac{\bar{f}^n(0)}{n} = \lim_{n\to\infty} \frac{n\alpha}{n} = \alpha.$$

EXAMPLE 2. Let f be a continuous monotonic map of the circle of degree one with a q-periodic point. Then the corresponding rotation number is a rational number of the form p/q for some p.

Indeed, the q-periodicity of a point x implies that

$$\bar{f}^q(x) - x = p$$

for some integer p. By induction we have

$$\bar{f}^{kq}(x) = x + kp.$$

Thus

$$r(\bar{f}) = \lim_{k\to\infty} \frac{\bar{f}^{kq}(x)}{kq} = \lim_{k\to\infty} \frac{x+kp}{kq} = \frac{p}{q}\,.$$

3.2. Periodic points; rational and irrational rotation numbers.

PROPOSITION 3.1. *Let f be a continuous monotonic map of the circle of degree one. Then f has a periodic point if and only if its rotation number is rational.*

PROOF. The necessity is shown by Example 2. We now prove sufficiency.

Let $\bar{f}$ be a lift of f with $r(\bar{f})$ equal to p/q. We will prove that the equation

$$\bar{f}^q(x) - x = p \tag{3.3}$$

has a root. If not, then

$$\bar{f}^q(x) - x > p \text{ or } \bar{f}^q(x) - x < p \text{ for any } x \in \mathbb{R}\,.$$

We only study the first possibility. The second may be brought to a contradiction in the same way.

Since $\bar{f}^q - \mathrm{id}$ is periodic, there exists a $c > 0$ such that

$$\bar{f}^q(x) - x \geqslant p + c \text{ for any } x \in \mathbb{R}\,.$$

Thus $\bar{f}^{kq}(x) - x \geqslant kp + kc$ and so

$$r(\bar{f}) \geqslant (p+c)/q,$$

which is a contradiction. Let $x \in \mathbb{R}$ be a root of (3.3); then $\Pi(x) \in S^1$ is a q-periodic point of f.

COROLLARY 3.1. *If for some x we have $p < \bar{f}^q(x) - x < s$, then*

$$p/q \leqslant r(\bar{f}) \leqslant s/q.$$

PROOF. We shall prove only the lower estimate for $r(\bar{f})$; the upper one is proved in the same way. Suppose that for some x_0 we have the equality $\bar{f}^q(x_0) - x_0 = p$. Then $r(\bar{f}) = p/q$ by Proposition 3.1. Suppose that this equality is false for any x. Then $\bar{f}^q(x) - x > p$ everywhere. Hence, $r(\bar{f}) > p/q$ as proved just above.

PROPOSITION 3.2. *Let f be a continuous monotonic map of the circle of degree one with irrational rotation number. Denote by $\omega(x)$ the set of ω-limit points of x and by Ω the set of nonwandering points. Then*

1. *$\omega(x) = \Omega$ for any $x \in S^1$.*
2. *$\omega(x)$ is a perfect set.*

PROOF. By definition it follows that the set $\omega(x)$ is a nonempty closed invariant set. Let $I = (u, v) \subset S^1$ be a gap of $\omega(x)$. Since f has no periodic points, the arcs $f^n(I)$, $n \in \mathbb{N}$, are pairwise disjoint. Therefore $I \cap \Omega = \varnothing$. This implies that $\Omega \subset \omega(x)$ for all $x \in S^1$. On the other hand, by the definition of ω-limits and nonwandering sets, we have $\omega(x) \subset \Omega$ for any x. Together with the previous inclusion, this implies statement 1.

We now prove 2. We claim that there are no isolated points in Ω. Assume the contrary and let $y \in \Omega$ be isolated. Let the open intervals $I = (u, y)$, $J = (y, v)$ be two gaps of Ω, that is, connected components of the complement of Ω. Since f has no periodic points, for any pair x, z and for n large enough, we have $f^n(x) \neq z$. Hence, none of the points y, u, v returns to the set $\{y, u, v\}$ after a large time. By the invariance of Ω, this implies that $f^n((u, v)) \cap (u, v) = \varnothing$ for n large enough. Hence, $y \notin \Omega$, a contradiction. This proves conclusion 2.

We now state a theorem due to Denjoy, which gives a sufficient condition for the equality $\Omega = S^1$.

THEOREM (Denjoy). *Let f be an orientation-preserving C^1 diffeomorphism of the circle with Df of bounded variation. If the rotation number $r(f)$ of f is irrational, then f is conjugated to the rigid rotation of the circle by the angle $2\pi r(f)$.*

For a proof of this theorem see, for example, [**A**].

3.3. Continuous and monotonic dependence of the rotation number on the map. Another important property of the rotation number is given by the following

PROPOSITION 3.3. *Let $f\colon \mathbb{R} \to \mathbb{R}$ be a monotonic continuous function satisfying $f(x+1) = f(x) + 1$. Then the number $r(f)$ depends continuously on f. More precisely, let $g\colon \mathbb{R} \to \mathbb{R}$ be another monotonic continuous function satisfying the relation $g(x+1) = g(x) + 1$. Then for any given $\varepsilon > 0$, there exists a $\delta > 0$ such that the inequality*

$$\|f - g\| \overset{\text{def}}{=} \sup_{x \in \mathbb{R}} |f(x) - g(x)| < \delta \tag{3.4}$$

implies $|r(f) - r(g)| < \varepsilon$.

PROOF. Given any $\varepsilon > 0$, take n so that $1/n < \varepsilon$. By the uniform continuity of f, there exists a $\delta > 0$ such that

$$(f + \delta)^n - f^n < 1.$$

The exponent here and below in this section is related to composition. If the argument is not specified, the inequality holds everywhere on $\mathbb{R}$.

The monotonicity of f implies that the above difference is nonnegative, and

$$(f + \delta)^{kn} - f^{kn} < k.$$

This is proved by induction in k. Hence,

$$\frac{(f + \delta)^{kn}}{kn} - \frac{f^{kn}}{kn} < \frac{1}{n} < \varepsilon.$$

Passing to the limit as $k \to \infty$, we get:

$$r(f + \delta) - r(f) \leqslant \frac{1}{n} < \varepsilon.$$

Similarly, $r(f) - r(f - \delta) < \varepsilon$. Hence

$$|r(f \pm \delta) - r(f)| < \varepsilon.$$

The same is true for any g from (3.4) substituted into the last displayed formula instead of $f \pm \delta$. Indeed, the monotonicity of f and g implies

$$(f - \delta)^n \leqslant g^n \leqslant (f + \delta)^n, \qquad r(f - \delta) \leqslant r(g) \leqslant r(f + \delta).$$

This proves the proposition.

The monotonicity of the rotation number $r(f)$ with respect to f is described by the following Lemma 3.2. It is important to note that, in the case when $r(f)$ is irrational, the rotation number as a function of the map strictly increases after a *local* increase of the map itself.

LEMMA 3.2. *Let f, g, h be three continuous monotonic maps of the circle of degree one. Let $r(g)$ be irrational. Suppose that the lifts $\bar{f}$, $\bar{g}$, $\bar{h}$ satisfy the inequalities $\bar{f} \leqslant \bar{g} \leqslant \bar{h}$. Moreover, suppose that for some point $x \in \mathbb{R}$ such that $\varphi = \Pi(x) \in \Omega_g$, we have $\bar{f}(x) < \bar{g}(x) < \bar{h}(x)$. Then*

$$r(\bar{f}) < r(\bar{g}) < r(\bar{h}).$$

PROOF. We shall prove the statement related to $\bar{f}$ only; the statement about $\bar{h}$ is proved in the same way.

Without loss of generality, we may assume that φ is not an end of a gap of Ω_g. Indeed, Ω_g is a perfect set, and the points with the desired property are dense in Ω_g. Hence we can find a point $\psi \in \Omega_g$ that belongs to the orbit of φ under g, lies close to φ to the right of it and has the following property: there exist two points x and $y > x$ and two positive integers p and q such that

$$\Pi(x) = \varphi, \quad \Pi(y) = \psi, \quad \bar{f}(y) < \bar{g}(x), \quad \bar{g}^q(x) = y + p.$$

Corollary 3.1, together with the relation $y > x$, implies that $r(\bar{g}) \geqslant p/q$. Since $r(\bar{g})$ is irrational, the equality is excluded. Hence, $r(\bar{g}) > p/q$.

On the other hand,

$$\bar{f}^q(\bar{g}(x)) \leqslant \bar{f}(\bar{g}^q(x)) = \bar{f}(y+p) = \bar{f}(y) + p < \bar{g}(x) + p.$$

Once more, by Corollary 3.1, $r(\bar{f}) \leqslant p/q$. This proves the lemma.

3.4. Structurally stable diffeomorphisms of the circle. Let g be a monotonic endomorphism of the circle of degree one with irrational rotation number. Denote by R_α the rigid rotation of the circle by the angle $2\alpha\pi$. Then $\bar{g}+\alpha$ is a lift of the composition $R_\alpha \circ g$. Now consider the function of α given by $h(\alpha) = r(\bar{g}+\alpha)$. By Proposition 3.3, $h(\alpha)$ is continuous. On the other hand, by Lemma 3.2, we have $h(\alpha) \neq h(0) = r(\bar{g})$ for $\alpha \neq 0$. Therefore for a given $\varepsilon > 0$ there exists $0 < \alpha < \varepsilon$ such that $h(\alpha)$ is rational, which, by Proposition 3.1, implies that $R_\alpha \circ g$ is not conjugate to g. Therefore, we have

PROPOSITION 3.4. *Any structurally stable orientation-preserving diffeomorphism of the circle has rational rotation number, and therefore possesses a periodic orbit.*

In fact, a stronger statement holds:

THEOREM 3.1. *Denote by $\mathrm{Diff}^r(S^1)$ the space of C^r diffeomorphisms of the circle. Then*

1. *$f \in \mathrm{Diff}^r(S^1)$, $r \geqslant 1$, is structurally stable if and only if f is a Morse–Smale system.*
2. *The set of Morse–Smale systems is open and dense in $\mathrm{Diff}^r(S^1)$.*

For the proof of the theorem above, see, for example, [**PM**].

Now we turn to one-parameter families of diffeomorphisms of the circle.

3.5. Sparkling rotation numbers in one-parameter families of diffeomorphisms of a circle. Let us begin with some notation. Denote by $\mathcal{A}^r$ the set of C^1 families of C^r maps $f_\varepsilon\colon \mathbb{R} \to \mathbb{R}$ such that

$$f_\varepsilon(x+1) = f_\varepsilon(x) + 1, \qquad \frac{\partial f_\varepsilon}{\partial x} > 0 \text{ for any } x \in \mathbb{R},\ \varepsilon \in [-1,1].$$

Then each family $f_\varepsilon \in \mathcal{A}^r$ can be regarded as the lift of some one-parameter family of orientation-preserving diffeomorphisms of the circle. A family $f_\varepsilon \in \mathcal{A}^r$ is said to be *monotonic* if $\partial f_\varepsilon/\partial\varepsilon > 0$ (or < 0) for any $x \in \mathbb{R}$, $\varepsilon \in [-1,1]$. Denote by $\mathcal{B}^r$ the set of monotonic families in $\mathcal{A}^r$. Then $\mathcal{B}^r$ is an open subset of $\mathcal{A}^r$.

PROPOSITION 3.5. *Let $f_\varepsilon \in \mathcal{B}^r$ be a monotonic family and let $r(\varepsilon) = r(f_\varepsilon)$ be the rotation number of f_ε. Then*

1. *$r(\varepsilon)$ is a monotonic continuous function of the parameter ε and takes any irrational value at one and only one point.*
2. *For any rational number p/q from the range of $r(\varepsilon)$, if ε_0 is an interior point of the set $r^{-1}(p/q)$, then f_{ε_0} has at least two periodic orbits on $\mathbb{R}/\mathbb{Z}$.*

PROOF. The first assertion is a consequence of Proposition 3.3 and Lemma 3.2. Now we prove the second assertion. If it were not true, f_ε would have a unique periodic orbit on $\mathbb{R}/\mathbb{Z}$ with rotation number p/q. Then this periodic orbit would be semistable. Now consider the family of functions

$$H_\varepsilon = f_\varepsilon^q(x) - x - p. \tag{3.5}$$

Then $H_{\varepsilon_0}(x)$ does not change sign for any $x \in \mathbb{R}$. On the other hand, H_ε is monotonic with respect to ε, hence there exists a half-neighborhood U of ε_0 such that for $\varepsilon \in U \setminus \{\varepsilon_0\}$, the equation $H_\varepsilon = 0$ has no roots, which means that f_ε has no periodic orbits with rotation number p/q. This is a contradiction with our assumption that ε_0 is an interior point of the set $r^{-1}(p/q)$.

LEMMA 3.3. *Let $f_\varepsilon \in \mathcal{A}^r$ and let $r(\varepsilon)$ be the rotation number function of f_ε. Let $\varepsilon_0 \in (-1,1)$ be a point such that*

1. *$r(\varepsilon_0) = p/q$ (a rational number).*
2. *$r(\varepsilon)$ is strictly monotonic at ε_0.*

Then $f = f_{\varepsilon_0}$ satisfies the functional equation

$$f^q(x) \equiv x + p. \tag{3.6}$$

PROOF. Let H_ε be the same as in (3.5). By assumption 1, $H_{\varepsilon_0}(x_0) = 0$ for some x_0. By the monotonicity of the family, we have

$$H_\varepsilon(x_0)(\varepsilon - \varepsilon_0) > 0 \quad \text{for } \varepsilon \neq \varepsilon_0.$$

By Corollary 3.1 and assumption 2, the same is true for x_0 replaced by any x. Therefore, $H_\varepsilon(x)$ changes sign for any x as ε passes through ε_0. Hence, $H_{\varepsilon_0} \equiv 0$.

3.6. Normal families.

DEFINITION 3.1. A family $f_\varepsilon \in \mathcal{A}^r$ is said to be *normal* if
(1) the function $r(\varepsilon) = r(f_\varepsilon)$ is monotonic;
(2) any irrational value of the interval $[r(-1), r(1)]$ is taken by the function $r(\varepsilon)$ at one and only one point;
(3) any rational value of the interval $(r(-1), r(1))$ is taken by $r(\varepsilon)$ on one segment of nonzero length. Moreover, when ε is an interior point of any such segment, f_ε has at least two periodic orbits on $\mathbb{R}/\mathbb{Z}$.

The following proposition gives a necessary and sufficient condition for a monotonic family to be normal.

PROPOSITION 3.6. *A monotonic family $f_\varepsilon \in \mathcal{B}^r$ is normal if and only if f_ε does not satisfy the functional equation* (3.6) *for any $\varepsilon \in (-1, 1)$, $q \in \mathbb{N}$, $p \in \mathbb{Z}$ with $p/q \in (r(-1), r(1))$.*

PROOF. Requirements (1), (2) of Definition 3.1 are fulfilled for any continuous monotonic family from $\mathcal{A}^r$ by virtue of Proposition 3.5. The second part of (3) follows from the first (also by Proposition 3.5). The violation of (3.6) is necessary and sufficient for the first part to hold, as proved below.

Necessity. Suppose f_ε is normal and there exist $\varepsilon_0 \in (-1, 1)$, $q \in \mathbb{N}$, $p \in \mathbb{Z}$ such that f_{ε_0} satisfies (3.6). Consider the function

$$H_\varepsilon(x) = f_\varepsilon^q(x) - x - p.$$

Then $H_{\varepsilon_0} \equiv 0$. From the assumption that f_ε is monotonic, we have one of the inequalities $\partial H_\varepsilon/\partial\varepsilon > 0\,(< 0)$. Thus

$$\min_{x\in\mathbb{R}} |H_\varepsilon(x)| > 0 \quad \text{for } \varepsilon \neq \varepsilon_0,$$

which implies

$$r(\varepsilon) = r(f_\varepsilon) \neq p/q \quad \text{for } \varepsilon \neq \varepsilon_0.$$

This contradicts the assumption that f_ε is normal.

Sufficiency. This is a corollary of Lemma 3.3.

3.7. Entire functions and normal families.

LEMMA 3.4. *Let $g(z)$ be a polynomial in z. If an entire function $f(z)$ satisfies the functional equation*

$$f^q(z) = g(z), \qquad z \in \mathbb{R},$$

then $f(z)$ is a polynomial. In particular, the functional equation (3.6) *has a unique solution in the class of entire functions satisfying the relation $f(z+1) = f(z) + 1$. This solution is $f(z) = z + p/q$.*

PROOF. Since f^q and g are entire functions, we have

$$f^q(z) = g(z), \qquad z \in \mathbb{C}.$$

If f is not a polynomial, then ∞ is an essential singular point. According to the Picard theorem, there exists a constant $B \neq \infty$ and a sequence $z_k \to \infty$ such that $f(z_k) = B$. Thus $g(z_k) = f^{q-1}(f(z_k)) = f^{q-1}(B)$. Therefore the polynomial $g(z) - f^{q-1}(B)$ has an infinite number of roots. This is a contradiction with the Gauss theorem, which proves our first statement.

When $g(z) = z + p$, the first assertion implies that $f(z)$ is a polynomial of degree one. Using the method of undetermined coefficients, we obtain the unique solution $f(z) = z + p/q$ that satisfies $f(z+1) = f(z) + 1$.

3.8. Genericity of normal families. Now we pass to the main results of this section.

From Proposition 3.6 and Lemma 3.4, we have

THEOREM 3.2. *Let $f_\varepsilon \in \mathcal{B}^r$ be a monotonic family. Assume that for each fixed value $\varepsilon \in [-1, 1]$, f_ε is an entire function different from $x + r$ for any rational r. Then f_ε is normal.*

EXAMPLE 3. Consider the family of maps

$$f_{\varepsilon,\alpha}\colon x \mapsto x + \varepsilon + \alpha h(x), \qquad |\alpha| \ll 1,$$

where $h(x) = h(x+1)$ is a real trigonometric polynomial. Then for any small fixed α, the family $f_{\varepsilon,\alpha}$ is normal by Theorem 3.2.

THEOREM 3.3. *Generic monotonic families of diffeomorphisms of the circle of degree one are normal. Moreover, for any r, the set of normal families in $\mathcal{B}^r$ contains a residual subset (countable intersection of open dense subsets).*

PROOF. Consider the family $F = \{f_\varepsilon \mid \varepsilon \in [-1, 1]\}$ from Theorem 3.3. Fix a positive integer r and suppose that $f_\varepsilon \in C^r(S^1)$. Then for sufficiently large N, the family

$$G = \{g_\varepsilon = \text{identity plus partial sum of the Fourier series of } f_\varepsilon \\ (g_\varepsilon \text{ is a trigonometric polynomial of degree } N)\}$$

forms a family C^r close to F. It may be nonnormal: for some ε all the Fourier coefficients of f_ε up to degree N except for the constant term may vanish. Let $a_1(\varepsilon)$ and $b_1(\varepsilon)$ be the coefficients of g_ε at $\sin 2\pi x$ and $\cos 2\pi x$. Let (α, β) be a small vector that does not belong to the curve $\{(a_1(\varepsilon), b_1(\varepsilon)) \mid \varepsilon \in [-1, 1]\}$. Then no function in the family of maps $h_\varepsilon = g_\varepsilon - \alpha \sin 2\pi x - \beta \cos 2\pi x$ takes the form $x + r$. By Theorem 3.2, this family is normal.

Now let us prove that the set of normal families contains a residual subset. Denote by $\Lambda_{p/q}$ the set of the families from $\mathcal{B}^r$ for which assumption (3.6) is violated. The requirement $f_\varepsilon^q \not\equiv x + p$ for any $\varepsilon \in [-1, 1]$ is an open condition. Hence, $\Lambda_{p/q}$ is open. Its density was proved above. Hence, the set $\bigcap_{\mathbb{Q}} \Lambda_{p/q}$ is a residual subset that consists of normal families. Theorem 3.3 is proved.

§4. Bifurcations on a noncritical homoclinic torus of a generic saddlenode family

Summarizing the persistence theorem of §2 and the normality theorem of §3, we obtain a description of the bifurcations that occur on the homoclinic torus of a saddlenode cycle when the cycle itself disappears.

4.1. Homoclinic tori and global cross-sections. By the persistence theorem, a saddlenode family with noncritical homoclinic torus has a family of invariant tori that depend smoothly on the parameter. We can identify all these tori and regard the restrictions of the vector fields of the family to the homoclinic tori as

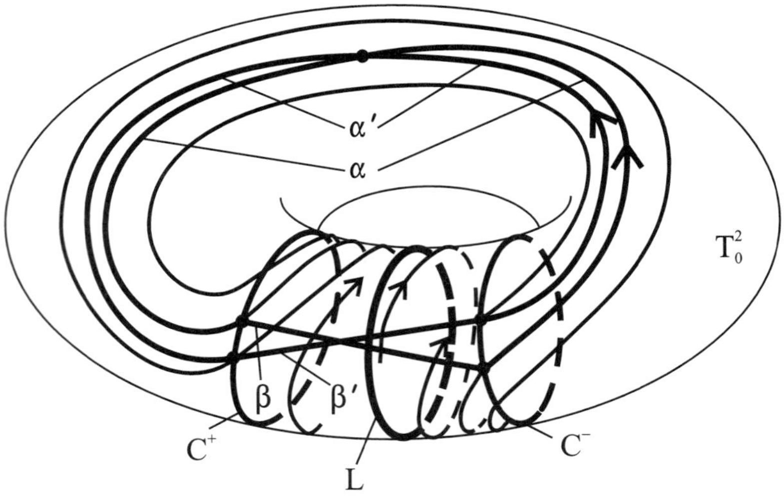

FIGURE 5.4. Construction of the global cross-section of the vector field X_0 on its homoclinic torus.

a family of vector fields on one and the same torus. We shall construct a global cross-section for the family of these vector fields, and check that the corresponding family of the Poincaré maps is, in general, normal.

THEOREM 4.1. *In a generic one-parameter family of vector fields in* $\mathbb{R}^n$, $n \geqslant 3$, *a vector field corresponding to the zero parameter value with the following properties may occur*:

1. *The vector field has a saddlenode cycle with multiplicity two.*
2. *The union of the cycle and its homoclinic orbits is a noncritical* 2*-torus.*

Then all the nearby vector fields of the family have an invariant finitely smooth 2*-torus. The restriction of these vector fields to these tori has a global cross-section smoothly depending on the parameter. The Poincaré map defined on this cross-section forms a normal family. The rotation number of these maps is zero to the left of* 0 *and takes nonzero values to the right of* 0.

The theorem is proved in this and the next three subsections. Denote by $\{X_\varepsilon\}$ the family of vector fields in Theorem 4.1 and by T_ε^2 the corresponding homoclinic tori.

Consider the torus T_0^2. It contains the saddlenode cycle L. All the other orbits of X_0 on this torus are homoclinic to L. Consider a narrow annular neighborhood U of L bounded by two circles C^+ and C^- transversal to X_0. All the orbits of X_0 except for L enter U through C^+ and exit through C^-. Take an arbitrary arc α of a phase curve of X_0 that starts on C^-, lands on C^+, and does not cross L. It may be completed to a piecewise smoothly embedded circle by adding an arc $\beta \subset U$ transversal to X_0 (see Figure 5.4).

Suppose that the saddlenode cycle L crosses the oriented arc β from the right to the left. In the opposite case "right" and "left" in this paragraph should be interchanged. Take a narrow flow box that contains α and has its top on C^+ and bottom on C^-. Take a "diagonal" α' of this flow box transversal to X_0 and oriented from C^- to C^+ in such a way that the orbits of X_0 cross α' from the right to the left. Add an arc β' close to β that, together with α', forms a smoothly embedded circle γ transversal to X_0. This is the desired cross-section.

As was explained above, all the tori may be identified and the restrictions of X_ε to T^2_ε may be regarded as a smooth family on the torus T^2_0. Then γ is the desired cross-section.

The family of the Poincaré maps $\mathcal{P}_\varepsilon : \gamma \to \gamma$ is of the very kind that was studied in 3.5–3.8. We will prove that this family is generically normal. To do this, we study the global correspondence map constructed for $\varepsilon > 0$ in the next subsection.

4.2. Two Poincaré maps of the same flow and their rotation numbers. The homoclinicity assumption of Theorem 4.1 implies that there exists a diffeomorphism $\Delta_0^{\text{reg}} : C^- \to C^+$ along the orbits of the vector field X_0. The nearby diffeomorphism $\Delta_\varepsilon^{\text{reg}} : C^- \to C^+$ is well defined for any small value of ε. On the other hand, a map $\Delta_\varepsilon^{\text{sing}} : C^+ \to C^-$ along the orbits of X_ε located in U is well defined for $\varepsilon > 0$. Hence, the family of maps

$$\mathcal{Q}_\varepsilon = \Delta_\varepsilon^{\text{sing}} \circ \Delta_\varepsilon^{\text{reg}} : C^- \to C^-$$

is well defined for small $\varepsilon > 0$. Below we shall prove that this family is monotonic for small positive ε. Now we describe the relation between the rotation numbers of the diffeomorphisms $\mathcal{P}_\varepsilon$ and $\mathcal{Q}_\varepsilon$.

PROPOSITION 4.1. *The rotation numbers of the maps $\mathcal{P}_\varepsilon$ and $\mathcal{Q}_\varepsilon$ are reciprocals.*

PROOF. The intersection index of the cycles γ and C^- equals 1. Hence, the cycles themselves are the generators of the group $H_1(T^2)$. Therefore, the torus T^2 may be represented as $T^2 = \mathbb{R}^2/\mathbb{Z}^2$ with the projection $\pi : \mathbb{R}^2 \to T^2$. Moreover, this projection brings the lines $\mathbb{R}^1 \times \{0\}$ and $\{0\} \times \mathbb{R}^1$ to loops homological to γ and C^- respectively. The vector field X_ε may be lifted to $\mathbb{R}^2$. The phase curves of this field have asymptotic directions. The tangent of the angle between this direction and the horizontal line is equal to $r(\mathcal{Q}_\varepsilon)$. The analogous tangent for the vertical direction is $r(\mathcal{P}_\varepsilon)$. These tangents are reciprocals. This proves the proposition.

4.3. Normal form of the family of singular correspondence maps.

LEMMA 4.1. *There exist coordinates φ and φ^- on C^+ and C^- depending smoothly on ε on some segment with left extremity 0 such that the singular correspondence map takes the form*

$$\Delta_\varepsilon^{\text{sing}} = \varphi - T(\varepsilon), \tag{4.1}$$

where

$$T(\varepsilon) = \frac{2}{\sqrt{\varepsilon}} \arctan \frac{\delta}{\sqrt{\varepsilon}}, \tag{4.2}$$

and δ depends on the choice on the charts φ and φ^-.

PROOF. Let L be a saddlenode cycle of X_0, S a cross-section transversal to L, and P_ε the Poincaré map of X_ε.

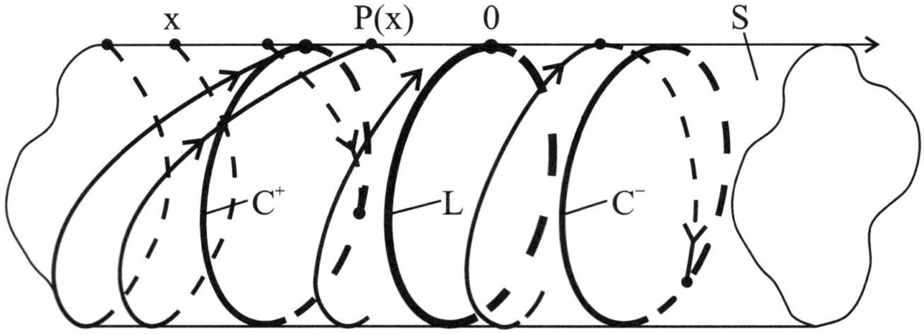

FIGURE 5.5. Construction of the special charts on the cross-sections C^-, C^+.

The genericity assumption of §1 implies that $\{P_\varepsilon\}$ is a generic saddlenode local family. By the partial embedding theorem (Theorem 2.5.4), the family $\{P_\varepsilon\}$ for $\varepsilon \geqslant 0$ is finitely smooth equivalent to a time one transformation of the equation

$$\dot{x} = (x^2 + \varepsilon)(1 + a(\varepsilon)x)^{-1}. \tag{4.3}$$

The local dynamics prescribes special charts on C^+ and C^-, generated by the chart rectifying vector field (4.3). The latter charts are equal to

$$t_\varepsilon^\pm = \int_{\pm\delta}^{x} \frac{1 + a(\varepsilon)y}{y^2 + \varepsilon}\, dy,$$

where δ is a small positive number. Both charts are defined in the neighborhood of zero for $\varepsilon > 0$. The chart t_ε^- is defined to the left, and t_ε^+ to the right of 0 for $\varepsilon = 0$. Note that the rectifying chart is well defined up to a shift. This shift is specified by the choice of δ.

Denote by φ and φ^- the ε-dependent coordinates on C^+ and C^- generated by $t_\varepsilon^\pm$ in the following way. Let π_ε^+ be the projection of $S \cap \{x < 0\}$ to C^+ along the orbits of X_ε, well defined for $\varepsilon \geqslant 0$. Let π_ε^- be the analogous projection of $S \cap \{x > 0\}$ to C^- (see Figure 5.5). Take

$$\varphi(\pi^+(x)) = t_\varepsilon^-(x) \pmod 1, \qquad \varphi^-(\pi^-(x)) = t_\varepsilon^+(x) \pmod 1.$$

For a point $p \in C^+$ there are several points x with $\pi^+(x) = p$. But they belong to the same orbit of the Poincaré map that preserves the functions $t_\varepsilon^\pm(x) \pmod 1$:

$$t_\varepsilon^\pm(P_\varepsilon(x)) = t_\varepsilon^\pm(x) \pmod 1.$$

The map $\Delta_\varepsilon^{\text{sing}}\colon C^+ \to C^-$ in these charts takes the form (4.1) with

$$T(\varepsilon) = \int_{-\delta}^{\delta} \frac{1 + a(\varepsilon)x}{x^2 + \varepsilon}\, dx = \frac{2}{\sqrt{\varepsilon}} \arctan \frac{\delta}{\sqrt{\varepsilon}}.$$

This proves the lemma.

4.4. Normality of the family $\{\mathcal{P}_\varepsilon\}$. In order to prove that the family $\{\mathcal{P}_\varepsilon\}$ is normal for small ε, we will prove that the family $\{\mathcal{Q}_\varepsilon\}$ is normal for small positive ε.

For small $\varepsilon \leqslant 0$, we have $r(\mathcal{P}_\varepsilon) = 0$ because $\mathcal{P}_\varepsilon$ has a fixed point.

LEMMA 4.2. *The family $\mathcal{Q}_\varepsilon$ is normal for small $\varepsilon > 0$ and for any generic family $\{\Delta_\varepsilon^{\mathrm{reg}}\}$.*

PROOF. We will prove the lemma by making use of Proposition 3.6 rather than Theorem 3.3.

To check the assumptions of this proposition, let us first prove monotonicity. By Lemma 4.1, $\mathcal{Q}_\varepsilon = \Delta_\varepsilon^{\mathrm{reg}} - T(\varepsilon)$. Hence,

$$\frac{\partial}{\partial \varepsilon}\mathcal{Q}_\varepsilon = \frac{\partial}{\partial \varepsilon}\Delta_\varepsilon^{\mathrm{reg}} - T'(\varepsilon).$$

By (4.2), $T'(\varepsilon) \to -\infty$ as $\varepsilon \to 0^+$. Hence, $\partial\mathcal{Q}_\varepsilon/\partial\varepsilon > 0$ for small positive ε.

The inequality $\mathcal{Q}_\varepsilon^q \not\equiv \varphi + p$ is fulfilled for any generic family $\Delta_\varepsilon^{\mathrm{reg}}$ and arbitrary integers p and q. This is proved in the same way as the genericity statement in Theorem 3.3.

Hence, by Proposition 3.6, the family $\mathcal{Q}_\varepsilon$ is normal. The conjugate (for $\varepsilon > 0$) family $\mathcal{P}_\varepsilon$ is also normal because normality (as opposed to monotonicity) is an invariant property.

On the other hand, generic families of vector fields correspond to generic families of regular maps. This is proved by standard gluing arguments, and we will not go into details.

Lemmas 4.1 and 4.2, with the last remark, prove Theorem 4.1.

§5. The blue sky catastrophe on the Klein bottle

In this section we will study the global bifurcations on a Klein bottle having a nonhyperbolic periodic orbit at the critical parameter value. The Klein bottle itself is the union of this orbit and its homoclinic curves.

5.1. Existence of the blue sky catastrophe. In Chapters 3 and 4 we described two ways in which the periodic orbit can disappear via a nonlocal bifurcation. The first bifurcation occurs when the cycle tends to the homoclinic loop of a saddle. The second is the same effect with the homoclinic orbit of a saddlenode. These two bifurcations have a common property: the period of the parameter-dependent periodic orbit tends to infinity, but the length of the orbit is bounded as the parameter tends to its critical value. The reason is that the cycle approaches the homoclinic orbit of the critical vector field, and the orbits run very slowly near singular points. In this section we will study another kind of disappearance of the periodic orbit, the so-called *blue sky catastrophe*. Let us begin with its definition.

DEFINITION 5.1. A one-parameter family of vector fields on a compact manifold has a *blue sky catastrophe* if for all parameter values on one side of a critical value, the field of the family has a periodic orbit with the following properties. The orbit continuously depends on the parameter; its period tends to infinity as the parameter tends to the critical value; the distance from the orbit to the singular points of the field has a positive lower bound.

The name stresses that the periodic orbit, becoming longer and longer, disappears not on a polycycle, but "into the clear blue sky".

Palis and Pugh [**PP**] stated the following problem: Can a blue sky catastrophe occur in generic one-parameter families of vector fields? The following theorem gives an affirmative answer to this question.

THEOREM 5.1. *Let a one-parameter family of vector fields on a Klein bottle contain a field having a nonhyperbolic semistable cycle with multiplier* 1. *Let all the other orbits of this field be homoclinic for this semistable cycle. Then, under certain genericity assumptions, a blue sky catastrophe occurs in this family at the moment when the semistable cycle vanishes.*

Genericity assumption. The semistable cycle bifurcates in a typical way; see fundamental assumptions 1, 2 in 1.1, taking $n = 2$, $u = s = 0$.

PROOF. We begin with some topological remarks. Denote by L and X_0 the semistable cycle and the corresponding vector field from Theorem 5.1. The multiplier of L equals 1. Hence, the corresponding Poincaré map preserves orientation. This implies that L is a coorientable curve on the Klein bottle. That is, it has a neighborhood homeomorphic to an annulus.

The crucial property of the Klein bottle is that any closed curve that has a unique point of transversal intersection with a coorientable curve is, in turn, not coorientable. This means that it has a neighborhood homeomorphic to a Möbius band.

Consider an annular neighborhood U of the semistable cycle L bounded by two curves transversal to the vector field X_0, as it was done in 4.3. Denote by C^+ one of the curves through which the orbits enter the neighborhood, and by C^- another one through which they exit. The homoclinicity assumption of Theorem 5.1 implies that there exists a diffeomorphism $\Delta_0^{\text{reg}}: C^- \to C^+$ along the orbits of the field X_0. Any arc of the orbit of X_0 emanating from C^- and terminating on C^+ may be closed by adding an arc having only one intersection with L. The loop γ thus obtained is not coorientable. The circles C^- and C^+ have an agreeing orientation induced by the representation of the annulus U as the Cartesian product of an oriented circle L by a segment. This orientation is reversed by the map Δ_0^{reg}, because the loop γ has a Möbius band as a neighborhood.

For nearby vector fields X_ε, the smooth family of maps $\Delta_\varepsilon^{\text{reg}}: C^- \to C^+$ along the orbits of X_ε is well defined.

By the genericity assumption, the cycle L splits into two on one side (say, to the left) of the critical (zero) value of the parameter ε, and vanishes for small positive parameter values. For the latter values the map $\Delta_\varepsilon^{\text{sing}}: C^+ \to C^-$ is well defined. It is orientation-preserving. The map

$$\Delta_\varepsilon = \Delta_\varepsilon^{\text{sing}} \circ \Delta_\varepsilon^{\text{reg}}: C^- \to C^-,$$

which is well defined for $\varepsilon > 0$, is orientation-reversing. Fixed points of this map correspond to cycles that are subject to the blue sky catastrophe. The following proposition proves the existence of these points.

PROPOSITION 5.1. *An orientation-reversing diffeomorphism of the circle has exactly two fixed points.*

PROOF. The proof is shown in Figure 5.6.

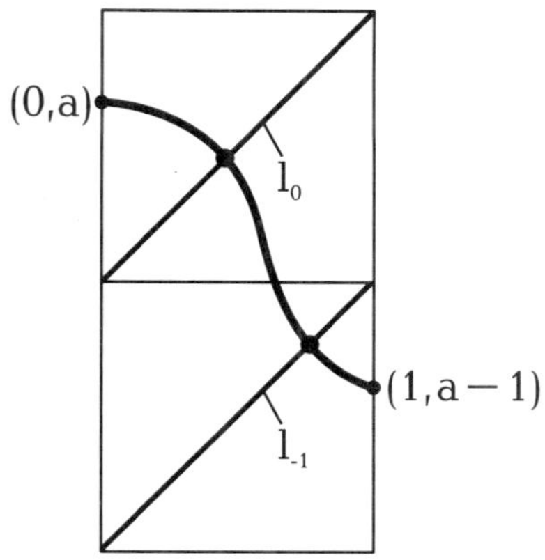

FIGURE 5.6. Fixed points of an orientation-reversing diffeomorphism of the circle.

Let f be the map in Proposition 5.1. Consider the graph of the covering map $\tilde{f}$ of the line corresponding to f:

$$\tilde{f}(t+1) = \tilde{f}(t) - 1, \qquad \tilde{f}' < 0.$$

Without loss of generality, this map may be chosen so that $\tilde{f}(0) = a \in [0,1)$. Fixed points of f satisfy the equation $\tilde{f}(t) = t \pmod{\mathbb{Z}}$. Figure 5.6 shows that the graph $\tau = \tilde{f}(t)$ intersects each of the lines l_0: $\tau = t$, l_{-1}: $\tau = t-1$, exactly once and never intersects any line l_n: $\tau = t + n$ for $n \neq 0, -1$. Indeed, the endpoints of the graph are $(0, a)$, $(1, a-1)$. They lie on different sides of the lines l_0, l_{-1}, and on one side of any other line l_n.

Proposition 5.1 proves the existence of two cycles that pass through fixed points of Δ_ε. The time length of the arc of these cycles connecting C^+ and C^- in the annular neighborhood U of the cycle L (that has already vanished) tends to infinity as ε tends to 0^+. Hence, the period of these cycles tends to infinity. When $\varepsilon = 0$, the cycles disappear, while a semistable cycle, distant from the previous ones, is born. Note that vector fields of the family have no singular points at all. Hence, the described evolution is really a blue sky catastrophe in the sense of Definition 5.1. This proves Theorem 5.1.

5.2. Sparkling flip cycles. It turns out that in generic families from Theorem 5.1, sparkling flip cycles occur. The family must satisfy some extra genericity assumptions.

THEOREM 5.2. *For generic families of vector fields on the Klein bottle satisfying assumptions of Theorem* 5.1, *there exists a sequence of parameter values tending to the critical value zero such that for any of them the corresponding vector field has a flip cycle, that is, a periodic orbit with multiplier* -1. *The above sequence may be split into an even number of sequences* $\{\varepsilon_n^i \mid n \in \mathbb{Z}^+, n > n_0\}$ *with* n_0 *sufficiently large so that*

$$\varepsilon_n^i = \frac{c}{(n+a_i)^2}(1 + o(1/n)), \qquad c \neq 0. \tag{5.1}$$

Here the numbers a_i are some constants depending on the vector field corresponding to the critical value of the parameter.

The origin of these constants will be explained when we state the genericity assumptions below.

5.3. Genericity assumptions. We will formulate the genericity assumptions by making use of Lemma 4.1. Let φ and φ^- be the charts on C^+ and C^- given by this lemma. In these charts the map $\Delta_\varepsilon^{\text{reg}}$ takes the form

$$\Delta_\varepsilon^{\text{reg}}(\varphi^-) = -\varphi^- + h_\varepsilon(\varphi^-), \tag{5.2}$$

where h_ε is a 1-periodic function: $h_\varepsilon(t+1) = h_\varepsilon(t)$. Now the genericity assumption for Theorem 5.2 may be stated, in addition to those of Theorem 5.1.

The function $h = h_0$ has nondegenerate critical points only.

REMARK. Before the charts φ and φ^- are introduced, the function h, together with the previous condition, has no invariant meaning. Only in charts defined in an invariant way is the function h well defined (up to a shift in the image and preimage).

Now we can specify the values a_j in Theorem 5.2. Let α_j be the critical points of the function h. In the proof of Theorem 5.2 below it is shown that

$$a_i = -2\alpha_i + h(\alpha_i) + 2/\delta. \tag{5.3}$$

5.4. Existence of sparkling flip cycles. The equation of the cycles has the form

$$\Delta_\varepsilon(t) = t.$$

Substituting (5.2) into the definition of Δ_ε, then into the last equation, we get

$$-\varphi + h_\varepsilon(\varphi) - T(\varepsilon) = \varphi. \tag{5.4}$$

Now a transparent heuristic explanation of the existence of flip cycles may be given. Suppose that h_ε does not really depend on ε. Then the graph of Δ_ε is just the vertically shifted graph of $f = \Delta_\varepsilon^{\text{reg}}$. The shift length $T(\varepsilon)$ tends to infinity as $\varepsilon \to 0$. The graph of f has the slope -1 at isolated points. Any time when the shifted graph intersects a line l_n: $\tau = \varphi + n$ at a point where the graph has the slope -1, a flip cycle occurs. As $T(\varepsilon) \to \infty$, there is a countable number of such intersections. This gives a countable number of values of the parameter corresponding to flip cycles.

Now we proceed to the rigorous proof of Theorem 5.2. The equation of the flip cycles has the form:

$$\Delta_\varepsilon(\varphi) = \varphi, \qquad \Delta'_\varepsilon(\varphi) = -1.$$

It is equivalent to

$$-\varphi + h_\varepsilon(\varphi) - T(\varepsilon) = \varphi \pmod{\mathbb{Z}}, \qquad h'_\varepsilon(\varphi) = 0. \tag{5.5}$$

Denote, as before, the critical points of h by α_i, $i = 1, \dots, 2k$ (if all the critical points of the function on the circle are nondegenerate, then their number is even). For any i, the equation $h'_\varepsilon(\varphi) = 0$, $\varphi(0) = \alpha_i$ determines a smooth function $\varphi_i(\varepsilon)$. Indeed, the implicit function theorem is applicable, because $h''(\alpha_i) \neq 0$.

Drop the index i for simplicity. Note that $T(\varepsilon) \to +\infty$ as $\varepsilon \to 0^+$. System (5.5) takes the form

$$g(\varepsilon) + n = T(\varepsilon), \qquad n \in \mathbb{Z}, \tag{5.6}$$

where $g(\varepsilon) = -2\varphi(\varepsilon) + h_\varepsilon(\varphi(\varepsilon))$ is a finitely smooth function.

PROPOSITION 5.2. *For any interval $B \ni 0$, equation (5.6) with C^1 smooth function g has a countable number of solutions*

$$\varepsilon_n = \frac{\pi^2}{(n+a)^2}\Big(1 + o\Big(\frac{1}{n}\Big)\Big), \tag{5.7}$$

where $a = g(0) + 2/\delta$, and n is large.

PROOF. An easy calculation shows that

$$T(\varepsilon) = \frac{\pi}{\sqrt{\varepsilon}} - \frac{2}{\delta} + o(1).$$

Suppose that, as in Proposition 5.2, $a = g(0) + 2/\delta$. Then equation (5.6) takes the form $a + o(1) + n = \pi/\sqrt{\varepsilon}$. It has the solution

$$\varepsilon_n = \frac{\pi^2}{(n+a)^2}\Big(1 + o\Big(\frac{1}{n}\Big)\Big),$$

which belongs to B for a sufficiently large n. This proves the proposition and hence the theorem.

5.5. Sparkling semistable cycles. In the last part of this section we will show that a semistable cycle may occur in generic families from Theorem 5.1 under some extra genericity assumptions. The following theorem repeats almost verbatim Theorem 5.2, with the only difference that flip cycles are replaced by saddlenode (semistable) ones.

THEOREM 5.3. *For generic families of vector fields on the Klein bottle satisfying the conditions of Theorem 5.1, there exists a sequence of parameter values tending to the critical value zero such that for any of them the corresponding vector field has a semistable cycle, that is, a periodic orbit with multiplier 1. The above sequence may be split into an even number of sequences $\{\varepsilon_n^i \mid n \in \mathbb{Z}\}$ so that*

$$\varepsilon_n^i = \frac{c}{(n+b_i)^2}\Big(1 + o\Big(\frac{1}{n}\Big)\Big), \qquad c \neq 0. \tag{5.8}$$

Here the b_i are some constants depending on the vector field corresponding to the critical value of the parameter.

5.6. Heuristic proof. A fixed point with multiplier 1 cannot occur for the map Δ_ε because this map is orientation-reversing. Yet it can occur for the map $\Delta_\varepsilon^{[2]}$. The corresponding parameter value is given by the system

$$\begin{cases} \Delta_\varepsilon^{[2]}(\varphi) = \varphi, \\ \partial\Delta_\varepsilon^{[2]}(\varphi)/\partial\varphi = 1. \end{cases}$$

By the definition of the map Δ_ε, this system is equivalent to

$$\begin{cases} h_\varepsilon(-\varphi + h_\varepsilon(\varphi) - T(\varepsilon)) = h_\varepsilon(\varphi), \\ \dfrac{\partial}{\partial\varphi} h_\varepsilon(-\varphi + h_\varepsilon(\varphi) - T(\varepsilon)) = \dfrac{\partial}{\partial\varphi} h_\varepsilon(\varphi)\Big/\Big(\dfrac{\partial}{\partial\varphi} h_\varepsilon(\varphi) - 1\Big). \end{cases} \tag{5.9}$$

In order to give the heuristic proof, we suppose that the function $h_\varepsilon = h_0 = h$ does not depend on ε. Then (5.9) becomes

$$\begin{cases} h(-\varphi + h(\varphi) - T(\varepsilon)) = h(\varphi), \\ h'(-\varphi + h(\varphi) - T(\varepsilon)) = h'(\varphi)/(h'(\varphi) - 1). \end{cases} \tag{5.10}$$

Now consider two closed curves in the (x, y)-plane:

$$\begin{aligned} &\gamma\colon\ x = h(\varphi),\ y = h'(\varphi), && \varphi \in \mathbb{R}, \\ &\widetilde{\gamma}\colon\ x = h(\varphi),\ y = h'(\varphi)/(h'(\varphi) - 1), && \varphi \in \mathbb{R}. \end{aligned} \tag{5.11}$$

Recall that the function h is 1-periodic. Hence, γ and $\widetilde{\gamma}$ are two closed curves. Suppose that no intersection point of these curves is at the same time the self-intersection point of any of them. Any point of the curve γ or $\tilde{\gamma}$ that is not a self-intersection point corresponds to the same parameter value, well defined modulo $\mathbb{Z}$. The equation (5.10) takes the form:

$$\gamma(-\varphi + h(\varphi) - T(\varepsilon)) = \widetilde{\gamma}(\varphi).$$

The way of solving system (5.9) will be the following. We take some intersection point p of the curves γ and $\widetilde{\gamma}$. Suppose it corresponds to the parameter values ψ (mod $\mathbb{Z}$) on γ and φ (mod $\mathbb{Z}$) on $\widetilde{\gamma}$. Let φ be a solution of (5.10). Then p is, at the same time, the right- and left-hand side of (5.10), $p = \gamma(\psi) = \widetilde{\gamma}(\varphi)$, and

$$\psi = -\varphi + h(\varphi) - T(\varepsilon) \pmod{\mathbb{Z}}.$$

This equation has the form (5.6), i.e.,

$$b + n = T(\varepsilon), \qquad n \in \mathbb{Z},$$

where $g(\varepsilon) \equiv \text{const} = -\varphi - \psi + h(\varphi)$.

This is an equation of type (5.6) with left-hand side not depending on ε. This equation has a sequence of solutions (5.7) for $a = b + 2/\delta$, by Proposition 5.2. This proves Theorem 5.3 for h_ε not depending on ε, modulo the statement that the number of the sequences (5.7) is even. The latter statement is proved in 5.8 below. The general proof follows the same lines.

5.7. Genericity assumptions in Theorem 5.3. We now specify the genericity assumptions which must be satisfied by the family X_ε in Theorem 5.3. They include the assumptions from 5.2, which will be repeated for convenience. As before, we assume that the Poincaré map for the semistable cycle L is a generic saddlenode family. Let h_ε be the same family of periodic functions as in 5.2, $h = h_0$. Suppose as before that h has nondegenerate critical points only. Let γ and $\widetilde{\gamma}$ be the same curves as in (5.11). Suppose that:

(1) Each of the curves γ and $\widetilde{\gamma}$ has at most a finite number of self-intersection points, none of them coinciding with their intersection point.
(2) The curves γ and $\widetilde{\gamma}$ have transversal intersections outside the x-axis.
(3) If $\alpha \in \mathbb{R}$ is a critical point of h, then

$$3h''(\alpha) + 2h'''(\alpha) \neq 0.$$

The last assumption describes the intersection of γ and $\widetilde{\gamma}$ lying on the x-axis.

This intersection is nonempty. Indeed, by assumption of 5.2, h has a nonempty finite set of critical points on the circle. Any such point provides a parameter value, which determines the same point $(h(\varphi), 0)$ on the curves γ and $\tilde{\gamma}$.

5.8. Intersections of γ and $\widetilde{\gamma}$.

LEMMA 5.2. *Under assumptions* (1)–(3) *the curves* (5.11) *have an even number of intersection points both on and outside the x-axis.*

PROOF. We begin with the points on the x-axis. By definition (5.11), the curves γ and $\widetilde{\gamma}$ intersect the x-axis at the points corresponding to the parameter values equal to the critical points of h. By the assumptions of 5.7 and 5.2, the function h has an even number of critical points in the interval $[0,1)$. Let $\alpha_1, \dots, \alpha_{2r} \in [0,1)$, $r \geqslant 1$, denote these critical points. Then,

$$\{P_i = (h(\alpha_i), 0)\}_{1 \leqslant i \leqslant 2r} = \gamma \cap \widetilde{\gamma} \cap \{y = 0\}.$$

Now let us prove that *the curves γ and $\tilde{\gamma}$ have cubic tangency at the intersection points lying on the x-axis.* Hence, the first of them passes at this point from one side of the second curve to the other: the curves cross over. Indeed, the critical points of h are nondegenerate. Hence, the y-coordinate may be taken as a parameter on both curves near P_i, i.e., γ: $x = f_i(y)$, $\tilde{\gamma}$: $x = \tilde{f}_i(y)$, where

$$f_i(h'(\varphi)) = h(\varphi), \tag{5.12}$$

$$\tilde{f}_i(h'(\varphi)/(1 - h'(\varphi))) = h(\varphi), \tag{5.13}$$

and $P_i = (f_i(0), 0) = (\tilde{f}_i(0), 0)$. Drop the subscript i for simplicity. By differentiating (5.12) and (5.13) at zero, we can easily get

$$f(0) = \tilde{f}(0) = h(\alpha), \qquad f'(0) = \tilde{f}'(0) = 0, \qquad f''(0) = \tilde{f}''(0) = h''(\alpha)^{-1},$$

and

$$f'''(0) = -2h'''(\alpha)h''(\alpha)^{-3}, \qquad \tilde{f}'''(0) = 6h''(\alpha)^{-1} + 2h'''(\alpha)h''(\alpha)^{-3}.$$

Therefore, by assumption (3),

$$\tilde{f}'''(0) - f'''(0) = 2h''(\alpha)^{-3}(3h''(\alpha)^2 + 2h'''(\alpha)) \neq 0.$$

This implies that γ and $\tilde{\gamma}$ have the desired cubic tangency and cross over at the points of the x-axis.

Now we can finish the proof of Lemma 5.2. For each self-intersection point $p \in \gamma$ (or $\widetilde{\gamma}$) we change the curve γ (resp. $\widetilde{\gamma}$) in a small neighborhood of p in the way shown in Figure 5.7.

Then the curve γ (resp. $\widetilde{\gamma}$) becomes the union of a finite number of disjoint simple closed curves γ_i (resp. $\widetilde{\gamma}_j$). On the other hand, by the argument above and by assumption (2), γ_i and $\widetilde{\gamma}_j$ cross over each other at each intersection point. Therefore, the number of intersection points of γ_i and $\widetilde{\gamma}_j$ for any i, j is even. The lemma now follows, because

$$\#\{\gamma \cap \widetilde{\gamma}\} = \sum_{i,j} \#\{\gamma_i \cap \widetilde{\gamma}_j\}.$$

5.9. Proof of the theorem. In order to solve the system of equations (5.9), we now consider two closed curves depending on ε on the (x,y)-plane defined by (5.11) with h replaced by h_ε:

$$\gamma_\varepsilon : \ x = h_\varepsilon(t), \ y = h'_\varepsilon(t), \qquad t \in \mathbb{R},$$

$$\widetilde{\gamma}_\varepsilon : \ x = h_\varepsilon(t), \ y = h'_\varepsilon(t)/(h'_\varepsilon(t) - 1), \quad t \in \mathbb{R}.$$

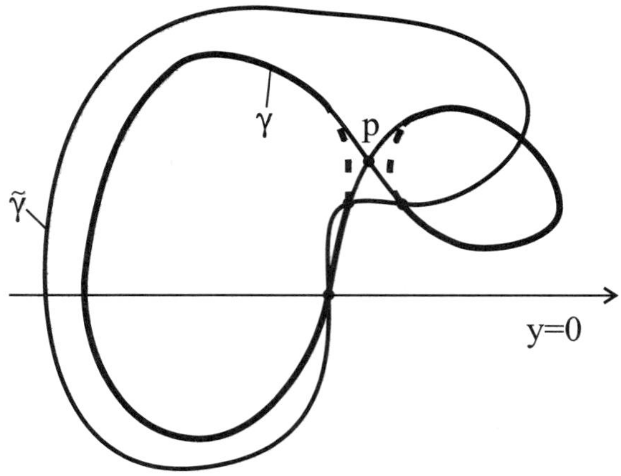

FIGURE 5.7. Even number of intersections of the curves γ and $\tilde{\gamma}$.

These curves satisfy assumptions (1)–(3) with h replaced by h_ε for small ε. Hence, by Lemma 5.2, they have an even number of intersections outside the x-axis. Denote these points by $p_1, \ldots, p_{2s}$. by assumption (2), the intersections are transversal. Hence, for small ε any point p_i defines a smooth map P_i and smooth functions ψ_i, φ_i such that

$$P_i(\varepsilon) \in \gamma_\varepsilon \cap \tilde{\gamma}_\varepsilon \cap \{y \neq 0\}, \qquad P_i(\varepsilon) = \gamma_\varepsilon(\psi_i(\varepsilon)) = \tilde{\gamma}_\varepsilon(\varphi_i(\varepsilon)).$$

A point (φ, ε) is a solution of (5.9) with $h_\varepsilon'(\varphi) \neq 0$ if and only if for some i

$$\varphi = \varphi_i(\varepsilon), \qquad \psi_i(\varepsilon) = -\varphi_i(\varepsilon) + h_\varepsilon(\varphi_i(\varepsilon)) - T(\varepsilon) \pmod{\mathbb{Z}}.$$

This equation has the form (5.6) with $g(\varepsilon) = -\varphi_i(\varepsilon) - \psi_i(\varepsilon) + h_\varepsilon(\tilde{\varphi}_i(\varepsilon))$. Proposition 5.2 now gives $2s$ sequences of solutions as in Theorem 5.3.

Now let us consider the intersection points of $\gamma_\varepsilon \cap \tilde{\gamma}_\varepsilon$ lying on the x-axis. They correspond to the critical values of h_ε. A pair (φ, ε) with $h_\varepsilon'(\varphi) = 0$ satisfies (5.9) if and only if $-2\varphi + h_\varepsilon(\varphi) = T(\varepsilon) \pmod{\mathbb{Z}}$. This is the case already investigated in 5.4. The corresponding flip cycle transforms to a semistable one when it is run over twice. Reference to Theorem 5.2 completes the proof of Theorem 5.3.

§6. Generalized Smale horseshoe existence theorem

In this section we shall generalize the Smale horseshoe existence theorem of Chapter 2 to the so-called *partially hyperbolic case*. The results of this section will be used to study the bifurcation of several noncritical homoclinic tori or Klein bottles in the next section. The main theorem of this section, stated in 6.2, will be proved in Chapter 8.

6.1. Generalized cone condition, annular solids and statement of the theorem. In the next section we will study the bifurcation of several homoclinic tori or Klein bottles by considering the Poincaré map of the nonhyperbolic cycle which is a saddle in its hyperbolic variables. The Poincaré map of such a cycle is not a hyperbolic but a partially hyperbolic one. Roughly speaking, it is hyperbolic in the direction of the hyperbolic variables but exhibits neutral growth in the direction

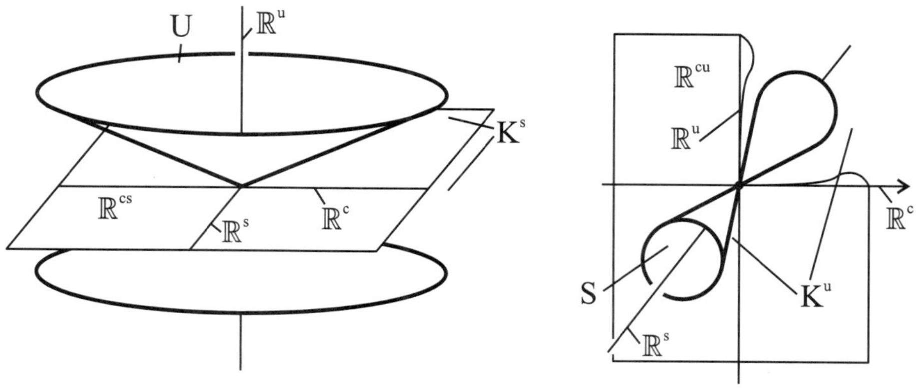

FIGURE 5.8. Cone family in the theory of partially hyperbolic maps.

of the center manifold. In order to study this kind of mapping, we must generalize the concept of hyperbolicity to the case where "center variables" exist.

In this subsection and later, we suppose that μ_h, μ_v are two positive constants with $\mu_h\mu_v < 1$. We split the tangent space at any point p of $\mathbb{R}^n = \mathbb{R}^s \oplus \mathbb{R}^c \oplus \mathbb{R}^u$ into $T_p\mathbb{R}^s \oplus T_p\mathbb{R}^c \oplus T_p\mathbb{R}^u$. Tangent vectors will be denoted by ξ, η and so on, and split to the components: $\xi = (\xi^s, \xi^c, \xi^u)$, according to the previous splitting of the tangent space. For brevity denote

$$\xi^{cs} = (\xi^s, \xi^c), \quad \xi^{cu} = (\xi^c, \xi^u), \quad |\xi^{cs}|^2 = |\xi^c|^2 + |\xi^s|^2, \quad |\xi^{cu}|^2 = |\xi^c|^2 + |\xi^u|^2.$$

Now we define the (μ_h, μ_v) cone family as the following tuple:

$$K_p^s = \{\xi \in T_p\mathbb{R}^n : |\xi^u| \leqslant \mu_h|\xi^{cs}|\}, \qquad K_p^u = \{\xi \in T_p\mathbb{R}^n : |\xi^s| \leqslant \mu_v|\xi^{cu}|\}$$

(see Figure 5.8).

Denote by U_p and S_p the complements of K_p^s and K_p^u respectively. For any map f and p from the domain of $f, \xi \in T_p\mathbb{R}^n$, denote by η the vector $df(p)\xi$. Recall that η is split into the components (η^s, η^c, η^u) according to the initial splitting of $\mathbb{R}^n$.

DEFINITION 6.1. Let $D \subset \mathbb{R}^s \oplus \mathbb{R}^c \oplus \mathbb{R}^u$ be a closed region, and let $f\colon D \to \mathbb{R}^s \oplus \mathbb{R}^c \oplus \mathbb{R}^u$ be a diffeomorphism. We say that f satisfies the *generalized* (μ_h, μ_v) *cone condition* on D if there exists a constant $\lambda > 1$ such that

1. $df(K_p^u) \subset K_{f(p)}^u$.
2. $df^{-1}(K_{f(p)}^s) \subset K_p^s$.
3. For any $\xi \in U_p$, $|\eta^u| \geqslant \lambda|\xi^u|$.
4. For any $\eta \in S_{f(p)}$, $|\xi^s| \geqslant \lambda|\eta^s|$.

REMARKS. 1. Assumptions 1 and 2 of the cone condition imply that there are two extra invariant families of cones:

1) $df(U_p) \subset U_{f(p)}$;
2) $df^{-1}(S_{f(p)}) \subset S_p$.

Indeed, if a linear isomorphism maps one cone into another, then the inverse isomorphism maps the complement of the second cone into the complement of the first.

2. For the case $c = 0$, replace the space T^c by the single point 0. Then the generalized cone condition becomes just the ordinary one defined in Chapter 2.

6.2. A sufficient condition for the existence of a generalized horseshoe. In this subsection we denote the c-dimensional torus by $T^c = \mathbb{R}^c/\mathbb{Z}^c$. Denote by B^s and B^u balls of arbitrary radii centered at zero in $\mathbb{R}^s$ and $\mathbb{R}^u$ respectively. The product $\mathcal{B} = B^s \times T^c \times B^u$ is called a *standard solid*, and the sets

$$\partial_h \mathcal{B} = B^s \times T^c \times \partial B^u, \qquad \partial_v \mathcal{B} = \partial B^s \times T^c \times B^u$$

are called its *horizontal* and *vertical boundaries* respectively.

DEFINITION 6.2. A *generalized μ_h-horizontal surface H* in $\mathbb{R}^s \times T^c \times \mathbb{R}^u$ is defined as the graph of a μ_h-Lipschitz map $h\colon D \to \mathbb{R}^m$, where D is the closure of a connected domain in $\mathbb{R}^s \times T^c$. The graph of the restriction of h to the boundary of D is called the *boundary* of H. Similarly, a *generalized μ_v-vertical surface V* in $\mathbb{R}^s \times T^c \times \mathbb{R}^u$ and its boundary ∂V are defined; only μ_h is replaced by μ_v, "horizontal" by "vertical", while s and u change places.

Let $\mathcal{B} = B^s \times T^c \times B^u$ be a standard annular solid. The graph of a μ_h-Lipschitz map $h\colon B^s \times T^c \to B^u$ is a *generalized μ_h-horizontal surface in $\mathcal{B}$*. A *generalized μ_v-vertical surface in $\mathcal{B}$* is defined similarly.

DEFINITION 6.3. A closed region $D \subset \mathbb{R}^s \times T^c \times \mathbb{R}^u$ is said to be a (μ_h, μ_v) *annular solid* if there exists a homeomorphism $F\colon \mathcal{B} \to D$ of a standard solid with $D = F(\mathcal{B})$ such that:

1. For any $z \in B^u$, the set $F(H_z)$, $H_z = B^s \times T^c \times \{z\}$, is a generalized μ_h-horizontal surface.
2. For any $y \in B^s$, the set $F(V_y)$, $V_y = \{y\} \times T^c \times B^u$ is a generalized μ_v-vertical surface.

The sets $\partial_h D = F(\partial_h \mathcal{B})$ and $\partial_v D = F(\partial_v \mathcal{B})$ are called the *horizontal* and *vertical parts* of the boundary of D respectively.

Obviously, a standard solid is a (μ_h, μ_v) annular solid for any positive constants μ_h, μ_v with $\mu_h \mu_v < 1$.

DEFINITION 6.4. A (μ_h, μ_v) annular solid D is called *vertically cylindric* in a standard solid $\mathcal{B}$ if $D \subset \mathcal{B}$ and $\partial_v D \in \partial_v \mathcal{B}$. It is called *horizontally cylindric* if the same holds for v replaced by h in the previous formula, i.e., if $D \subset \mathcal{B}$, $\partial_h D \in \partial_h \mathcal{B}$.

The main content of this section is the theorem below.

In this theorem $f^l D$ stands for the image of the maximal subset of D in which the iterate f^l is well defined.

Denote by Σ^N the set of all the sequences of N symbols with the standard metric recalled in §7.1; σ denotes the Bernoulli shift on the space Σ^N.

THEOREM 6.1 (Generalized Horseshoe Existence Theorem). *Let $\mathcal{B} \subset \mathbb{R}^s \times T^c \times \mathbb{R}^u$ be a standard solid, and $S = \{1, \dots, N\}$. Let D_i be pairwise disjoint (μ_h, μ_v) annular solids vertically cylindric in $\mathcal{B}$, and let D_i' be pairwise disjoint (μ_h, μ_v) annular solids horizontally cylindric in $\mathcal{B}$, $i \in S$. Let $\mathbf{D} = \bigcup_{i=1}^N D_i$, $\mathbf{D}' = \bigcup_{i=1}^N D_i'$, and let $f\colon \mathbf{D} \to \mathbf{D}'$ be a diffeomorphism with the following properties:*

1. *f satisfies the generalized (μ_h, μ_v) cone condition.*

2. $f(D_i) = D_i'$, $f(\partial_h D_i) = \partial_h D_i'$, $f^{-1}(\partial_v D_i') = \partial_v D_i$.
Then there exists a map Φ *of*

$$\Lambda = \bigcap_{l=-\infty}^{\infty} f^l(\mathbf{D})$$

onto Σ^N *such that the diagram*

$$\begin{array}{ccc} \Lambda & \xrightarrow{f} & \Lambda \\ {\scriptstyle\Phi}\downarrow & & \downarrow{\scriptstyle\Phi} \\ \Sigma^N & \xrightarrow{\sigma} & \Sigma^N \end{array} \tag{6.1}$$

is commutative.

Moreover, there exists a homeomorphism $\Lambda \to \Sigma^N \times T^c$. *The fiber of the map* Φ *is a Lipschitz torus* T^c.

REMARKS. 3. The map Φ is just the map: point $\mapsto$ fate. Namely, $\varphi(x) = \dots\omega_0\dots\omega_n\dots$, $\omega_n = j$ iff $f^n(x) \in D_j$. See §2.3 or 8.4.3 and 8.5.5.

4. For the case $c = 0$, the theorem above is just the Smale horseshoe existence theorem stated in Chapter 2.

Theorem 6.1 will be proved in Chapter 8.

6.3. Properties sufficient for the linear operator to satisfy the generalized (μ_h, μ_v) **cone condition.** Consider Euclidean space $\mathbb{R}^n$ with the splitting

$$\mathbb{R}^n = \mathbb{R}^s \oplus \mathbb{R}^c \oplus \mathbb{R}^u, \qquad \mathbb{R}^{cu} = \mathbb{R}^c \oplus \mathbb{R}^u, \qquad \mathbb{R}^{cs} = \mathbb{R}^s \oplus \mathbb{R}^c.$$

In the statement of the following lemma we refer to the formulas of 4.2.6. In §7, where these formulas are established, they are repeated.

The generalized (μ_h, μ_v) cone condition for a linear operator A is obtained from Definition 6.1 by substituting A, A^{-1} for df, df^{-1}, taking $\eta = A\xi$ and dropping the subscripts p and $f(p)$. Namely, the condition claims that there exists a $\lambda > 1$ such that

1. $AK^u \subset K^u$.
2. $A^{-1}K^s \subset K^s$.
3. For any $\xi \in U$, $|\eta^u| \geqslant \lambda|\xi^u|$.
4. For any $\eta \in S$, $|\xi^s| \geqslant \lambda|\eta^s|$.

LEMMA 6.1. *Consider two invertible linear operators* H *and* Σ *of* $\mathbb{R}^n$ *into itself. Let* H *satisfy conditions* (4.2.1).

a) *Consider the splitting*

$$\mathbb{R}^n = \mathbb{R}^{cs} \oplus \mathbb{R}^u, \tag{6.2}$$

and the corresponding splitting of H *and* Σ *into blocks*:

$$H = \begin{pmatrix} a & b \\ c & d \end{pmatrix}, \qquad \Sigma = \begin{pmatrix} \Lambda & B \\ D & M \end{pmatrix}. \tag{6.3}$$

Let assumptions (4.2.2), (4.2.3), (4.2.13), (4.2.15)–(4.2.17) *hold.*

b) *Consider the splitting*

$$\mathbb{R}^n = \mathbb{R}^s \oplus \mathbb{R}^{cu}, \tag{6.4}$$

and the corresponding splitting of H *and* Σ *into blocks:*

$$H = \begin{pmatrix} a' & b' \\ c' & d' \end{pmatrix}, \qquad \Sigma = \begin{pmatrix} \Lambda' & B' \\ D' & M' \end{pmatrix}. \tag{6.5}$$

Let the assumptions (4.2.2), (4.2.13)–(4.2.16), (4.2.18) *hold with* Λ, M, B, D, c, *and* d *replaced by* Λ', M', B', D', c', *and* d'.

Then the operator $A = H\Sigma$ *satisfies the generalized* (μ_h, μ_v) *cone condition with* $\lambda = 2$. *In more details, for any sufficiently large* μ_v *depending on* L, *any* $\mu_h < 1/\mu_v$ *and any sufficiently small* δ *depending on* L, μ_v, *and* μ_h, *the operator* A *satisfying the above assumptions will, at the same time, satisfy the generalized* (μ_h, μ_v) *cone condition.*

PROOF. Assumption 2 in the generalized (μ_h, μ_v) cone condition may be replaced by the condition

1′. $AU \subseteq U$,

because U is the complement of K^s. Similarly, assumption 1 takes the form

2′. $A^{-1}S \subseteq S$.

Lemma 6.1 now follows from Proposition 4.2.3. Indeed, condition a) of the lemma allows us to apply Proposition 4.2.3 in the case a). This application yields assumptions 1′ and 3 of the generalized (μ_h, μ_v) cone condition for the operator A. Condition b) together with Proposition 4.2.3, case b), implies assumptions 2′ and 4.

This completes the proof of Lemma 6.1.

§7. Several noncritical homoclinic tori or Klein bottles of a nonhyperbolic cycle

In this section we will consider the bifurcation of a nonhyperbolic periodic orbit having $N > 1$ noncritical homoclinic tori or Klein bottles. By Theorem 2.1, for all values of the parameter near the critical one, these smooth tori or Klein bottles will persist. The bifurcations occurring on these surfaces have been studied in §4 and §5. The result of this section will show that except for the bifurcations considered in §4 and §5, a complicated invariant set containing an infinite number of tori or Klein bottles is created as the result of the disappearance of the cycle.

7.1. The bifurcation creating a partially hyperbolic invariant set: statement of the main theorem.

THEOREM 7.1. *Let* X_ε *be a generic one-parameter family of smooth vector fields in* $\mathbb{R}^n$, $n \geqslant 4$, *satisfying the fundamental assumptions from* §1 *so that* X_0 *has* $N > 1$ *pairwise disjoint noncritical homoclinic cylinders* H_i $(i = 1, \dots, N)$ *of a nonhyperbolic cycle* L. *Suppose that the unions* $H_i \cup L$ *are of type* $(1, \pm 1)$. *Then there exists a neighborhood* U *of* $\bigcup_{i=1}^N H_i \cup L$ *and a neighborhood* V *of* $\varepsilon = 0$ *such that for the parameter values on one side of* $\varepsilon = 0$ *in* V, *the field* X_ε *has two hyperbolic periodic orbits tending to* L *as* $\varepsilon \to 0$ *and no other nonwandering orbits in* U. *For the parameter* ε *on the other side of zero in* V, X_ε *has a compact invariant set* Ω_ε *in* U *tending to* $\bigcup_{i=1}^N H_i \cup L$ *as* $\varepsilon \to 0$ *with the following properties:*

1. Ω_ε *has an uncountable number of path-connected components each of which is a two dimensional Lipschitz cylinder, torus or Klein bottle.*
2. *The compact path-connected components,* i.e., *tori or Klein bottles, are dense in* Ω_ε.

3. *There exists an invariant cylinder which is a dense subset of* Ω_ε.

We will prove this theorem in the next eight subsections.

7.2. Normal form for the parameter-depending Poincaré map of a saddlenode cycle. Let Γ be a hypersurface transversal to the vector field X_0 at a point of L. Consider the Poincaré map P_ε of X_ε defined in some neighborhood Γ_0 of $L \cap \Gamma$ on Γ. The fundamental assumption 1 in §1 allows us to apply the theorem on the local normal form for the perturbation of a saddlenode (see §5 of Chapter 2). A finite smooth change of coordinates and parameter brings the local family of Poincaré maps P_ε to the form

$$\begin{gathered} P_\varepsilon(x,y,z) = (x',y',z'), \\ x' = f_\varepsilon(x) = x + (\varepsilon + x^2)\, b(x,\varepsilon), \quad y' = A(x,\varepsilon)\, y, \quad z' = B(x,\varepsilon)\, z, \\ (x,y,z) \in (\mathbb{R}^{1+s+u}, 0), \quad \varepsilon \in (\mathbb{R}, 0), \quad b(0,0) > 0, \end{gathered} \tag{7.1}$$

where $f_\varepsilon(x)|_{\varepsilon \geqslant 0}$ is the time one map of the equation

$$\dot{x} = w_\varepsilon(x), \qquad w_\varepsilon(x) = (x^2 + \varepsilon)(1 + a(\varepsilon)\, x)^{-1}, \tag{7.2}$$

and we have

$$\|A(x,\varepsilon)\| < \lambda < 1, \qquad \|B(x,\varepsilon)^{-1}\| < \lambda < 1 \tag{7.3}$$

for some constant λ. By (7.1), the planes $\{x = \pm\sqrt{-\varepsilon}\}$ are invariant under the action of the Poincaré map for $\varepsilon < 0$. Therefore, the orbits beginning on the left side of these invariant planes cannot go to their right side. This implies that there are no other nonwandering orbits of X_ε except for the periodic orbits corresponding to the fixed points $(\pm\sqrt{-\varepsilon}, 0, 0)$ of P_ε. This proves Theorem 7.1 for $\varepsilon < 0$. In what follows we assume that $\varepsilon \geqslant 0$.

7.3. Singular correspondence map. Take a tubular neighborhood U of a saddlenode cycle L of the vector field X_0. Let (x,y,z) be the ε-dependent normalizing chart for the Poincaré map (see §7.2). Extend the function x from the section Γ_0 to U in such a way that

- the plane $x = 0$ is invariant for X_0 and contains L;
- the function x monotonically increases in U along the orbits of any vector field X_ε for small $\varepsilon > 0$;
- the same holds for X_0 in $U \setminus \{x = 0\}$.

For small positive ε let us define a singular correspondence map $\Delta_\varepsilon^{\text{sing}}$ from a solid torus in $\{x = -\delta\}$ to one in $\{x = \delta\}$, and describe its normal forms and normalizing coordinates. These coordinates are generated by the normalizing coordinates of the Poincaré map as described below.

DEFINITION 7.1. The *singular correspondence map* $\Delta_\varepsilon^{\text{sing}}$ is the map of the intersection $U \cap \{x = -\delta\}$ to $U \cap \{x = \delta\}$ along the orbit of the vector field X_ε in U, $\varepsilon > 0$, defined whenever possible.

REMARK. As will be shown below, the domain of $\Delta_\varepsilon^{\text{sing}}$ is an annular solid, very narrow in the z-direction. This is obvious for the model family of vector fields X_ε giving rise to the following equation in U:

$$\dot{\varphi} = 1, \quad \dot{x} = x^2 + \varepsilon, \quad \dot{y} = (\log\lambda)\, y, \quad \dot{z} = (\log(1/\lambda))\, z, \qquad \lambda \in (0,1).$$

The general picture is similar to this example.

7.4. Normal form for the singular correspondence map. In the following notation for the ε-depending coordinates, the subscript ε is dropped for simplicity. The functions φ and φ_- range over the real axis. The actual angle coordinate near L is $\varphi \pmod{\mathbb{Z}}$.

LEMMA 7.1. *The neighborhood $U' \subset U$ of the cycle L and the ε-depending coordinates (Y, φ, Z) on $\Gamma^+ = U' \cap \{x = -\delta\}$ and (Y^-, φ^-, Z^-) on $\Gamma^- = U' \cap \{x = \delta\}$ may be chosen so that the singular correspondence map will have the form*

$$\Delta_\varepsilon^{\text{sing}}\colon (Y, \varphi, Z) \mapsto (Y^-, \varphi^-, Z^-),$$
$$\varphi^- = \varphi - T(\varepsilon), \quad \text{where } T(\varepsilon) = \int_{-\delta}^{\delta} \frac{dx}{w_\varepsilon(x)}, \tag{7.4}$$
$$Y^- = \mathbf{A}(\varphi, \varepsilon) Y, \qquad Z^+ = \mathbf{B}(\varphi, \varepsilon) Z,$$
$$\text{where } \|\mathbf{A}(\varphi, \varepsilon)\| \to 0, \ \|\mathbf{B}^{-1}(\varphi, \varepsilon)\| \to 0 \ \text{as } \varepsilon \to 0. \tag{7.5}$$

The domains Γ^+, Γ^- are standard solids in these coordinates.

REMARK. The domains Γ^+, Γ^- should be regarded as two different copies of the standard solid $\mathcal{B}_0 = B^s \times T^1 \times B^u$, where B^s and B^u are the unit balls in $\mathbb{R}^s$ and $\mathbb{R}^u$ respectively.

PROOF. We define coordinates on the domains Γ^+ and Γ^- together with the domains themselves. We will describe only the first; the second is constructed in the same way (see Figure 5.9).

Take the segment $I = [-\delta, f_\varepsilon(-\delta)]$ on the x axis in Γ_0. Let G be a neighborhood of I in Γ_0. The projection $\pi\colon G \to \{x = -\delta\}$ along the orbits of the vector field X_ε is well defined provided that the orbits are considered in U only. Note that the map π is not one-to-one. Points transformed one to another by the Poincaré map of the cycle L are projected to the same point by π. We define $(Y, \varphi, Z)(\pi p)$ via $(x, y, z)(p)$ in such a way that

$$(Y, \varphi, Z)(\pi p) = (Y, \varphi, Z)(\pi P_\varepsilon p). \tag{7.6}$$

Let

$$\varphi_\varepsilon(\pi p) = \int_{-\delta}^{x(p)} \frac{d\zeta}{w_\varepsilon(\zeta)}. \tag{7.7}$$

Then $\varphi_\varepsilon(\pi p) \pmod 1$ is a coordinate on $S^1 = \mathbb{R}/\mathbb{Z}$. Let $\psi(\varphi)$ be a smooth nonnegative function in a neighborhood of $[0,1]$ on $\mathbb{R}$ with values in $[0,1]$, identically equal to 1 near 0 and 0 near 1. Let $\Psi(p) = \psi(\varphi_\varepsilon(p))$. For any $p \in G$ let

$$Y \circ \pi(p) = A^{\Psi(p)}(x(p), \varepsilon)\, y(p), \tag{7.8}$$
$$Z \circ \pi(p) = B^{\Psi(p)}(x(p), \varepsilon)\, z(p). \tag{7.9}$$

The functions Y and Z defined by (7.8), (7.9) have property (7.6). Hence, (Y, φ, Z) are really smooth coordinates on πG. Now choose G in such a form that the coordinates (Y, φ, Z) map $\Gamma^+ = \pi G$ to a standard solid $B_y \times S^1 \times B_z$ where B_y, B_z are balls centered at zero in $\mathbb{R}^s$ and $\mathbb{R}^u$ respectively. Rescaling Y and Z transforms the domain Γ^+ into the standard solid $\mathcal{B}_0$ from Lemma 7.1.

Proceed with the same construction for the section $\{x = \delta\}$ and the segment $I^- = [\delta, f_\varepsilon(\delta)]$. This will produce a neighborhood G^- of I^-, the domain $\Gamma^- = \{x = \delta\}$ with coordinates (Y^-, φ^-, Z^-) that map Γ^- to the same standard solid

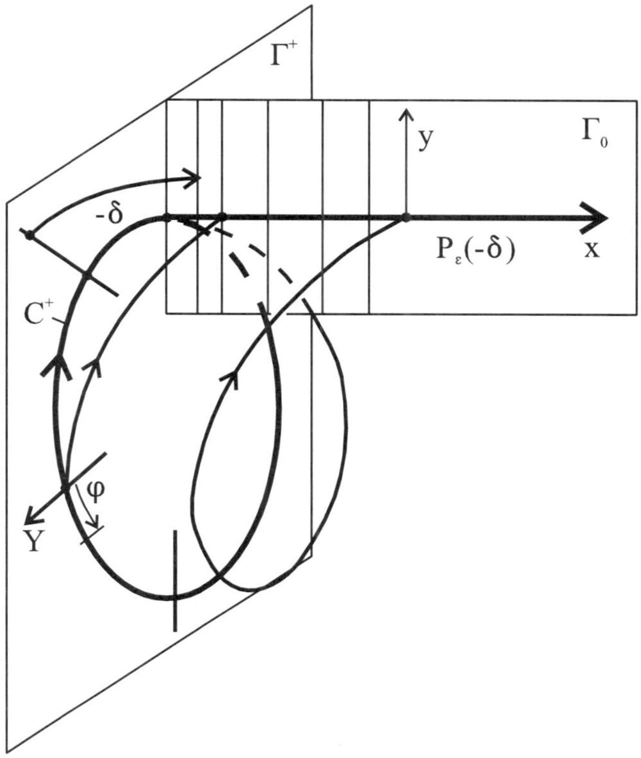

FIGURE 5.9. Coordinates (Y, φ) on Γ^+. The picture shows the case without unstable variables: $u = 0$.

$\mathcal{B}_0$, and a projection $\pi^- : G^- \to \Gamma^-$ along the orbits of X_ε. The domains Γ^-, Γ^+ and the normalizing coordinates on them are defined. Now take a neighborhood U' of L such that Γ^- and Γ^+ are the intersections from Lemma 7.1. In fact, the source and target of the map $\Delta_\varepsilon^{\text{sing}}$ are much smaller than Γ^+, Γ^- and degenerate at $\varepsilon = 0$. But it is important to have a coordinate system on larger domains that have a smooth limit as ε tends to 0.

Let us now prove the formula for $\Delta_\varepsilon^{\text{sing}}$ in the domain where it is defined.

Let $a \in \Gamma^+$, $b = \Delta_\varepsilon^{\text{sing}} a \in \Gamma^-$, $p \in G$, $\pi(p) = a$, $q \in G^-$, $\pi^-(q) = b$. Then

$$q = P_\varepsilon^k p \text{ for some } k > 0. \tag{7.10}$$

Let

$$t_\varepsilon^\pm(x) = \int_{\pm\delta}^{x} \frac{d\zeta}{w_\varepsilon(\zeta)}\,. \tag{7.11}$$

Then $t_\varepsilon^+(p) - t_\varepsilon^-(p) = t_\varepsilon^+(q) - t_\varepsilon^-(q) \pmod 1$ by (7.10) and because $t_\varepsilon \circ P_\varepsilon = t_\varepsilon + 1$ by (7.1). On the other hand, by (7.11), $t_\varepsilon^+ - t_\varepsilon^- \equiv T(\varepsilon)$. By (7.7), $\varphi(a) = t^-(p)$. Similarly, $\varphi^-(b) = t^+(q)$. This proves (7.4).

To prove (7.5), let $p = p_0$, $p_{i+1} = P_\varepsilon(p_i)$, $q = p_k$, where p, q, k are the same as in (7.10). Let $x(p_i) = x_i$, $A(x_i, \varepsilon) = A_i$. By (7.2), $\|A_i\| \leqslant \lambda < 1$. We will prove in full detail the first formula in (7.5); the second can be proved in the same way.

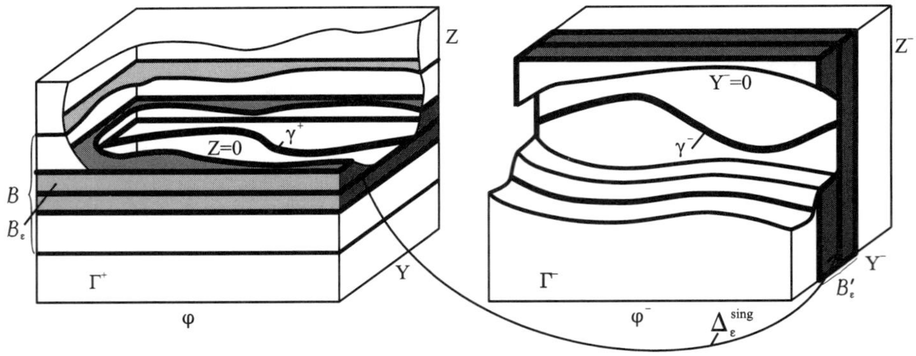

FIGURE 5.10. The domains $\mathcal{B}_\varepsilon$ and $\mathcal{B}'_\varepsilon$; $\mathcal{B}_\varepsilon$ is much narrower in the Z-direction, and $\mathcal{B}'_\varepsilon$ in the Y-direction than shown in the picture. The circles are shown as segments parallel to horizontal edges parametrized by φ, φ^-. Subsets of the solid torus Γ^+ (left) and Γ^- (right) are obtained by gluing together all the pairs of points $\{0\}\times p$ and $\{1\} \times p$.

Let $\Psi^-(q) = \psi(\varphi^-(q))$. By (7.1), (7.10), $y(q) = A_{k-1} \circ \cdots \circ A_0 y(p)$. By (7.8)

$$Y(a) = A_0^{\Psi(p)} y(p), \qquad Y^-(b) = A_k^{\Psi^-(q)} y(q).$$

Hence

$$Y^-(b) = \mathbf{A}(\varphi,\varepsilon) Y(a), \qquad \mathbf{A}(\varphi,\varepsilon) = A_k^{\Psi^-(q)} \circ A_{k-1} \circ \cdots \circ A_0^{1-\Psi(p)} Y(a). \quad (7.12)$$

Similarly, by (7.9),

$$Z^-(b) = \mathbf{B}(\varphi,\varepsilon) Z(a), \qquad \mathbf{B}(\varphi,\varepsilon) = B_k^{1+\Psi^-(q)} \circ B_{k-1} \circ \cdots \circ B_0^{1-\Psi(p)} Z(a). \quad (7.13)$$

As $\varepsilon \to 0$, $T(\varepsilon) \to \infty$, hence $k \to \infty$. Together with (7.3), this proves (7.5), and Lemma 7.1.

REMARK. All the factors in formula (7.12) for $\mathbf{A}(\varphi,\varepsilon)$ have bounded derivatives of any order restricted by the finite smoothness of the normalizing coordinates. This will be used below in 6.2.1.

Now let us describe the domains $\mathcal{B}_\varepsilon \subset \Gamma^+$ and $\mathcal{B}_\varepsilon^- \subset \Gamma^-$, the source and target of $\Delta_\varepsilon^{\text{sing}}$. They are defined by

$$|Y| \leqslant 1, \qquad |\mathbf{B}(\varphi,\varepsilon) Z| \leqslant 1, \quad (7.14)$$

$$|Z^-| \leqslant 1, \qquad |\mathbf{A}^{-1}(\varphi + T(\varepsilon),\varepsilon) Y| \leqslant 1 \quad (7.15)$$

respectively (see Figure 5.10).

7.5. Principal correspondence map, its source and target. The construction of this subsection is shown in Figures 5.10–5.12.

Let $H_i \cup L$ be the same torus or Klein bottle as in Theorem 7.1. Let $\gamma_i^\pm = H_i \cap \Gamma^\pm$. Note that $\gamma_i^\pm$ are pairwise disjoint. Below for any $i = 1, \ldots, N$ we will

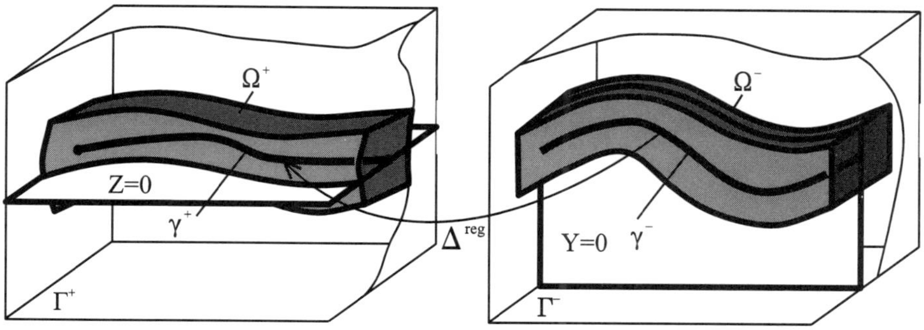

FIGURE 5.11. Domains of the regular correspondence map. The circles $T^1 \times \{0\}$ in Γ^+ and Γ^- are shown as segments. The same convention as for Figure 5.10 holds.

define a map Δ_ε from some neighborhood $D_i \subset \Gamma^+$ of γ_i^- into a domain $D_i' \subset \Gamma^+$. The map $f = \Delta_\varepsilon \colon \mathbf{D} = \bigcup D_i \to \mathbf{D}' = \bigcup D_i'$ will be subject to the application of Theorem 6.1. This will prove Theorem 7.1, as explained in 7.9.

For simplicity, we omit the subscript i and write H, $\gamma^\pm$, D, D' instead of H_i, $\gamma_i^\pm$, D_i, D_i'.

The surface H is homoclinic. Hence, the orbits of X_0 that start on γ^- come to γ^+ going outside the slice $U \cap \{-\delta \leqslant x \leqslant \delta\}$. This defines a regular correspondence map $\Delta_\varepsilon^{\mathrm{reg}} \colon \gamma^- \to \gamma^+$ along the orbits of X_0. By the smooth dependence of solutions on initial conditions and parameters, this map may be extended to a diffeomorphism $\Delta_\varepsilon^{\mathrm{reg}}$ of neighborhoods of γ^- and γ^+ along the orbits of X_ε.

We identify Γ^+ and Γ^- with two copies of the standard solid $B^s \times T^1 \times B^u$, with angular coordinates φ on Γ^+ and φ^- on Γ^-. Note that the curves γ^+ and γ^- belong to the unstable and stable sets of L respectively. Hence, $Z = 0$ on γ^+ and $Y = 0$ on γ^- (see Figure 5.10).

By the noncriticality and type assumptions, γ^+ and γ^- are the graphs of smooth maps $\tau^+ \colon T^1 \to B^s$ and $\tau^- \colon T^1 \to B^u$. Let

$$\Omega^- = \bigcup_{\varphi^- \in T^1} B_\rho^s \times Q^u(\varphi^-),$$

where

$$B_\rho^s = \{|y| \leqslant \rho\} \subset \mathbb{R}^s, \qquad Q^u(\varphi^-) = \{|z - \tau^-(\varphi^-)| \leqslant \rho\} \subset \mathbb{R}^u$$

(see Figure 5.11). Let ρ be so small that the map $\Delta_\varepsilon^{\mathrm{reg}}$ may be extended to Ω^- as a diffeomorphism to its image Ω^+, i.e., $\Omega^+ = \Delta_\varepsilon^{\mathrm{reg}} \Omega^-$. Let

$$V = \bigcup_{\varphi^- \in T^1} \{0\} \times Q^u(\varphi^-)$$

be the neighborhood of γ^- in the intersection $S_L^u \cap \Gamma^-$ (see Figure 5.12). Further let $B_\alpha^u = \{|z| \leqslant \alpha\} \subset B^u$ (see Figure 5.12). By the transversality assumption (item 4 in 1.1), for sufficiently small ρ and α the set $V_\varepsilon = \Delta_\varepsilon^{\mathrm{reg}}(V)$ contains the graph of

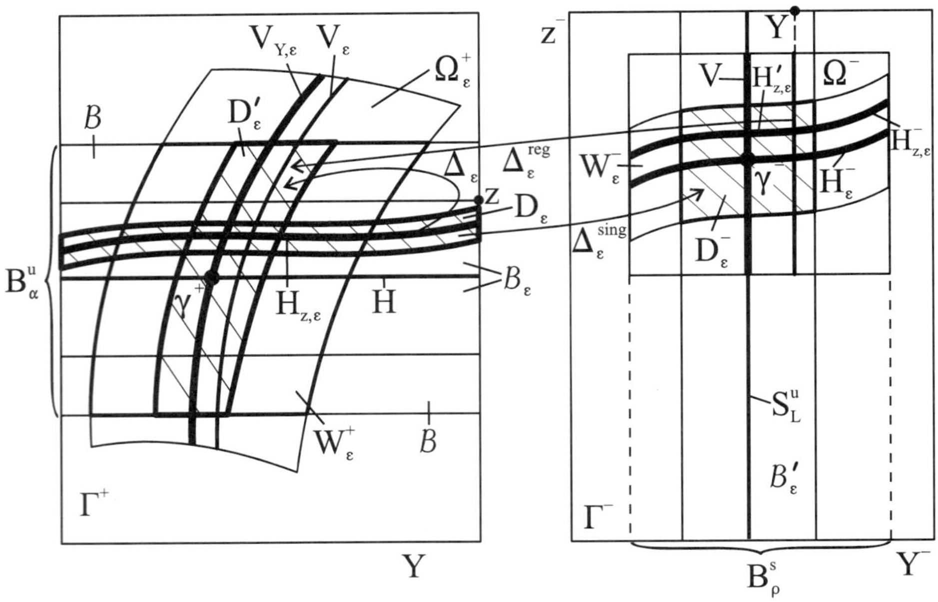

FIGURE 5.12. Domains of the correspondence map and their partitions. The sections $\varphi = \text{const}$ are drawn. In particular, γ^+ and γ^- are shown as single points.

the smooth map

$$v_\varepsilon\colon T^1 \times B^u_\alpha \to B^s$$

(see Figure 5.12). An extra assumption on the choice of α follows in 7.6.

Take $\mathcal{B} = B^s \times T^1 \times B^u_\alpha$ and $W^+_\varepsilon = \Omega^+ \cap \mathcal{B}$. Let $W^-_\varepsilon = (\Delta^{\mathrm{reg}}_\varepsilon)^{-1} W^+_\varepsilon$.

Recall that $\mathcal{B}_\varepsilon$ and $\mathcal{B}'_\varepsilon$ are the source and target of $\Delta^{\mathrm{sing}}_\varepsilon$. They are given by (7.14), (7.15) (see Figure 5.10). Let

$$D^-_\varepsilon = \mathcal{B}^-_\varepsilon \cap W^-_\varepsilon, \quad D_\varepsilon = (\Delta^{\mathrm{sing}}_\varepsilon)^{-1} D^-_\varepsilon, \quad D'_\varepsilon = \Delta^{\mathrm{reg}}_\varepsilon D^-_\varepsilon.$$

Further let

$$\Delta_\varepsilon = \Delta^{\mathrm{reg}}_\varepsilon \circ \Delta^{\mathrm{sing}}_\varepsilon\colon D_\varepsilon \to D'_\varepsilon$$

(see Figure 5.12).

This construction is carried out for any $i = 1, \dots, N$. The source and target of the principal correspondence map is the union of all the domains D_ε, respectively, D'_ε constructed for all i. The map itself coincides on any D_ε with the map Δ_ε constructed above. The desired domains and maps are constructed.

7.6. Structure of the generalized (μ_h, μ_v)-annular solid on D_ε, D'_ε.

LEMMA 7.2. *There exist positive constants μ_h, μ_v with $\mu_h \mu_v < 1$, such that for small ε the sets D_ε, D'_ε are (μ_h, μ_v) annular solids, horizontally and vertically cylindric in $\mathcal{B}$ respectively.*

PROOF. The proof follows the same lines as that of Lemma 4.3.2 (see Figure 5.12).

Let us construct the auxiliary generalized vertical and horizontal surfaces. We write φ, Y instead of φ^-, Y^- for brevity. Let

$$V_{Y,\varepsilon} = \Delta_\varepsilon^{\mathrm{reg}}\Big(\bigcup_{\varphi\in T^1}\{Y\}\times Q^u(\varphi)\Big).$$

For $|Y|$ and $|\alpha|$ sufficiently small, $V_{Y,\varepsilon}\cap\mathcal{B}$ is the graph of a smooth map:

$$v_{Y,\varepsilon}\colon T^1\times B^u_\alpha\to B^s,\qquad v_{0,\varepsilon}=v_\varepsilon.$$

This is the extra assumption on the choice of α mentioned above. Take μ_v greater than the maximal Lipschitz constant of the maps $v_{Y,\varepsilon}$ for small Y, ε. Let $H = B^s\times T^1\subset\Gamma^+$. The transversality assumption (item 4 in 1.1) implies that

$$H_\varepsilon^- = \{|Y|\leqslant\rho\}\cap(\Delta_\varepsilon^{\mathrm{reg}})^{-1}H$$

is the graph of a smooth map $h_\varepsilon^-\colon B^s_\rho\times T^1\to B^u$ for small ρ. We define

$$H^-_{Z,\varepsilon}\subset\Gamma^- \text{ as } H^-_{Z,\varepsilon} = (\Delta_\varepsilon^{\mathrm{reg}})^{-1}(B^s\times T^1\times\{Z\}),\qquad Z\in B^u_\alpha.$$

This surface is a graph of a smooth map $h^-_{Z,\varepsilon}\colon B^s_\rho\times T^1\to B^u$, $Z\in B^u_\alpha$, for sufficiently small ρ and α.

The surface $V_{Y,\varepsilon}$ is a manifold with boundary $\partial V_{Y,\varepsilon}$, and

$$\partial V_{Y,\varepsilon}\subset\partial_h\mathcal{B}. \tag{7.16}$$

The surface $H'_{Z,\varepsilon} = H^-_{Z,\varepsilon}\cap\mathcal{B}'_\varepsilon$ is a manifold with boundary $\partial H'_{Z,\varepsilon}$ and we have $\partial H'_{Z,\varepsilon}\subset\partial_v\mathcal{B}'_\varepsilon$. Now define $H_{Z,\varepsilon} = (\Delta_\varepsilon^{\mathrm{sing}})^{-1}H'_{Z,\varepsilon}$. Then

$$\partial H_{Z,\varepsilon}\subset\partial_v\mathcal{B}_\varepsilon\subset\partial_v\mathcal{B}. \tag{7.17}$$

The surface $H_{Z,\varepsilon}$ is the graph of the following map:

$$h_{Z,\varepsilon}\colon B^s\times T^1\to B^u_\alpha,\qquad h_{Z,\varepsilon}(Y,\varepsilon) = \mathbf{B}_\varepsilon^{-1}(\varphi,\varepsilon)\circ h^-_{Z,\varepsilon}(\mathbf{A}(\varphi,\varepsilon)Y,\varphi - T(\varepsilon)).$$

The Lipschitz constant of this map tends to zero as $\varepsilon\to 0$ by (7.4), (7.5). Hence it becomes smaller than an arbitrarily small μ_h. Fix some $\mu_h\in(0,\mu_v^{-1})$.

Let

$$\begin{aligned} F_\varepsilon\colon B^s\times T^1\times B^u_\alpha\to D_\varepsilon,&\quad (Y,\varphi,Z)\to(Y,\varphi,h_{Z,\varepsilon}(Y,\varphi)),\\ F'_\varepsilon\colon B^s\times T^1\times B^u_\alpha\to D'_\varepsilon,&\quad (Y,\varphi,Z)\to(v_{Y',\varepsilon}(\varphi,Z),\varphi,Z),\quad Y'=\mathbf{A}(\varphi,\varepsilon)Y. \end{aligned}$$

The images of the fibers $Y=\text{const}$ and $Z=\text{const}$ under F_ε are subsets of $Y=\text{const}$ and $H_{Z,\varepsilon}$ respectively. Hence they are generalized μ_v-vertical and μ_h-horizontal respectively. Consequently, by (7.17), D_ε is a generalized (μ_h,μ_v) annular solid vertically cylindric in $\mathcal{B}$. The images of the fibers $Y=\text{const}$ and $Z=\text{const}$ under F'_ε are subsets of $V_{Y',\varepsilon}$ and $Z=\text{const}$ respectively. Hence they are generalized μ_v-vertical and μ_h-horizontal respectively. Consequently, by (7.16), D'_ε is a generalized (μ_h,μ_v) annular solid horizontally cylindric in $\mathcal{B}$.

This proves Lemma 7.2.

Now recall the convention from 7.5 about dropping the index i. In fact, for any $i \in \{1, \dots, N\}$ we defined a horizontally cylindric solid D_i and a vertically cylindric solid D'_i in $\mathcal{B}$ and a map $\Delta_\varepsilon \colon D_i \to D'_i$ such that

$$\Delta_\varepsilon(\partial_h D_i) = \partial_h D'_i, \qquad \Delta_\varepsilon(\partial_v D_i) = \partial_v D'_i.$$

The domains D_i as well as D'_i are pairwise disjoint, because the curves $\gamma_i^\pm$ are. Let $\mathbf{D} = \bigcup D_i$, $\mathbf{D}' = \bigcup D'_i$. We must prove that the map

$$\Delta_\varepsilon \colon \mathbf{D} \to \mathbf{D}'$$

satisfies all the assumptions of Theorem 6.1.

In this theorem take the same $\mathcal{B}$ as in 7.5. By Lemma 7.2, D_i and D'_i are (μ_h, μ_v) annular solids horizontally and vertically cylindric in $\mathcal{B}$ respectively. It remains to prove the cone condition for Δ_ε. To do this, we must check the conditions of Lemma 6.1 for $\Sigma = d\Delta_\varepsilon^{\text{sing}}$ and $H = d\Delta_\varepsilon^{\text{reg}}$ in the domain of Δ_ε. This will be done in the two following subsections.

7.7. Verifying the cone condition: derivative of the regular map. The tangent spaces to the standard solid $\mathbb{B}$ are split into the sum:

$$\mathbb{R}^n = \mathbb{R}^s \oplus \mathbb{R}^c \oplus \mathbb{R}^u, \qquad \mathbb{R}^{cs} = \mathbb{R}^s \oplus \mathbb{R}^c, \qquad \mathbb{R}^{cu} = \mathbb{R}^c \oplus \mathbb{R}^u. \tag{7.18}$$

In Theorem 7.1, $c = 1$.

The assumptions of Lemma 6.1 on the matrix H are (4.2.1) and (4.2.2). They require the existence of a number $\mathbf{L}$ such that

$$\|H\| \leqslant \mathbf{L}, \quad \|H^{-1}\| \leqslant \mathbf{L}, \quad \|d^{-1}\| \leqslant \mathbf{L}, \quad \|(d')^{-1}\| \leqslant \mathbf{L}. \tag{7.19}$$

Here d and d' are the lower right blocks of the matrix H corresponding to the splitting (6.2) and (6.4) respectively. We take $\mathbf{L}$ for L in Lemma 6.1 in order to avoid confusion with the periodic orbit L.

LEMMA 7.3. *There exists an $\mathbf{L} > 0$ such that for any $p \in \Omega^-$ and ε sufficiently small the matrix $H = d\Delta_\varepsilon^{\text{reg}}(p)$ satisfies conditions* (7.19).

PROOF. The first two estimates in (7.19) follow from the regularity of $\Delta_\varepsilon^{\text{reg}}$ in Ω^-. The third and fourth follow, roughly speaking, from noncriticality and transversality (assumption 4 in 1.1).

In more detail, inequalities (7.19) may be checked on γ^- only, then extended to Ω^- by continuity, assuming that ρ is small enough. Recall that the choice of a small Ω^- was at the basis of the construction of domains and maps in 7.5, 7.6.

Let $p \in \gamma^+$, $q \in \gamma^-$, $p = \Delta_\varepsilon^{\text{reg}}(q)$. We have

$$T_p\mathbb{R}^n = \mathbb{R}^s \oplus \mathbb{R}^1 \oplus \mathbb{R}^u, \qquad T_q\mathbb{R}^n = \mathbb{R}^s \oplus \mathbb{R}^1 \oplus \mathbb{R}^u,$$

where different copies of the same space are denoted by the same symbol. Let

$$\pi_s \colon T_p\mathbb{R}^n \to \mathbb{R}^1 \oplus \mathbb{R}^u, \qquad \pi_{cs} \colon T_p\mathbb{R}^n \to \mathbb{R}^u$$

be the natural projections. By definition,

$$d' = \pi_s \circ H|\mathbb{R}^{cu} \colon \mathbb{R}^{cu} \to \mathbb{R}^{cu}, \qquad d = \pi_{cs} \circ H|\mathbb{R}^u \colon \mathbb{R}^u \to \mathbb{R}^u.$$

Let $S = T_p\gamma^+$, $S^- = T_q\gamma^-$, $S = HS^-$. Then, by the noncriticality assumption,

$$T_p\mathbb{R}^n = \mathbb{R}^s \oplus S \oplus \mathbb{R}^u, \qquad T_q\mathbb{R}^n = \mathbb{R}^s \oplus S^- \oplus \mathbb{R}^u.$$

By the transversality condition,

$$T_p\mathbb{R}^n = \mathbb{R}^s \oplus H(S^- \oplus \mathbb{R}^u) = \mathbb{R}^s \oplus H(\mathbb{R}^{cu}).$$

Hence d' is an isomorphism.

On the other hand, by the noncriticality assumption,

$$T_p\mathbb{R}^n = \mathbb{R}^s \oplus H(S^- \oplus \mathbb{R}^u) = \mathbb{R}^s \oplus S \oplus H\mathbb{R}^u = \mathbb{R}^{cs} \oplus H\mathbb{R}^u.$$

Hence, d is an isomorphism.

By the compactness of γ^-, the operators d and d' have uniformly bounded inverses. This proves the lemma.

7.8. Verifying the cone condition: derivative of the singular map.

LEMMA 7.4. *For any $p \in \mathcal{B}_\varepsilon$, $q \in \Omega^-$ and sufficiently small ε, the matrices $\Sigma = d\Delta_\varepsilon^{\text{sing}}(p)$ and $H = d\Delta_\varepsilon^{\text{reg}}(q)$ satisfy the conditions of Lemma* 6.1 *uniformly in p, q.*

PROOF. According to the splitting (7.18), the matrix Σ is split into blocks. For $p = (Y, \varphi, Z)$, we have

$$\Sigma(p) = \begin{pmatrix} \mathbf{A}(\varphi,\varepsilon) & \mathbf{A}_\varphi(\varphi,\varepsilon)Y & 0 \\ 0 & 1 & 0 \\ 0 & \mathbf{B}_\varphi(\varphi,\varepsilon)Z & \mathbf{B}(\varphi,\varepsilon) \end{pmatrix}$$

(see Lemma 7.1). Consequently, for the splitting a in Lemma 6.1, we have

$$\Lambda = \begin{pmatrix} \mathbf{A} & \mathbf{A}_\varphi Y \\ 0 & 1 \end{pmatrix}, \quad M = \mathbf{B}, \quad B = \begin{pmatrix} 0 \\ 0 \end{pmatrix}, \quad D = (0, \mathbf{B}_\varphi Z). \tag{7.20}$$

For the splitting b,

$$\Lambda' = \mathbf{A}, \quad M' = \begin{pmatrix} 1 & 0 \\ \mathbf{B}_\varphi & \mathbf{B} \end{pmatrix}, \quad B' = (\mathbf{A}_\varphi Y, 0), \quad D' = \begin{pmatrix} 0 \\ 0 \end{pmatrix}. \tag{7.21}$$

Conditions (4.2.1), (4.2.2) required by Lemma 6.1, were checked previously (in 7.7).

Moreover, for the splitting a, the following conditions are required by Lemma 6.1: (4.2.3), (4.2.17), (4.2.13), (4.2.15), (4.2.16). Namely,

$$\begin{gathered} \|M^{-1}\| < \delta, \quad \|\Lambda - BM^{-1}D\| \leqslant \mathbf{L}, \quad \|BM^{-1}\| < \mathbf{L}, \\ \|M^{-1}D\| < \delta, \quad \|cBM^{-1}\| < \delta. \end{gathered} \tag{7.22}$$

The third and the fifth inequalities from (7.22) become trivial: $0 < \delta$, because in (7.20) we had $B = 0$. By (7.20), relation (7.22) takes the form

$$\|\mathbf{B}^{-1}\| < \delta, \qquad \left\| \begin{matrix} \mathbf{A} & \mathbf{A}_\varphi Y \\ 0 & 1 \end{matrix} \right\| < \mathbf{L}, \qquad \|\mathbf{B}^{-1}\mathbf{B}_\varphi Z\| < \delta. \tag{7.23}$$

The first inequality in (7.23) for small ε follows from (7.5), which holds for any $p \in \mathcal{B}$.

The third inequality makes use of the assumption $p \in \mathcal{B}_\varepsilon$. Indeed, in (7.13) let

$$B_0^{1-\psi(\varphi)} = C_0(\varphi), \qquad B_k^{\psi(\varphi)} = C_k(\varphi), \qquad B_j = C_j, \quad j = 1, \dots, k-1.$$

The norms $\|C_{j\varphi}\|$ are bounded uniformly in $\varphi \in S^1$, $\varepsilon \in [0, \varepsilon_0]$ because of the smoothness of the normal form (7.1) up to $\varepsilon = 0$: $\|C_{j\varphi}\| < C$. Moreover, we have

$\mathbf{B}^{-1}\mathbf{B}_\varphi Z = \mathbf{B}^{-1}\mathbf{B}_\varphi\mathbf{B}^{-1}\mathbf{B}Z$. But $\|\mathbf{B}Z(p)\| \leqslant 1$ because $p \in \mathcal{B}_\varepsilon$ (see (7.14)). On the other hand,

$$\mathbf{B}^{-1}\mathbf{B}_\varphi\mathbf{B}^{-1} = \sum_{j=1}^{k} C_0^{-1}\cdots C_j^{-1}C_{j\varphi}C_j^{-1}\cdots C_k^{-1}.$$

Therefore, $\|\mathbf{B}^{-1}\mathbf{B}_\varphi Z\| \leqslant Ck\lambda^k < \delta$ for sufficiently small ε, because $k = k(\varphi,\varepsilon) \to \infty$ as $\varepsilon \to \infty$. This proves the third inequality in (7.23).

The same calculation proves that $\|\mathbf{A}_\varphi\| \to 0$ as $\varepsilon \to 0$. This implies the second inequality in (7.23).

For the splitting b, the following conditions are required by Lemma 6.1: (4.2.18), (4.2.13)–(4.2.16), with primes added. Namely,

$$\begin{gathered}\|(M')^{-1}\| < \mathbf{L}, \quad \|B'(M')^{-1}\| < \mathbf{L}, \quad \|\Lambda' - B'(M')^{-1}D'\| < \delta, \\ \|(M')^{-1}D'\| < \delta, \quad \|c'B'(M')^{-1}\| < \delta.\end{gathered} \tag{7.24}$$

By (7.21), these inequalities take the form

$$\begin{gathered}\left\|\begin{matrix} 1 & 0 \\ -\mathbf{B}^{-1}\mathbf{B}_\varphi Z & \mathbf{B}^{-1}\end{matrix}\right\| < \mathbf{L}, \quad \|\,\mathbf{A}_\varphi Y \quad 0\,\| < \mathbf{L}, \\ \|\mathbf{A}\| < \delta, \quad \|c'(\,\mathbf{A}_\varphi Y \quad 0\,)\| < \delta.\end{gathered} \tag{7.25}$$

The first inequality in (7.25) follows from the last one in (7.23) and (7.5). The second and third follow from (7.5) and the fact that $\|\mathbf{A}_\varphi\| \to 0$ as $\varepsilon \to 0$, which is already proved. The last one follows from (7.5) and (7.19).

Thus we have checked all the assumptions of Lemma 6.1 for Σ and H, and Lemma 7.4 is proved.

7.9. End of the proof of Theorem 7.1. By the discussion in 7.5–7.8, the principal correspondence map Δ_ε for positive small ε satisfies all the conditions of the generalized Smale horseshoe existence theorem (Theorem 6.1). Therefore there exists an invariant set Λ of Δ_ε in $\mathcal{B}$ which is a collection of Lipschitz circles, and a map $\phi\colon \Lambda \to \Sigma^N$ such that the diagram

$$\begin{array}{ccc} \Lambda & \xrightarrow{\ \Delta_\varepsilon\ } & \Lambda \\ {\scriptstyle\phi}\downarrow & & \downarrow{\scriptstyle\phi} \\ \Sigma^N & \xrightarrow{\ \sigma\ } & \Sigma^N \end{array} \tag{7.26}$$

is commutative. Here σ is the Bernoulli shift of the space Σ^N of bi-infinite sequences of N symbols.

The orbits of the initial vector field X_ε passing through the points of Λ form an invariant set denoted by Ω_ε. By the construction of the map Δ_ε, each orbit of Δ_ε corresponds to a connected component of Ω_ε. Finally, all properties of the set Ω_ε listed in Theorem 7.1 may be obtained from the diagram (7.26) and Theorem 8.1.1.

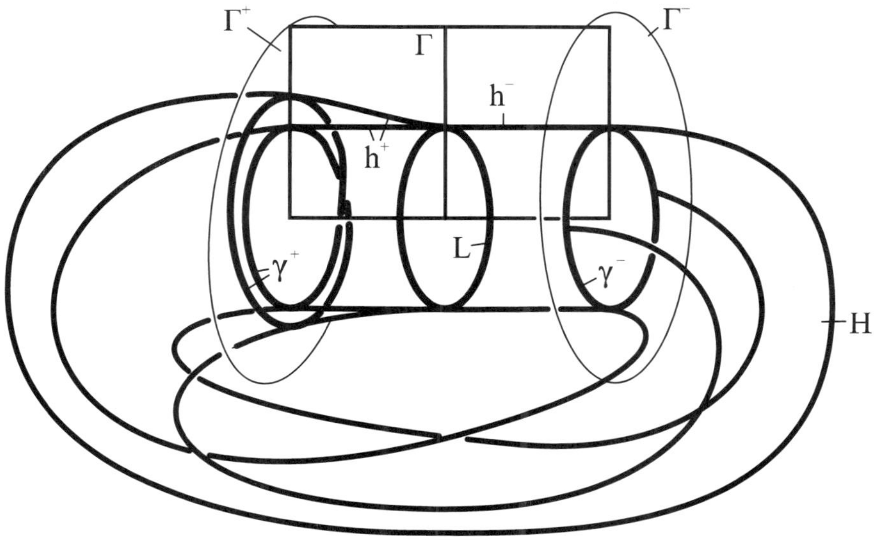

FIGURE 5.13. A homoclinic surface of type $(2, 1)$.

§8. Generation of a strange attractor via the bifurcation of a twisted homoclinic surface

In this subsection the bifurcation of a homoclinic surface of a semihyperbolic cycle of type $(m, 1)$ for $|m| > 1$ is discussed. This surface is shown in Figure 5.13.

8.1. Description of the family and main result. We say that two invariant sets of two vector fields are *topologically equivalent* if there exists a homeomorphism of one set onto another that conjugates the corresponding flows restricted to these sets. The same notion is defined for maps with evident modifications.

Consider the following map of the three-dimensional solid torus

$$\mathbf{T}^3 = D \times T^1, \quad T^1 = \mathbb{R}/\mathbb{Z}, \quad D = \{z \in \mathbb{C} : |z| \leqslant 1\}$$

to itself:

$$F\colon (z, \varphi) \mapsto (1/2e^{i\varphi} + z/2m, m\varphi), \qquad z \in D, \quad \varphi \in T^1, \quad |m| > 1. \tag{8.1}$$

It is called the *standard solenoid map of type m*.

THEOREM 8.1. *In a typical one-parameter family, a saddlenode cycle, a stable node in hyperbolic variables, with a homoclinic surface of type $(m, 1)$, $|m| > 1$, may occur. The parameter values on one side of the critical value, say, $\varepsilon < 0$, correspond to a vector field with a nonwandering set in the neighborhood of a homoclinic surface that consists of two hyperbolic cycles. On the other side, say $\varepsilon > 0$, the parameter values correspond to a hyperbolic maximal attractor topologically equivalent to the standard Smale–Williams solenoid.*

8.2. Genericity assumptions. The family of vector fields X_ε in Theorem 8.1 must satisfy the following genericity assumptions.

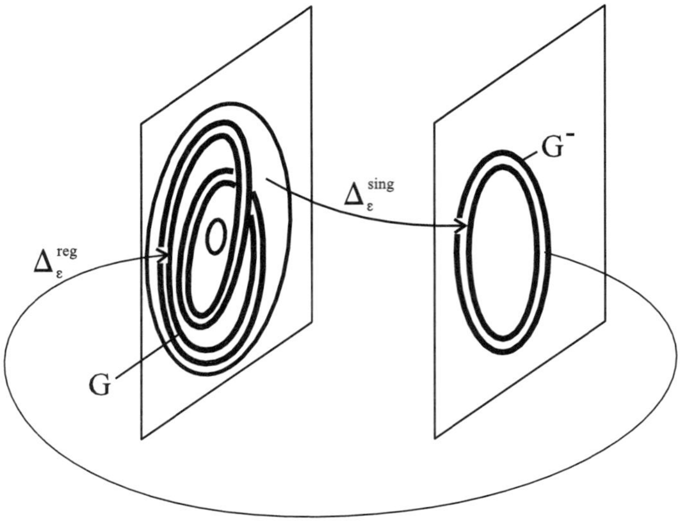

FIGURE 5.14. Singular, regular and global maps. The latter has a Smale–Williams solenoid as attractor.

1. *Fundamental assumption in* 1.1. It implies that the saddlenode cycle L has a Poincaré map defined on a section Γ_0 transversal to L, and this map has the following property. It is finitely smoothly equivalent to the normal form (7.1)–(7.3) with some modification: the unstable part of the spectrum and the unstable variable z are deleted. This normal form is:

$$P_\varepsilon(x,y) = (x',y'), \quad x' = f_\varepsilon(x) = x + (\varepsilon + x^2)b(x,\varepsilon), \quad y' = A(x,\varepsilon)y,$$
$$\|A\| < \lambda < 1, \quad b(0,0) > 0.$$

2. The regular map from $h^- = \Gamma_0 \cap H \cap \{x > 0\}$ to $h^+ = \Gamma_0 \cap H \cap \{x < 0\}$ along the orbits of X_0 is expanding in the following sense. Let π_0 be the projection $(x,y) \mapsto x$, $p \in h^+$ an arbitrary point, and $q \in h^-$ a point on the same orbit of X_0 as p. Let $\Delta\colon (h^-,p) \to (h^+,q)$ be the germ of a map along the orbits of X_0. Denote $g = \pi_0 \circ \Delta$. Let ω_ε be the same as in (7.2), and $\omega_0 = \omega_\varepsilon|_{\varepsilon=0}$. Then a vector field w_0 on $\mathbb{R}^1$ near p is mapped to a proportional one near $g(p)$: $g_* w_0(x) = \nu(x) w_0(g(x))$. The *expansion property means that* $\nu(x) > 1$. The vector field w_0 on the x-axis is invariant under the Poincaré map P_0. Hence the choice of q is irrelevant to the statement of the expansion property.

8.3. Reformulation of the expansion hypothesis. The previous hypothesis becomes especially simple in the coordinates (Y,φ), (Y^-,φ^-) on the sections Γ^+, Γ^- introduced in 7.4. Denote by $\gamma^\pm = H \cap \Gamma^\pm$ the intersections of the homoclinic surface with the Poincaré sections $\Gamma^\pm$. Since the unstable variables are absent, the unstable set in Γ^- coincides with γ^-. Therefore $\gamma^- = \{Y^- = 0,\ \varphi^- \in T^1\}$. Let π be the projection $\Gamma^+ \to \gamma^+$, $(Y,\varphi) \mapsto \varphi$. Let $\Delta_\varepsilon^{\text{reg}}$ be a diffeomorphism of γ^- onto γ^+ along the orbits of the vector field X_ε. It may be extended to a diffeomorphism of a neighborhood G^- of γ^- onto a neighborhood G of γ^+. The assumption is that the map $f\colon \pi\circ\Delta_\varepsilon^{\text{reg}}\colon \gamma^-\to\gamma^+$ is expanding in the charts φ^-, φ for $\varepsilon=0$, hence for any small ε.

Let us now give a coordinate description of $\Delta_\varepsilon^{\text{reg}}$ based on assumption 2. This assumption implies

$$\begin{aligned}\Delta_\varepsilon^{\text{reg}}\colon (Y^-,\varphi^-)\mapsto(Y,\varphi),\quad &(Y,\varphi)=(f_\varepsilon(\varphi^-),\Phi_\varepsilon(\varphi^-))+F_\varepsilon,\\ \Phi'_\varepsilon>1,\quad F_\varepsilon(0,\varphi^-)=0,\quad &\Phi_\varepsilon(\varphi^-+1)=\Phi_\varepsilon(\varphi^-)+m.\end{aligned}\tag{8.2}$$

The last equality follows from the assumption that the surface H is of type $(m,1)$.

The following three subsections contain the proof of Theorem 8.1.

8.4. Correspondence map and limit maps. A singular map $\Delta_\varepsilon^{\text{sing}}$ of a subdomain of Γ^+ to its image in Γ^- is defined in the same way as in 7.4 (see Figure 5.14). Formulas (7.4), (7.5) hold with evident changes:

$$\begin{aligned}\Delta_\varepsilon^{\text{sing}}\colon (Y,\varphi)\mapsto(Y^-,\varphi^-),\quad &\varphi^-=\varphi-T(\varepsilon),\quad Y^-=\mathbf{A}(\varphi,\varepsilon)Y,\\ T(\varepsilon)=\int_{-\delta}^{\delta}\frac{dx}{w_\varepsilon(x)}\to\infty,\quad &\|\mathbf{A}\|\to 0\ \text{as}\ \varepsilon\to 0.\end{aligned}\tag{8.3}$$

Now let us define the principal correspondence map of some neighborhood D of γ^+ into itself by the formulas $\Delta_\varepsilon=\Delta_\varepsilon^{\text{reg}}\circ\Delta_\varepsilon^{\text{sing}}$, $D\to D'=\Delta_\varepsilon D$. The formulas above imply the following limit property of this map.

LEMMA 8.1. *Let $T(\varepsilon)$ be the same as in (8.3). For any $\tau\in[0,1]$ and $k\in\mathbb{N}$ define ε_k by setting $T(\varepsilon_k)=k+\tau$. Then the following limit exists:*

$$\Delta(\tau)=\lim_{k\to\infty}\Delta_{\varepsilon_k},\quad \textit{where } \Delta(\tau)\colon\Gamma^+\to\gamma^+,\ (Y,\varphi)\mapsto(f(\varphi+\tau),\Phi(\varphi+\tau));\tag{8.4}$$

here $|\Phi'|>\mu>1$, $(f,\Phi)=(f_\varepsilon,\Phi_\varepsilon)$ for $\varepsilon=0$.

PROOF. Formulas (8.4) immediately follow from (8.2), (8.3).

In what follows, we only need to know that any principal correspondence map is a small perturbation of some limit map from Lemma 8.1.

8.5. Hyperbolicity of the attractor. The map Δ_ε brings the neighborhood Γ^+ of γ^+ strictly into itself. Hence, the maximal attractor

$$A=\bigcap_{n=0}^{\infty}(\Delta_\varepsilon)^n\Gamma^+$$

is nonempty. We will prove that it is hyperbolic in the sense of Definition 8.1.1. To do this, we need to prove only the following

LEMMA 8.2. *For any map $\Delta(\tau)$ (see (8.4)), there exist μ_h, μ_v, $\mu_h\mu_v<1$, such that any diffeomorphic C^1 small perturbation of this map satisfies the (μ_h,μ_v) cone condition with the family of cones K^-, K^+ related to the splitting $(\xi^-,\xi^+)=(dy,d\varphi)$. Namely*

$$K^-=\{|\xi^+|\leqslant\mu_h|\xi^-|\},\qquad K^+=\{|\xi^-|\leqslant\mu_v|\xi^+|\}.\tag{8.5}$$

PROOF. Recall the (μ_h, μ_v) cone condition for a linear operator A. We must have

1. $AK^+ \subset K^+$;
2. $A^{-1}K^- \subset K^-$.

Let $A\xi = \eta$. Then there exists a $\lambda > 1$ such that

3. for any $\xi \in K^+$, $|\eta^+| > \lambda|\xi^+|$;
4. for any $\eta \in K^-$, $|\xi^-| > \lambda|\eta^-|$.

By (8.4), $d\Delta(\tau) = \begin{pmatrix} 0 & f'(\varphi) \\ 0 & \Phi'(\varphi) \end{pmatrix}$. Then $d\Delta_\varepsilon = \begin{pmatrix} \alpha & f'_\varepsilon(\varphi) \\ \beta & \Phi'_\varepsilon(\varphi) \end{pmatrix}$ with f_ε, Φ_ε C^1-close to f, Φ; α and β small.

LEMMA 8.3. *Let* $A = \begin{pmatrix} \alpha & a \\ \beta & b \end{pmatrix}$ *be the block matrix related to the splitting*

$$\mathbb{R}^n = \mathbb{R}^s \oplus \mathbb{R}, \quad \|\alpha\|, \|\beta\| \ll 1, \quad |b| > \lambda > 1.$$

Then there exist δ *and* μ_h, μ_v, $\mu_h\mu_v < 1$, *such that* A *satisfies the* (μ_h, μ_v) *cone condition for the cones* (8.5), *provided that* $\|\alpha\| < \delta$, $\|\beta\| < \delta$.

PROOF. 1. Let $\widetilde{K}^+ = \{|\xi^+| \geqslant \mu_h|\xi^-|\}$. We shall prove that $A\widetilde{K}^+ \subset K^+$. This will prove assumptions 1 and 2 of the (μ_h, μ_v) cone condition at the same time, because $\widetilde{K}^+$ contains K^+ and is the complement of K^-.

Let $\xi \in \widetilde{K}^+$. Then $|\xi^+| \geqslant \mu_h|\xi^-|$, by the definition of $\widetilde{K}^+$. We want to prove that

$$|\eta^-| \leqslant \mu_v|\eta^+|. \tag{8.6}$$

We have

$$|\eta^-| = |\alpha\xi^- + a\xi^+| \leqslant (|\alpha|/\mu_h + |a|)|\xi^+| = C^-|\xi^+|, \tag{8.7}$$

$$|\eta^+| = |\beta\xi^- + b\xi^+| \geqslant (|b| - |\beta|/\mu_h)|\xi^+| = C^+|\xi^+|. \tag{8.8}$$

Take μ_v so large that $\mu_v(|b| - 1/2) > |a| + 1/2$. After that take μ_h, α, β so small that

$$\mu_v\mu_h < 1, \quad |\alpha|/\mu_h < 1/2, \quad |\beta|/\mu_h < 1/2.$$

Then we have $C^- < \mu_v C^+$. This inequality, together with (8.7) and (8.8), proves (8.6).

2. Let us check the third statement of the cone condition. By (8.8),

$$|\eta^+| \geqslant C^+|\xi^+| \text{ with } C^+ = |b| - |\beta|/\mu_h.$$

For $|\beta| < \mu_h(|b| - \lambda)$ this implies assumption 3 of the (μ_h, μ_v) cone condition.

3. Let us check assumption 4 of the (μ_h, μ_v) cone condition. We want to prove that $|\xi^-| \geqslant \lambda|\eta^-|$ provided that $\eta \in K^-$, i.e., $|\eta^+| \leqslant \mu_h|\eta^-|$. We have $\xi \in K^-$. Then

$$|\xi^+| \leqslant \mu_h|\xi^-|, \qquad \lambda|\eta^-| \leqslant \lambda(|\alpha| + |a|\mu_h)|\xi^-| = C_0|\xi^-|.$$

Take $|\alpha|$ and μ_h so small that $C_0 < 1$. Then assumption 4 holds. Thus, the map f satisfies all the assumptions of the (μ_h, μ_v) cone condition. Lemma 8.3, together with Lemma 8.2, is proved.

8.6. Homotopy to the standard solenoid map. By Lemma 8.3, the maximal attractor of Δ_ε is hyperbolic. Any C^1 small perturbation of the map Δ_ε is topologically equivalent to Δ_ε in some neighborhood of the attractor (by Theorem 8.1.1). Moreover, any continuous family of maps of some domain into itself that contains Δ_ε and satisfies the (μ_h, μ_v) cone condition, consists of maps topologically equivalent to each other near their maximal attractors (by the same theorem).

Indeed, for all small perturbations of any map of the family, its maximal attractor is hyperbolic, and the maps in the neighborhood of the attractor are topologically equivalent.

Let us apply this argument to the map Δ_ε. The topological equivalence of maps is understood below as equivalence in some neighborhood of the maximal attractors of these maps. For small ε and $\tau = \{T(\varepsilon)\}$, the map Δ_ε is C^1 close to the map $\Delta(\tau)$ by Lemma 8.1. Hence, it has the form

$$\Delta_\varepsilon(Y, \varphi) = (g_\varepsilon(\varphi), \Psi_\varepsilon(\varphi)) + A_\varepsilon(\varphi)Y + F_\varepsilon(Y, \varphi).$$

Here Ψ_ε, g_ε are close to $\Phi \circ (\mathrm{id} + \tau)$, $f \circ (\mathrm{id} + \tau)$ respectively (see (8.4)). Further, $A_\varepsilon \colon \mathbb{R}^u \to \mathbb{R}^u \oplus \mathbb{R}^1$ is a linear operator with a very small norm, $F_\varepsilon = O(|Y|^2)$. Taking the neighborhood U of γ^+ small enough, we can assume, without loss of generality, that the map $\Delta_\varepsilon - sF_\varepsilon$ is a diffeomorphism for any $s \in [0, 1]$. Hence, Δ_ε is topologically equivalent to

$$\Delta_A \colon (Y, \varphi) \mapsto \gamma_\varepsilon(\varphi) + A_\varepsilon(\varphi)(Y).$$

Let $Y = (z, Y')$, $z \in \mathbb{C}$, $|z| \leqslant \delta$, $Y' \in \mathbb{R}^{s-2}$. Let γ be a map similar to the restriction of (8.1) to the circle $z = 0$:

$$\gamma(\varphi) = (z', Y', \varphi') = (1/2e^{i\varphi}, 0, m\varphi).$$

The linear homotopy of the map Δ_ε to

$$(Y, \varphi) \mapsto \gamma(\varphi) + A_\varepsilon(\varphi)Y \tag{8.9}$$

consists of topologically equivalent maps of Γ_+ into itself. The maps (8.9) and (8.1) are topologically equivalent on their attractors by the standard encoding arguments; see [**KH**]. This proves Theorem 8.1.

CHAPTER 6

Homoclinic Torus of a Nonhyperbolic Periodic Orbit: Semicritical Case

In this chapter we study the generation of a strange attractor via the bifurcation of a nonsmooth homoclinic torus of a nonhyperbolic periodic orbit, for the case in which the Poincaré map of this orbit is a saddlenode of a special type, i.e., a node in its hyperbolic variables. This means that one multiplier of the Poincaré map is equal to one, and all the others lie inside the unit circle. Moreover, the union of all homoclinic curves of the periodic orbit with the orbit itself is assumed to form a compact surface homeomorphic to a two-dimensional torus, which is not a smooth submanifold of the phase space. The principal tools used in this chapter are two-fold: normal forms of the perturbations of a saddlenode germ of a map and the theory of endomorphisms of a circle. The latter theory is presented in the last two sections; in the first two the breaking of the homoclinic torus itself is described.

§1. Theorem on the generation of a strange attractor

In this section the statement of the main theorem and the general outline of its proof are given.

1.1. Statement of the theorem. Consider a set of vector fields on the boundary of a Morse–Smale domain having the following properties.

Any vector field of the set has a saddlenode cycle L, which is a node in its hyperbolic variables. The homoclinic orbits of this cycle fill a cylinder H, a homoclinic surface of type $(1,1)$ in the sense of §5.1. The union $H \cup L = T^2$ is a nonsmooth (i.e., critical) two-torus.

The class of these vector fields is denoted by CHT (for critical homoclinic torus). As explained in Chapter 1, vector fields of class CHT occur in a nonavoidable manner in one-parameter families.

THEOREM 1.1 (Main Theorem). *In the space of CHT families an open dense set is formed by families with the following properties. After a suitable reparametrization, a CHT vector field corresponds to the zero value of the parameter. Vector fields of the family $\{X_\varepsilon\}$ corresponding to $\varepsilon < 0$ have two hyperbolic periodic orbits (one of them is stable) and have no other nonwandering points in some neighborhood U of the homoclinic torus T^2. For systems corresponding to positive parameter values the following holds.*

1. *There is a set S of parameter values for which the corresponding vector field has a transversal homoclinic orbit of a hyperbolic cycle with two-dimensional unstable manifold. The set S has the following density property:*

$$\liminf_{\varepsilon \to 0} \frac{1}{\varepsilon} \operatorname{mes}(S \cap [0, \varepsilon]) > 0. \tag{1.1}$$

Here mes *denotes Lebesgue measure on* $\mathbb{R}$.

2. *The neighborhood U is an absorbing domain, and for any $\varepsilon in S$ the maximal attractor of X_ε in U is strange, i.e., is not a finite union of submanifolds of the phase space.*

3. *There is a sequence $\varepsilon_k \in S$ of values of the parameter tending to 0 such that for any k the field X_{ε_k} has a homoclinic tangency of stable and unstable manifolds of some hyperbolic periodic orbit.*

The entire chapter, starting with 1.3, is devoted to the proof of this theorem.

Recall that the maximal attractor of a vector field X_ε was defined in §2.1 by the formula

$$A_{\max} = \bigcap_{t>0} g^t_{X_\varepsilon} U.$$

1.2. Summary of results: line 6 of the main table. The theorem above implies all the results mentioned in the line 6 of the main table of Chapter 1. For the definition of accessibility and families leading out of the Morse–Smale set, see 1.6.2.

COROLLARY 1.1. *The class of boundary points of a Morse–Smale set consisting of vector fields with nonhyperbolic periodic orbit having the type of a node in its hyperbolic variables with multiplier one and nonsmooth homoclinic two-torus (the class CHT) is accessible from the left and nonaccessible from the right.*

PROOF. The main theorem above implies that all vector fields of a typical one-parameter family crossing the class CHT that correspond to small negative parameter values are of Morse–Smale type in the neighborhood U, and therefore, the class is accessible from the left.

Statement 3 of the main theorem provides a sequence of parameter values tending to zero from the right such that the corresponding vector fields have homoclinic tangency of the mustaches of some periodic orbit. These vector fields are not of Morse–Smale type. Therefore, the class CHT is nonaccessible from the right.

COROLLARY 1.2. *Typical one-parameter unfoldings of vector fields of the class CHT lead out of the set of Morse–Smale systems.*

PROOF. Statement 1 of Theorem 1.1 implies the existence of vector fields in the family with transversal intersection of stable and unstable manifolds of some hyperbolic periodic orbit, which, therefore, are not of Morse–Smale type. The set of such systems is open, hence its complement is not dense.

COROLLARY 1.3. *The unfolding in Corollary 1.2 generates an infinite number of hyperbolic cycles and a nontrivial hyperbolic invariant set.*

PROOF. This is a consequence of statement 1 of Theorem 1.1 and of the Birkhoff–Smale theorem.

Now we pass to the formal proof of the main theorem.

1.3. Singular and regular correspondence maps. Global return map. The singular map in the present context is the very same as in 5.8.4. Namely, the constructions of 5.7.4, 5.8.4 provide two cross-sections Γ^+ and Γ^- in the tubular neighborhood of the saddlenode cycle L which contains L in between (any of the cross-sections $\Gamma^\pm$ divides U in two parts). Moreover, there are coordinates (Y, φ)

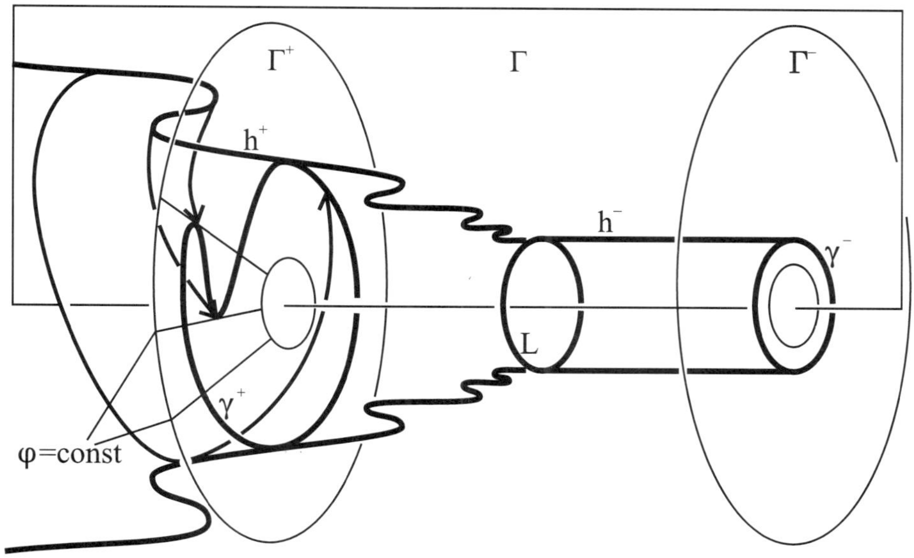

FIGURE 6.1. Nonsmooth part of the critical homoclinic torus and the curve γ^+.

on Γ^+ and (Y^-, φ^-) on Γ^- such that the singular correspondence map $\Delta_\varepsilon^{\mathrm{sing}}$ of Γ^+ to its image in Γ^- has the form (5.8.3),

$$(Y, \varphi) \mapsto (Y^-, \varphi^-) = (\mathbf{A}(\varphi, \varepsilon)Y, \varphi - T(\varepsilon)), \qquad \|\mathbf{A}\| \to 0 \text{ as } \varepsilon \to 0, \qquad (5.8.3)$$

where the expression for $T(\varepsilon)$ is recalled in (1.2) below.

The intersection γ^- of the homoclinic torus H with the exit cross-section Γ^- is the same as in 5.8.3: it is defined by the equation $Y^- = 0$. A regular map along the orbits of X_0 that brings $\gamma^- \in \Gamma^-$ to the curve $H \cap \Gamma^+ = \gamma^+ \subset \Gamma^+$ is well defined. This curve is homotopically equivalent to the cycle L in U. This follows from the assumption of Theorem 1.1 that the homoclinic surface H is of type (1.1); see 5.1.3 for the definition. But the curve γ^+ is quite different from that of 5.8.3, not only because of a different homotopy type. The difference is between the critical and noncritical cases. In the present (critical) case, the curve γ^+ is tangent to some fibers $\varphi = \text{const}$ in Γ^+ (see Figure 6.1).

Indeed, the curve γ^+ is smooth by the transversality assumption of Theorem 1.1. Let $h^+ = H \cap \Gamma_0 \cap \{x < 0\}$, the same notation as in 5.8.2. Let $\pi \colon h^+ \to \gamma^+$ be the projection along the orbits of X_0 located in U. If γ^+ is transversal to the fibers $\varphi = \text{const}$, then h^+ is transversal to the fibers $x = \text{const}$. Hence the union $H \cup L$ is smooth by Theorem 5.1.1. This contradicts the assumption of Theorem 1.1 that the closure $H \cup L$ is nonsmooth.

Therefore, γ^+ is really tangent to some fibers $\varphi = \text{const}$. Let

$$T^1 = \{(Y, \varphi) \mid Y = 0, \varphi \in \mathbb{R} \pmod 1\},$$
$$T^1_- = \{(Y^-, \varphi^-) \mid Y^- = 0, \varphi^- \in \mathbb{R} \pmod 1\}$$

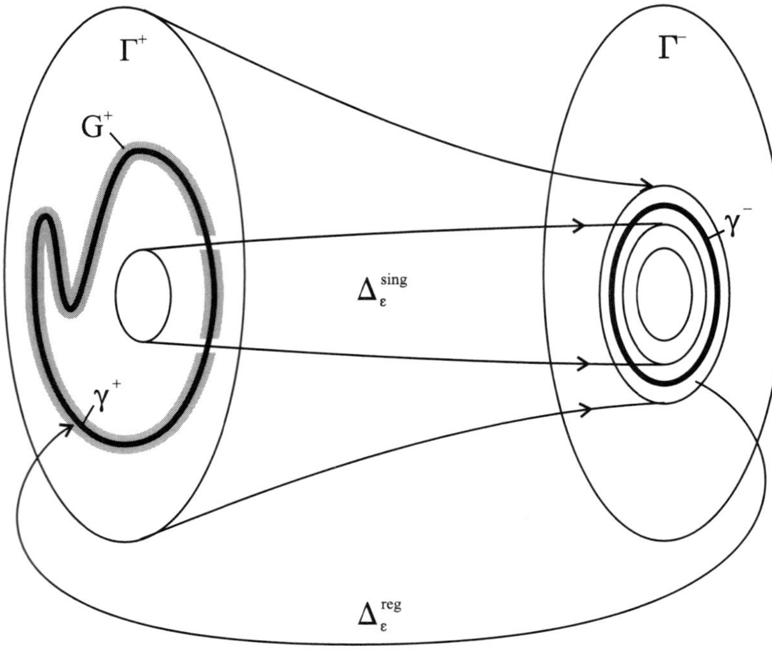

FIGURE 6.2. Singular, regular and global maps.

be circles in Γ^+, Γ^- respectively. Then the projection $f\colon \gamma^+ \to T^1$, $(Y,\varphi) \mapsto \varphi$, is an endomorphism of circles of degree 1.

The regular map $\Delta_\varepsilon^{\text{reg}}$ of the same neighborhood $\Gamma^- \supset \gamma^-$ to a neighborhood $\Gamma^+ \supset \gamma^+$ along the orbits of the ε-dependent vector fields of the family from Theorem 1.1 is well defined. Hence, a global correspondence map Δ_ε of Γ^+ to a neighborhood G^+ of the curve γ^+ is well defined:

$$\Delta_\varepsilon = \Delta_\varepsilon^{\text{reg}} \circ \Delta_\varepsilon^{\text{sing}} \colon \Gamma^+ \mapsto G^+$$

(see Figure 6.2).

Let $\mu \in [0,1)$ and define a sequence ε_k for large k by the equation $T(\varepsilon_k) = k+\mu$, where $T(\varepsilon)$ is the same as in 5.5.3:

$$T(\varepsilon) = \int_{-\delta}^{\delta} \frac{dx}{w_\varepsilon(x)}, \qquad w_\varepsilon(x) = (x^2+\varepsilon)(1+a(\varepsilon)x)^{-1}. \tag{1.2}$$

The following lemma describes the limit map Δ_{ε_k} for $k \to \infty$. It is similar to Lemma 5.8.1. Let

$$\pi^+\colon \Gamma^+ \to T^1, \qquad (Y,\varphi) \to (0,\varphi)$$

be the projection along the fibers $\varphi = \text{const}$. Define the map

$$R_\mu\colon T^1 \to T^1, \qquad \varphi \mapsto \varphi - \mu.$$

LEMMA 1.1. *For the sequence ε_k defined above, we have*

$$\Delta_{\varepsilon_k} \to \Lambda_\mu = \Delta_0^{\text{reg}} \circ R_\mu \circ \pi^+ \colon \Gamma^+ \to \gamma^+$$

in the C^2 topology.

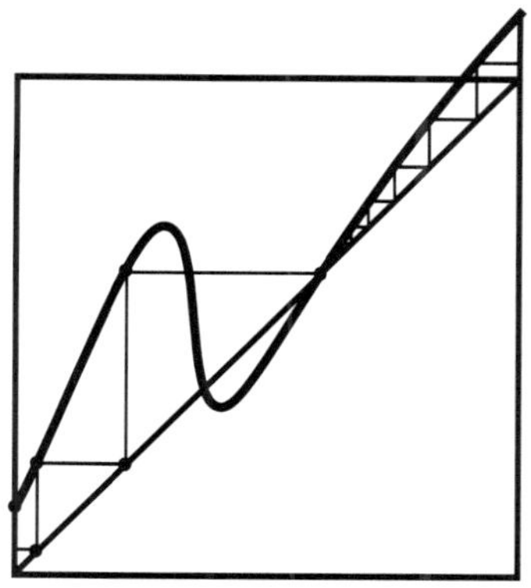

FIGURE 6.3. A transversal homoclinic orbit of a repelling fixed point.

The main part of the proof is the same as for Lemma 5.8.1. The details are carried out in §2.

The limit map $\Lambda_\mu\colon \Gamma^+ \to \gamma^+$ plays a crucial role in what follows. Its properties are determined by its restriction to γ^+. Let

$$L_\mu = \Lambda_\mu|\gamma^+ : \quad L_\mu = \Delta_0^{\mathrm{reg}} \circ R_\mu \circ \pi^+ \colon \gamma^+ \to \gamma^+. \tag{1.3}$$

It is convenient to study another map, $F_\mu\colon T^1 \to T^1$, conjugated to L_μ:

$$F_\mu = \pi^+ \circ \Delta_0^{\mathrm{reg}} \circ R_\mu \colon T^1 \to T^1. \tag{1.4}$$

The conjugating diffeomorphism is $H_\mu = \Delta_0^{\mathrm{reg}} \circ R_\mu \colon T^1 \to \gamma^+$. The conjugacy equality $H_\mu \circ F_\mu = L_\mu \circ H_\mu$ is evident. Note that conjugated maps have the same geometric properties.

The maps F_μ and L_μ are not one-to-one: the projection $\pi^+\colon \gamma^+ \to T^1$ has, in general, fold points. Hence, these maps are not diffeomorphisms, but endomorphisms.

1.4. Endomorphisms of the circle: periodic and homoclinic orbits. Let f be a continuous map of degree 1 of the circle to itself which is not one-to-one. We shall call these maps *endomorphisms* and denote the corresponding space with the C topology by $\mathrm{End}(S^1)$.

A characteristic property of endomorphisms is that the inverse image of a point is not necessarily well defined.

DEFINITION 1.1. An *orbit* of a point x under the endomorphism f is a sequence

$$\{x_n \mid n \in \mathbb{Z}\} \text{ such that } x_0 = x,\ x_{n+1} = f(x_n).$$

The point x determines its future orbit $\{x_n \mid n \geqslant 0\}$ but not, in general, the whole orbit. For instance, a periodic point of f may have a nonperiodic past orbit. See Figure 6.3, which shows this effect for a fixed point (periodic orbit of period 1).

DEFINITION 1.2. A *q-periodic* point ($q > 0$) of an endomorphism f of the circle is a point that returns to itself after q iterates of f, but no earlier. A *short periodic*

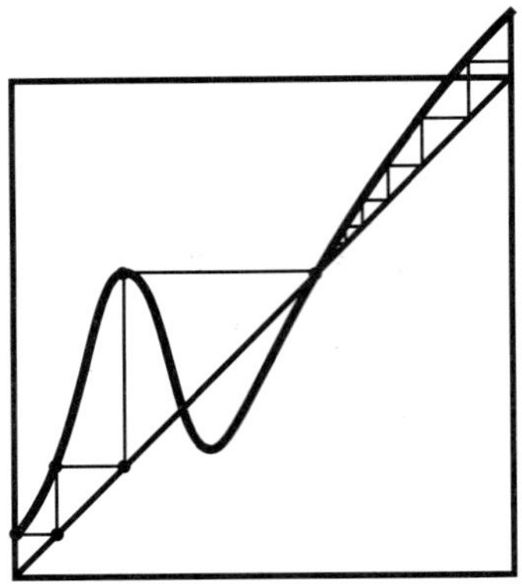

FIGURE 6.4. A tangent homoclinic orbit of a repelling fixed point.

orbit of a q-periodic point x is the set

$$s = \{x, f(x), \ldots, f^{q-1}(x)\}.$$

It corresponds to the actual periodic orbit

$$\{x_n \mid n \in \mathbb{Z}\} \text{ with } x_0 = x,\ x_n = f^k(x) \text{ for } k = 0, \ldots, q-1,\ n = k \pmod q.$$

A nonperiodic orbit may contain a short periodic orbit (see Figure 6.3 for $q = 1$).

DEFINITION 1.3. A q-periodic point of the endomorphism of the circle is *hyperbolic* (together with the corresponding short periodic orbit) if the derivative of f^q at any point of the periodic orbit differs from 1. This derivative is the same for any point of the short periodic orbit. It is called the *multiplier* of the orbit. The periodic orbit is *repelling* if this multiplier is greater than 1.

REMARK. The inequality $(f^q)'(x) \neq 1$ for at least one point of a short periodic orbit implies the hyperbolicity of the orbit by the chain rule.

DEFINITION 1.4. Consider a repelling q-periodic point $\mathbf{p}$ of an endomorphism of the circle. The maximal interval $W \ni \mathbf{p}$ having no other fixed points of f^q and such that $f|_W$ is a diffeomorphism is said to be the *local unstable manifold* of $\mathbf{p}$. The *global unstable manifold* of $\mathbf{p}$ is defined as

$$W^u_{\mathbf{p}} = \bigcup_{m \geqslant 0} f^m W.$$

REMARK. The global unstable manifold of an endomorphism of the circle may coincide with the entire circle. For a fixed point (a periodic orbit of period 1) this may be seen in Figure 6.3.

DEFINITION 1.5. Let s be a short periodic orbit of an endomorphism of the circle. The orbit $\{x_n\}$ is called *homoclinic to s* if $x_n \in s$ for all sufficiently large n, and

$$\operatorname{dist}(x_n, s) \to 0 \text{ as } n \to -\infty.$$

The example of a homoclinic orbit of a fixed point is shown in Figure 6.3.

DEFINITION 1.6. A homoclinic orbit of a short periodic orbit of an endomorphism of the circle f is called *transversal* if it contains no critical points of f, and *tangent* in the opposite case (see Figures 6.3, 6.4).

LEMMA 1.2. *There exists an open dense set A in the space of endomorphisms of the circle in the C^2 topology such that all the maps $f \in A$ have the following properties.*

1. *There exists a segment $J \subset [0,1]$ such that for any $\mu \in J$ the map*

$$f_\mu = f \circ R_\mu$$

has a repelling short periodic orbit s_μ and an orbit transversally homoclinic to s_μ. The global unstable manifold of s_μ contains the whole circle.

2. *There exists a point $\nu \in J$ such that the short periodic orbit s_ν has a tangent homoclinic orbit.*

APPENDIX TO LEMMA 1.2. *The set A in Lemma 1.2 can be chosen so that it possesses the following additional properties.*

1. *The orbit s_μ in statement 1 of Lemma 1.2 depends continuously on μ. That is, there exists a periodic point $p_\mu \in s_\mu$ continuously depending on μ.*

2. *Let $\bar{f}$, $\bar{p}_\mu$ be the lifts of f, p_μ, and suppose $g_\mu = \bar{f}_\mu^q - p$ has a fixed point $\bar{p}_\mu$. Then for any neighborhood $W \ni \bar{p}_\mu$ there exists a finite union of segments $I_\mu \subset W$ continuously depending on μ with respect to the Hausdorff distance, a segment $K \subset J$, and integers N and l with the following properties.*

Let $\sigma_\mu = \max_{I_\mu} g_\mu^N$. Then the value σ_μ is taken inside I_μ, and the function

$$\mu \mapsto \sigma_\mu - \bar{p}_\mu - l$$

has different signs at the endpoints of K.

This lemma together with the appendix is proved in §4. Necessary material about endomorphisms of the circle with a new proof of the Block–Franke theorem is presented in §3.

Finally we can state the genericity assumption in the main theorem (Theorem 1.1).

1.5. Genericity assumptions. 1. The family X_ε in Theorem 1.1 satisfies the fundamental assumptions of 5.1.1 for the semihyperbolic cycle which is a node in its hyperbolic variables.

We do not recall these assumptions. They were already discussed in 1.1, and used afterwards.

The second assumption is related to the map F_0 (see (1.4)).

2. The map $f = F_0 = \pi^+ \circ \Delta_0^{\mathrm{reg}} \circ R_0 \colon T^1 \to T^1$ belongs to the set A described in Lemma 1.2 and the appendix to it.

1.6. Transversal homoclinic orbits of the global return map: proof of statement 1 of the main theorem. The proof will be given modulo Lemmas 1.1, 1.2, and Lemmas 1.3, 1.4 stated below. The map Δ_ε will be studied as a perturbation of the limit map Λ_μ for an appropriate $\mu(\varepsilon)$. Lemma 1.2 provides transversal and tangent homoclinic orbits for the endomorphisms $\mathrm{L}_\mu = \Lambda_\mu|_{\gamma^+}$. The degenerate Hadamard–Perron theorem below allows us to use this material to build transversal and tangent homoclinic orbits for the global return map (see Figure 6.5 for the transversal case).

Let us now pass to the detailed proof. Let $T(\varepsilon)$ be the same as in (1.2). Denote

$$k(\varepsilon) = [T(\varepsilon)], \qquad \mu(\varepsilon) = \{T(\varepsilon)\}.$$

Recall that $k(\varepsilon) \to \infty$ as $\varepsilon \to 0$. Let J be the interval from Lemma 1.2 for $f = F_0$. The lemma is applicable because $F_0 \in A$ by assumption 2. Let

$$S = \{\varepsilon \mid \mu(\varepsilon) \in J\}. \tag{1.5}$$

LEMMA 1.3. *The set S has the density property* (1.1).

We shall prove that for small $\varepsilon \in S$, the map Δ_ε has a hyperbolic periodic orbit with a transversal intersection of its stable and unstable manifolds. A similar statement for the homoclinic tangency will be proved in a similar way.

To do this we use the following result.

DEGENERATE VERSION OF THE HADAMARD–PERRON THEOREM. *Let F be a map of class C^2 with a hyperbolic fixed point, possibly degenere (no multipliers on the unit circle, but, possibly, some zero multipliers). Then F has the germs of stable and unstable manifolds of class C^2 at this point; the restriction of F to the stable manifold may be noninjective. If the sequence F_k tends to F in C^r, $r \geqslant 1$, then for all sufficiently large k the maps F_k have hyperbolic fixed points with stable and unstable manifolds tending in C^r to the stable and unstable manifolds of the initial fixed point respectively.*

The map F_0 belongs to the set A by assumption 2. Hence, by statement 1 of Lemma 1.2, for some positive integer q and for any $\mu \in J$, the map F^q_μ has a repelling fixed point with a transversal homoclinic orbit. The same is true for the conjugated endomorphisms L^q_μ. Denote the corresponding repelling fixed point of the latter map by p_μ. The map L^q_μ is the restriction to γ^+ of a limit map Λ^q_μ. Hence Λ^q_μ has a repelling fixed point p_μ with one-dimensional unstable manifold included in γ^+, and a shrinking stable manifold $S_\mu = (\pi^+)^{-1}(\pi^+ p_\mu)$ (see Figure 6.5).

Fix any $\mu \in J$. By Definitions 1.4 and 1.5, there exists an arc $I \subset \gamma^+$ that lies close to p_μ but does not contain it and a large m such that

$$I' = L^{qm}_\mu I \ni p_\mu$$

(see Figure 6.5). Moreover, the arc I' transversally intersects the shrinking stable manifold S_μ of the map Λ^q_μ at p_μ.

Let ε_k be the monotonic sequence defined by $\mu(\varepsilon_k) = \mu$. The maps

$$G_k = \Delta^q_{\varepsilon_k} : \Gamma^+ \to G_k\Gamma^+ \subset \Gamma^+$$

tend to Λ^q_μ in C^2. Hence, the maps G_k have a sequence of fixed points q_k tending to p_μ. The local stable and unstable manifolds of q_k under G_k, W^s_k, and W^u_k tend (in C^1) to the intersections of S_μ and γ^+ with some neighborhood of p_μ. Let $I_k \subset W^u_k$ be a sequence of intervals that tends to I. Then the interval $I'_k = \Delta^{qm}_{\varepsilon_k} I_k$ intersects the stable manifold W^s_k transversally at a point close to p_μ, for large k. This is the desired homoclinic intersection. Its existence proves statement 1 of the main theorem.

1.7. Tangent homoclinic orbit of the global return map: proof of statement 3 of the main theorem. Let T be the same function as in (1.2). For

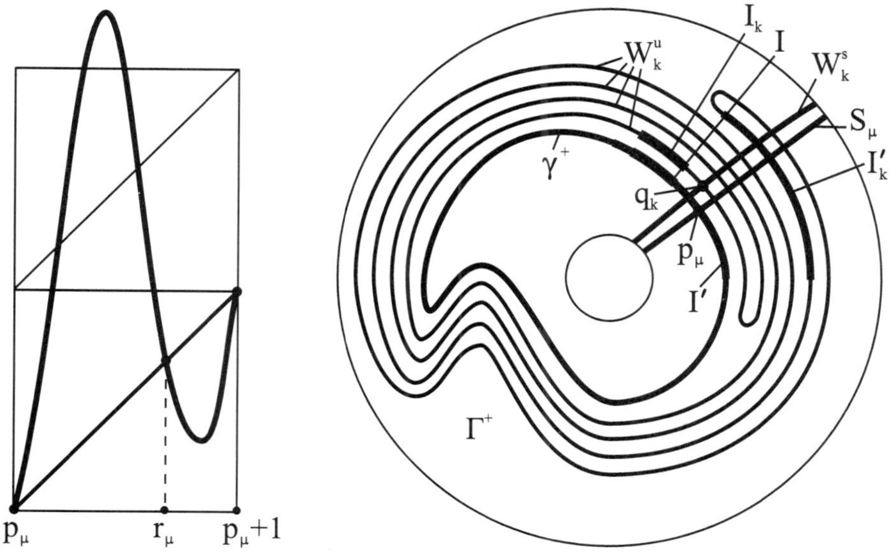

FIGURE 6.5. A homoclinic orbit of the Poincaré map.

any sufficiently large k, we will find a point $\mu_k \in [0,1)$ such that $\varepsilon_k = T^{-1}(k+\mu_k)$ fits statement 3 of Theorem 1.1.

We will apply appendix to Lemma 1.2 to the map $f = F_0 = \pi^+ \circ \Delta_0^{\mathrm{reg}} \circ R_0 : T^1 \to T^1$, more precisely, to the conjugated map $L_0 \colon \gamma^+ \to \gamma^+$ (see (1.3)). Let us formulate the consequences of the appendix for this map. The map L_μ has a q-periodic point p_μ (already considered in 1.6) depending continuously on μ. The point p_μ is not a fold point of the projection π^+, because it is a hyperbolic periodic point of the map L_μ.

The curve γ^+ has no parametrization given *a priori*. Let us take an arbitrary parametrization ψ of γ^+ that coincides with φ in some neighborhood $\mathbf{W}$ of p_μ and such that $\gamma^+ = \mathbb{R}/\mathbb{Z}$ with respect to this parametrization. Let $W = \mathbf{W} \cap \gamma^+$, $x = \varphi|W$.

Let $\bar\psi$ be the lift of ψ to $\mathbb{R}$, the universal covering over γ^+, and $\pi \colon \mathbb{R} \to \mathbb{R}/\mathbb{Z} = \gamma^+$ be the natural projection. Let $g_\mu = \bar L_\mu^q - p$ be the lift of L_μ^q with the fixed point $\bar p_\mu$ (this determines the choice of p). Then, by the appendix, there exists a finite union of segments $I_\mu \subset W$, a segment $K \subset J$, and positive numbers N and l with the following properties.

1. The function

$$\Phi_\mu \colon x \mapsto \bar\psi \circ g_\mu^N(x) - \bar\psi(\bar p_\mu) - l$$

has the maximum value $\lambda_\mu = \max_{I_\mu} \Phi_\mu$, which is taken in some *interior* point of I_μ.

Note that $\lambda_\mu(x) = \sigma_\mu(x) + \bar\psi(p_\mu) + l$, where σ_μ is the same as in the appendix, because the difference $\lambda_\mu - \sigma_\mu$ is constant on I_μ.

2. There exists a segment $K \subset J$ such that the function $\lambda \colon \mu \mapsto \lambda_\mu$ has different signs at the endpoints of K.

Let $\varepsilon(k,\mu) = T^{-1}(k+\mu)$. Take a universal covering $\bar{U}$ over some tubular neighborhood U of γ^+ in Γ^+ with the projection $\bar{U} \to U$ still denoted by π. Denote by Ψ the extension of $\bar{\psi}$ to $\bar{U}$ such that $\Psi(\operatorname{mod}\mathbb{Z})$ is constant on any fiber of π. Let $G(k,\mu)$ be the lift of the map $\Delta^q_{\varepsilon(k,\mu)}$ to $\bar{U}$ chosen in such a way that

$$G(k,\mu)|\gamma^+ \to g_\mu \text{ as } k \to \infty.$$

Now apply the degenerate version of the Hadamard–Perron theorem to the sequence $G(k,\mu)$. By this theorem, the map $G(k,\mu)$ for large k has a fixed point $q(k,\mu)$ with one-dimensional unstable manifold $W^u(k,\mu)$ and a stable manifold $W^s(k,\mu)$. Moreover, $q(k,\mu) \to \bar{p}(\mu)$ as $k \to \infty$, and $W^s(k,\mu)$, $W^u(k,\mu)$ tend to $W^s(\bar{p}_\mu)$, $W^u(\bar{p}_\mu)$ and the following holds:

1. In the neighborhood $\mathbf{W}$, the manifold $W^s(k,\mu)$ is given by an equation of the form $\Psi = h_{k,\mu}(y)$, where $h_{k,\mu}(y) \to \bar{\psi}(\bar{p}_\mu) = \text{const}$ in the C^1 topology.
2. There exists a sequence (still μ-dependent) $i_{k,\mu} \colon W \to W^u(k,\mu)$ such that $i_{k,\mu} \to \text{id}$ in the C^1 topology as $k \to \infty$.

Now we can find the homoclinic tangency for the map $G(k,\mu)$ corresponding to some $\mu = \mu_k \subset K$. Consider the function

$$\Phi_{k,\mu} = (\Psi - h_{k,\mu}(y)) \circ G^N_{k,\mu} \circ i_{k,\mu} - l.$$

It tends uniformly to $\Phi \colon (\mu,\varphi) \mapsto \Phi_\mu(\varphi)$, $\mu \in K$, $\varphi \in I_\mu$, by the above argument. Hence, the functions $\lambda_k \colon K \mapsto \mathbb{R}$, $\mu \mapsto \max_{I_\mu} \Phi_{k,\mu}$, converge uniformly on K to the function λ. This function changes sign on K; the same is true for λ_k for sufficiently large k. Hence, for some $\mu = \mu_k$, the function $\Phi_{k,\mu}$ takes a zero maximal value at some interior point $x \in I_\mu$. The point

$$y = \pi \circ G^N_{k,\mu} \circ i_{k,\mu}(x)$$

is the point of the homoclinic tangency of stable and unstable manifolds of the periodic point $q(k,\mu)$ of the map $\Delta_{\varepsilon(k,\mu)}$. This proves statement 3 of the main theorem.

1.8. Why is the maximal attractor of the global return map strange? The answer is given by the following

LEMMA 1.4. *The phase flow maps of the vector field X_ε corresponding to sufficiently large positive time values have an absorbing domain which is a neighborhood of the homoclinic 2-torus, and contract three-dimensional volumes in this domain.*

This lemma will be proved in 2.3. Hence, the maximal attractor of X_ε cannot contain three-dimensional manifolds. Indeed, by the appendix to Theorem 2.2.1, the Hausdorff dimension of this attractor is less than 3. But the Hausdorff dimension of a 3-manifold equals 3, a contradiction. On the other hand, the global return map Δ_ε for the values of $\varepsilon \in S$ considered in the previous subsection has a periodic orbit with one-dimensional unstable manifold and homoclinic intersection of moustaches. By the Smale–Birkhoff theorem, this map has an infinite number of periodic points with one-dimensional unstable manifolds. These manifolds give rise to two-dimensional invariant manifolds of the phase flow of X_ε, which belong to the maximal attractor of the phase flow transformation. Therefore, this attractor cannot be the finite union of embedded two-dimensional submanifolds of the phase space. This proves statement 2 of the main theorem.

§2. Lemmas on limit maps, density and volume contraction

In this section we prove Lemmas 1.1, 1.3 and 1.4.

2.1. Limit family of global return maps: proof of Lemma 1.1. The map Δ_{ε_k} from Lemma 1.1 is the composition of two maps, a regular and a singular one. The regular map tends to $\Delta_0^{\rm reg}$ with all its derivatives. It remains to prove that

$$\Delta_k = \Delta_{\varepsilon_k}^{\rm sing} \to R_\mu \circ \pi^+ \text{ as } k \to \infty.$$

Indeed, by (5.8.3), we have

$$\Delta_k(Y, \varphi) = (\mathbf{A}(\varphi, \varepsilon_k)Y, \varphi - \mu), \tag{2.1}$$

where $\mathbf{A}(\varphi, \varepsilon_k)$ is the product of $k+1$ factors. All these factors have bounded derivatives up to order 2 (see the remark at the end of 5.7.4). At least $k-1$ of them have norm smaller than some $\lambda < 1$. Hence, by the Leibniz rule, $\mathbf{A}(\varphi, \varepsilon_k)Y$ exponentially tends to zero together with all its derivatives as $k \to \infty$. Hence (2.1) implies the lemma.

2.2. Density of the parameter set corresponding to strange attractors. In this subsection, we prove Lemma 1.3, namely

LEMMA 1.3. *The set* (1.5)

$$S = \{\varepsilon > 0 \mid T(\varepsilon) \in J\},$$

with $T(\varepsilon)$ defined by (1.2), *has the density property* (1.1).

PROOF. Consider the function ε_k on $[0, 1)$ given by $\varepsilon_k(\mu) = T^{-1}(k + \mu)$. Let $J_k = \varepsilon_k(J)$.

PROPOSITION 2.1. *For the sequence of functions ε_k and sets J_k defined above, we have*

1. $\displaystyle\lim_{k\to\infty} \frac{\varepsilon_k(0)}{\varepsilon_{k+1}(0)} = 1;$
2. $\displaystyle\lim_{k\to\infty} \frac{|J_k|}{\varepsilon_k(0) - \varepsilon_{k+1}(0)} = |J|,$ *where* $|\cdot|$ *denotes the length of the interval.*

PROOF. The function T may be easily calculated by using (1.2):

$$T(\varepsilon) = \int_{-\delta}^{\delta} \frac{1 + a(\varepsilon)x}{\varepsilon + x^2} dx = \frac{2}{\sqrt{\varepsilon}} \arctan \frac{\delta}{\sqrt{\varepsilon}} = \frac{\pi}{\sqrt{\varepsilon}}(1 + O(\sqrt{\varepsilon})).$$

Hence

$$\varepsilon_k(0) = T^{-1}(k) = \frac{\pi^2}{k^2}\Big(1 + O\Big(\frac{1}{k}\Big)\Big),$$

which implies assertion 1. Now we prove assertion 2. It is implied by the fact that the greater is k, the more is ε_k like an affine function:

$$|J_k| = |\varepsilon_k(J)| = |\varepsilon_k'(\theta)||J| = T'(\varepsilon_k(\theta))|^{-1}|J| = \frac{2\pi^2}{k^3}(1 + o(1))|J|, \tag{2.2}$$

$$\varepsilon_k(0) - \varepsilon_{k+1}(0) = \varepsilon_k(0) - \varepsilon_k(1) = |\varepsilon_k'(\theta_1)| = |T'(\varepsilon_k(\theta_1))|^{-1} = \frac{2\pi^2}{k^3}(1 + o(1)),$$

which, together with (2.2), implies assertion 2.

Now we prove inequality (1.1). For $\varepsilon \in [\varepsilon_{k+1}(0), \varepsilon_k(0)]$, $k \gg 1$, we have

$$\begin{aligned}\frac{1}{\varepsilon}\operatorname{mes}(S \cap [0,\varepsilon]) &\geqslant \varepsilon_k(0)^{-1}\sum_{i>k}|J_i| > \frac{1}{2}\,\varepsilon_{k+1}(0)^{-1}\sum_{i>k}|J_i| \\ &= \frac{1}{2}\,\frac{\sum_{i>k}|J_i|}{\sum_{i>k}(\varepsilon_i(0)-\varepsilon_{i+1}(0))} > \frac{1}{3}\,|J|.\end{aligned}$$

The last inequality follows from assertion 2 of Proposition 2.1.

2.3. Contraction of three-dimensional volumes. By the construction in 1.3, the global return map Δ_ε takes some neighborhood W of the homoclinic torus to itself. Therefore, the maximal attractor of this map in W is well defined. In order to prove Lemma 1.4, it is sufficient to prove that the global correspondence map contracts two-dimensional volumes for sufficiently small ε. As we mentioned before, this map is the product of two factors (regular and singular). The regular map has a bounded distortion uniformly for all small $\varepsilon > 0$. Let us prove that the singular map contracts two-dimensional volumes with a coefficient that tends to 0 as $\varepsilon \to 0$.

Indeed,

$$D\Delta_\varepsilon^{\text{sing}} = \begin{pmatrix} \mathbf{A} & \mathbf{A}_\varphi \\ 0 & 1 \end{pmatrix}.$$

This map preserves the $d\varphi$-component and contracts all the vectors from the Y-plane. The contraction coefficient tends to 0 as $\varepsilon \to 0$, that is, as $k \to \infty$. This proves the lemma.

The main theorem is proved modulo Lemma 1.2 and the appendix to it. This will be done in the next two sections.

§3. Rotation sets and periodic points of circle endomorphisms

In this section we introduce and study the notion of rotation sets for maps of the circle of degree one. This notion extends the notion of Poincaré rotation number from diffeomorphisms to endomorphisms. After that we apply the theory of rotation sets to study the existence of periodic orbits of endomorphisms and give a proof of the Block–Franke theorem, which will be used in §4 to prove Lemma 1.3.

3.1. Rotation set for maps of the circle. We begin with the definition of the rotation set. Denote by $\operatorname{End}(S^1)$ the set of continuous not one-to-one maps of the circle of degree one onto itself and by $\overline{\operatorname{End}}(S^1)$ the set of nonbijective continuous maps $f\colon \mathbb{R} \to \mathbb{R}$ satisfying the relation

$$f(x+1) = f(x) + 1. \tag{3.1}$$

We use the standard C^0 topology on the spaces $\operatorname{End}(S^1)$ and $\overline{\operatorname{End}}(S^1)$.

DEFINITION 3.1. Let $f \in \overline{\operatorname{End}}(S^1)$. The *rotation number* of the f-orbit of the point x is the upper limit:

$$\varlimsup_{n\to\infty} \frac{f^{[n]}(x)}{n} = r_f(x).$$

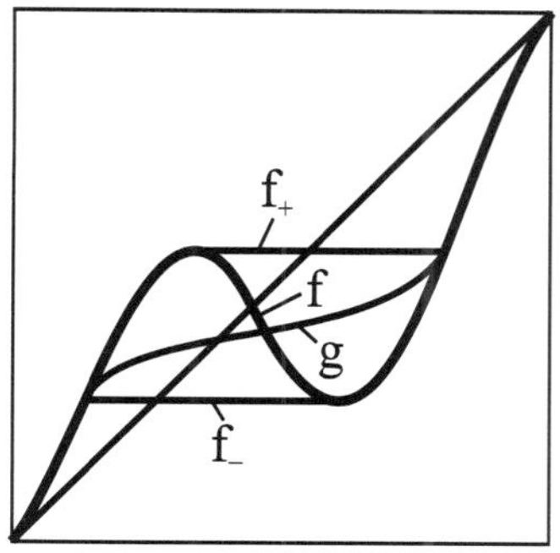

FIGURE 6.6. Graphs of the maps f_-, f_+ and the graph of the intermediate map g in the proof of the Block–Franke theorem.

DEFINITION 3.2. The *rotation set* $r(f)$ of a map $f \in \overline{\mathrm{End}}(S^1)$ is the closure of the set of the rotation numbers of all orbits:

$$r(f) = \mathrm{closure}\{r_f(x) \mid x \in \mathbb{R}\}.$$

For $f \in \mathrm{End}(S^1)$ with lift $\bar{f} \in \overline{\mathrm{End}}(S^1)$, we define the *rotation set* of f to be $r(\bar{f})$, which is unique up to the addition of an integer.

REMARK. It is easy to prove that the rotation set is a topological invariant: if two endomorphisms are topologically conjugated, then their rotation sets coincide modulo $\mathbb{Z}$.

If the endomorphism is in fact a homeomorphism, then its rotation set is a single point, equal to its rotation number. It turns out that if the lifted map is only monotone, possibly taking some segments to points, its rotation set still is a singleton (see §3 of Chapter 5). In the general case, the rotation set is always a segment depending continuously on the map. We prove in this section that its endpoints are rotation numbers for some monotonic maps naturally constructed from the initial one (see Figure 6.6). The most surprising fact is that a smooth nonhomeomorphic endomorphism of a circle never has a rotation set consisting of only one irrational point (this is the Block–Franke theorem below).

The following property of periodic orbits is evident: their rotation number is always rational. This simple remark is the key tool of the subsequent proofs. Suppose $f \in \overline{\mathrm{End}}(S^1)$. Let

$$f_-(x) = \min_{y \geqslant x} f(y), \qquad f_+(x) = \max_{y \leqslant x} f(y) \tag{3.2}$$

(see Figure 6.6).

Obviously, the maps $f_\pm$ are monotonic continuous functions satisfying (3.1) and

$$f_-(x) \leqslant f(x) \leqslant f_+(x). \tag{3.3}$$

The main result of this subsection is the following

THEOREM 3.1. *The rotation set of $f \in \overline{\mathrm{End}}(S^1)$ is the closed interval*

$$[r(f_-), r(f_+)],$$

where $r(f_\pm)$ denotes the rotation number of $f_\pm$.

Before proving this theorem, let us give some definitions and lemmas.

DEFINITION 3.3. Let $f \in \overline{\text{End}}(S^1)$. A point $x \in \mathbb{R}$ is called a *periodic point* of f if there exists an integer p and a natural number q such that $f^q(x) = x + p$, where f^q denotes the iterate $f \circ \cdots \circ f$ of f with itself q times. The rational number p/q is called the *rotation number* of x and of the corresponding periodic orbit.

LEMMA 3.1. *If the map $f \in \overline{\text{End}}(S^1)$ has no periodic point with rotation number p/q, $p \in \mathbb{Z}$, $q \in \mathbb{N}$, then $r(f)$ is contained in the set $\{x \in \mathbb{R} \mid x < p/q\}$ or in the set $\{x \in \mathbb{R} \mid x > p/q\}$.*

PROOF. Since $f^q(x) \neq x + p$ for any $x \in \mathbb{R}$, either $f^q(x) - x < p$ for all $x \in \mathbb{R}$, or $f^q(x) - x > p$ for all $x \in \mathbb{R}$. Notice that the function $f^q(x) - x$ is periodic in x. Hence there exists a positive constant c such that $f^q(x) - x < p - c$ or $f^q(x) - x > p + c$ for all $x \in \mathbb{R}$. Therefore $r(f) \subset \{x \in \mathbb{R} \mid x \leqslant (p-c)/q\}$ or $r(f) \subset \{x \in \mathbb{R} \mid x \geqslant (p+c)/q\}$.

COROLLARY 3.1. a) *Let $f \in \overline{\text{End}}(S^1)$, $a, b \in r(f)$, and $a \leqslant p/q \leqslant b$ for some rational number p/q. Then f has a periodic point with rotation number p/q and hence $p/q \in r(f)$.*

b) *The rotation set $r(f)$ is either a single point in $\mathbb{R}$ or a closed interval.*

PROOF. Statement a) of the corollary is a direct consequence of Lemma 3.1. Statement b) follows from a) and Lemma 3.1.

Denote by $r_-(f)$ and $r_+(f)$ the left and right end of the segment $r(f)$ respectively.

LEMMA 3.2. *Suppose $f \in \overline{\text{End}}(S^1)$. Then the functions $f \mapsto r_\pm(f)$ are continuous.*

PROOF. Notice that for any rational number p/q, the inequality $p/q < r_-(f)$ (or $p/q > r_+(f)$) is equivalent to $f^q(x) - x > p$ (resp. to $f^q(x) - x < p$) for all $x \in [0, 1]$; this is an "open condition", i.e., the set of $f \in \overline{\text{End}}(S^1)$ with $r_-(f) > p/q$ (resp. $r_+(f) < p/q$) is open. Finally, $p/q \in (r_-(f), r_+(f))$ if and only if for some large natural number N there are $x, y \in [0, 1]$ with

$$f^{Nq}(x) - x > Np + 1, \qquad f^{Nq}(y) - y < Np - 1.$$

Again, this condition is open. Hence $r_-(f)$ and $r_+(f)$ depend continuously on f.

DEFINITION 3.4. A C^2 map $f \in \overline{\text{End}}(S^1)$ is called *regular* if all its critical points are nondegenerate, i.e., the second derivative does not vanish at these points.

REMARKS. 1. Nondegenerate critical points of a map are isolated. Therefore, the critical set of a regular endomorphism of a circle is finite.

2. Regular endomorphisms form an open dense set in the space of all nondiffeomorphic C^2 endomorphisms of a circle.

3. The composition of regular endomorphisms is not necessarily regular itself. A counterexample for the line is $x^2 \circ x^2 = x^4$. Yet *any critical point of an iterate of a regular map is either a local minimum or maximum.*

Theorem 3.1 follows from Corollary 3.1b) and the following result.

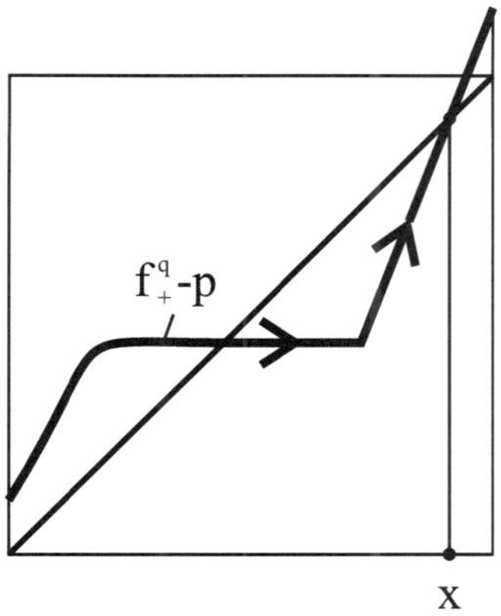

FIGURE 6.7. Graph of the map $f_+^q - p$.

LEMMA 3.3. *Let* $f \in \overline{\mathrm{End}}(S^1)$. *Then* $r_\pm(f) = r(f_\pm)$.

PROOF. From (3.3), we have

$$r(f_-) \leqslant r_-(f) \leqslant r_+(f) \leqslant r(f_+). \tag{3.4}$$

We begin with the principal case when $r(f_+) = p/q$ is a rational number. We will prove that there exists a periodic orbit of f with rotation number p/q which is, at the same time, the orbit of f^+. Consider the "noncoincidence set"

$$C = \{x \in \mathbb{R} \mid f_+(x) > f(x)\}.$$

We claim that there exists a periodic orbit of f that does not intersect C. This means that the orbits of x for the maps f and f_+ coincide. The point x is found in Figure 6.7 and in the next three paragraphs.

Let us first describe the periodic orbit of f_+ that intersects C; let z be a point of this orbit. Then z is an *interior point of the set* $C_q = \{x \in \mathbb{R} \mid Df_+^q(x) = 0\}$. This can be proved in the following way. Let $y \in \mathrm{orb}(z) \cap C$ and $y = f_+^s(z)$, $0 \leqslant s < q$. Then the equality $Df_+^{s+1} = 0$ holds in some neighborhood of z. The chain rule implies that the same holds for Df_+^q. Since C is open, a small perturbation of z keeps it in C_q.

Now let us find a periodic point x of f_+ that does not belong to $\mathrm{Int}(C_q)$. This will imply our claim.

Consider the function $f_+^q(x) - p$. Our assumption $r(f_+) = p/q$ implies that the graph Γ of this function has a nonempty intersection with the diagonal $\Delta = \{x = y\}$. The graph Γ has the orientation induced by the natural orientation of the x-axis. Its intersection with the diagonal Δ is closed and invariant under the shift

$$T\colon (x, y) \to (x + 1, y + 1), \qquad T(\Gamma \cap \Delta) = \Gamma \cap \Delta. \tag{3.5}$$

If the x-projection of an intersection point $u \in \Gamma \cap \Delta$ is an interior point of the set C_q, then the graph Γ passes from the left half-plane ($y > x$) to the right one ($y < x$); see Figure 6.7. The graph Γ must return to Δ, since otherwise u would be the last point of $\Gamma \cap \Delta$, which contradicts (3.5). The return point of Γ to Δ from the right half-plane is the required one. Its x-projection is the periodic point of f_+ whose orbit does not intersect C.

Let x be a periodic point of f_+ with rotation number $p/q = r(f_+)$ whose orbit does not intersect C. Then f has the same periodic orbit. This implies that $p/q \in r(f)$. By (3.4), $p/q = r_+(f)$. This proves Lemma 3.3 for the case in which $r(f_+)$ is rational.

In the case when $r(f_+)$ is irrational, we use the previous result, the density of rational numbers, and the continuity of the function r_+. In more detail, let us consider $\mu(\varepsilon) = r(f_+ + \varepsilon)$ as a function in ε. By Proposition 5.3.3, $\mu(\varepsilon)$ is continuous. By Lemma 5.3.2, $\mu(0) < \mu(\varepsilon)$ for $\varepsilon > 0$. Therefore, we can find a sequence $\varepsilon_n \to 0$ such that the numbers $r(f_+ + \varepsilon_n)$ are rational. Then we have

$$r_+(f) = \lim_{n\to\infty} r_+(f + \varepsilon_n) = \lim_{n\to\infty} r(f_+ + \varepsilon_n) = r(f_+).$$

We can prove that $r_-(f) = r(f_-)$ similarly. This proves Lemma 3.3 and Theorem 3.1.

3.2. Existence of periodic orbits: the Block–Franke theorem. Let f be a C^1 map of an interval. We say that a critical point of f is a *topological fold* if it is isolated and f' changes sign at this point. Note that a topological fold is not necessary a Whitney fold.

THEOREM 3.2 (Block–Franke). *Let $f \in \mathrm{End}(S^1)$ be a C^1 map with first derivative of bounded variation and critical set consisting of a finite number of topological folds. Then f has a periodic orbit.*

PROOF. Let A be the set $A = \{x \in \mathbb{R} \mid f_+(x) \geqslant f_-(x)\}$. By the assumption of the theorem, A is the finite union of closed intervals.

By Corollary 3.1, it is sufficient to prove that $r(f)$ cannot consist of a single irrational number. Suppose, on the contrary, that $r_-(f) = r_+(f) \notin \mathbb{Q}$. Then, by Theorem 3.1,

$$r(f_-) = r(f_+). \tag{3.6}$$

Now we can construct a C^1 diffeomorphism g of the circle whose first derivative is of bounded variation so that $f_-(x) \leqslant g(x) \leqslant f_+(x)$ (see Figure 6.6). The map g coincides with f outside of A. The bounded variation condition is easily satisfied on any closed interval of A. In the complement of A, we have $g = f$, and the above condition follows from the assumption of the theorem. The monotonicity of the rotation number, together with (3.6), implies

$$r(f_-) = r(g) = r(f_+) \notin \mathbb{Q}.$$

Hence, the Denjoy theorem may be applied to the map g. By this theorem, the nonwandering set of g coincides with the circle. Hence, the inequality $g < f_+$ holds on some interval of $\Omega(g)$. Therefore, Lemma 5.3.2 may be applied to g and f_+, yielding $r(g) < r(f_+)$, a contradiction. This proves the Block–Franke theorem.

§4. Homoclinic orbits of circle endomorphisms

In this section we prove Lemma 1.2 and the appendix to it. The proof will be split into three parts: Lemmas 4.1, 4.2 give explicit sufficient conditions for the assertions of Lemma 1.2 and the appendix to hold. Lemma 4.3 implies that these conditions are fulfilled for maps from a certain open dense subset in the space of degree one endomorphisms of the circle.

4.1. Transversal homoclinic orbits. Let R_μ be the same rotation as in 1.3: $R_\mu\colon x \to x - \mu$, $f_\mu = f \circ R_\mu$. Below we use the definitions from 1.4 and §3.

LEMMA 4.1. *Let f be a regular endomorphism of the circle such that*

1. *The upper rotation number $r_+(f)$ is irrational.*
2. *There exists a rational number $p/q \in r(f)$ such that all periodic points of f with rotation number p/q are hyperbolic.*

Then f has a transversal homoclinic orbit of some q-periodic point $\mathbf{p}$ with the rotation number p/q. The global unstable manifold of this point coincides with the whole circle:

$$W^u_{\mathbf{p}} = S^1. \tag{4.1}$$

REMARK. The point $\mathbf{p}$ belongs to the periodic orbit mentioned in assumption 2. The existence of a rational number in $r(f)$ follows from assumption 1 and the Block–Franke theorem. Assumption 2 implies the hyperbolicity of all the corresponding periodic orbits.

PROOF OF LEMMA 4.1. Let $\bar{f}$ be the lift of f, and $g = \bar{f}^q - p$. Then all the points of the periodic orbits of $\mathbf{p}$ with rotation number p/q lifted to $\mathbb{R}$ are fixed points of g. They are hyperbolic by assumption 2. On the other hand, $r(g) = r(\bar{f}) - p/q$. Hence, $r(g)$ is a segment with the point 0 inside.

Now the crucial difference between endomorphisms and diffeomorphisms of the circle enters the game. Namely, *there exists an interval between two fixed points of g whose length under the iterates of g becomes unbounded.* Indeed, suppose that for any interval I in the complement to the set of the fixed points of g, the sequence of lengths $|g^n I|$ is bounded. Then $r(g) = \{0\}$, a contradiction.

Now let I be an interval between two fixed points of g such that $|g^n I| \to \infty$ as $n \to \infty$. These fixed points are hyperbolic by assumption. At least one of them is a repeller; denote it by $\mathbf{p}$. The difference $g - \mathrm{id}$ keeps its sign on I. Hence, W^u contains I. On the other hand, there exists a k such that $|g^k I| > 1$, because the sequence $|g^n I|$ tends to infinity. Hence, $W^u_{\mathbf{p}} = S^1$.

Now let us find a transversal homoclinic orbit of $\mathbf{p}$ under g. By (4.1) there exist $y \in I$ and $k \in \mathbb{N}$ such that $g^k(y) = \mathbf{p} + 1$ or $g^k(y) = \mathbf{p} - 1$. Consider the first case. By the regularity of f, this y may be chosen to be a noncritical point of g^k. Indeed, by Remark 3 to Definition 3.4, all the critical points of g^k are either maxima or minima. The first point to the left at which the graph of g^k crosses over the horizontal line on the level $\mathbf{p} + 1$ corresponds to a noncritical value of g^k (see Figure 6.8). The x-coordinate of this point is y.

There exists an orbit $\{y_n\}$ of y under g^k with the following properties: $y_n = \mathbf{p}$ for $n > 0$; $\mathbf{p} \neq y_n \to \mathbf{p}$ as $n \to -\infty$ for $n \leqslant 0$. The same reasoning as above allows us to choose the points y_n, $n < 0$, as noncritical points of g^k.

Now we can construct the orbit of f claimed in Lemma 4.1. Let

$$x_{nq} = y_n + pn, \qquad x_{nq+m} = f^m(y_n + pn), \quad 0 < m < q.$$

This is a transversal homoclinic orbit of the short periodic orbit

$$\{\mathbf{p}, f(\mathbf{p}), \dots, f^{q-1}(\mathbf{p})\}.$$

Lemma 4.1 is proved.

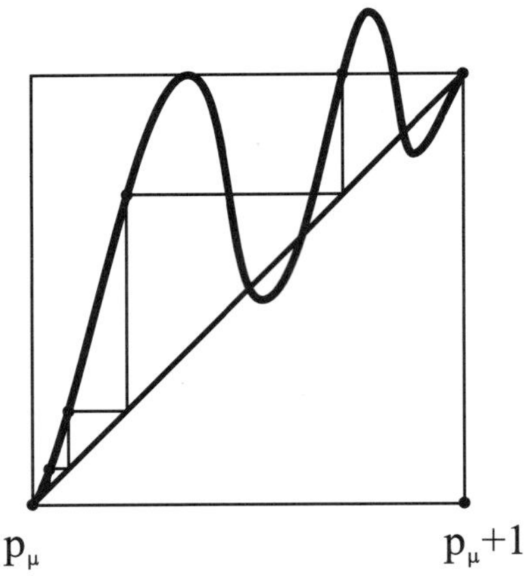

FIGURE 6.8. Finding a transversal homoclinic orbit.

4.2. Tangent homoclinic orbits. In what follows all the lifts of the maps and points of the circle continuously depending on the parameter are continuous in this parameter as well.

LEMMA 4.2. *Let f be the same as in Lemma 4.1, and $\bar{f}$ its lift. Then*

1. *There exist negative δ such that for any $\mu \in [\delta, 0]$ the map $f_\mu = f \circ (\mathrm{id} - \mu)$ has a hyperbolic periodic point p_μ with the property (4.1) continuously depending on μ. Moreover, there exists a point $\nu \in [\delta, 0]$ such that f_ν has a tangent homoclinic orbit of the periodic point p_ν.*

2. *For any neighborhood $W \ni \bar{p}_\mu, \mu \in [\delta, 0]$ there exist positive integers N and l and a finite union of segments $I_\mu \subset W$ depending on μ continuously (in the sense of the Hausdorff distance) such that the maximum value*

$$\sigma_\mu = \max_{I_\mu} \bar{f}_\mu^N \tag{4.2}$$

is taken at an interior point of I_μ. Moreover, the function $\lambda\colon \mu \mapsto \sigma_\mu - \bar{p}_\mu - l$ has different signs at the endpoints of $K = [\delta, 0]$:

$$\sigma_0 < \bar{p}_0 + l, \tag{4.3}$$

$$\sigma_\delta > \bar{p}_\delta + l. \tag{4.4}$$

Here $\bar{p}_\mu$ is the lift of p_μ.

PROOF. 1. The first statement follows from the second. The function λ is continuous because I_μ depends continuously on μ. It has different signs at 0 and δ. Hence, it vanishes at some point $\nu \in K_\delta$. Then the function $\bar{f}_\nu^N - \bar{p}_\nu - l$ has a zero critical value at some point $x \in I_\nu$. This point belongs to a tangent homoclinic orbit of p_ν because of property (4.1); see Figure 6.9. The figure is drawn for the case in which p_ν is a fixed point.

2. Let us now prove the second statement. This will be done in two steps.

Let the rotation number of the short periodic orbit of p_μ be p/q. As in 1.7, denote by $g_\mu = \bar{f}_\mu^q - p$ the map with the repelling hyperbolic fixed point $\bar{p}_\mu$. Relation (4.1) implies that for any neighborhood W of this point and some j one of the

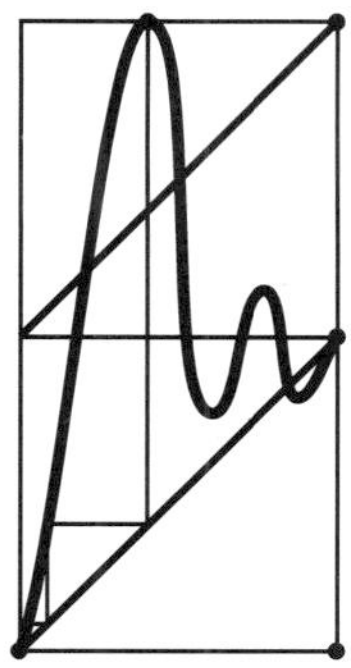

FIGURE 6.9. Finding a tangent homoclinic orbit

following inclusions hold: $g_\mu^j W \supset [\bar{p}_\mu, \bar{p}_\mu \pm 1]$. Without loss of generality we assume that the inclusion with the $+$ sign holds.

PROPOSITION 4.1. *Let $g(\mu)$ be the regular endomorphism of the circle with a repelling fixed point p_μ. Let $g(\mu)$ be continuous in μ with two first derivatives, and let p_μ be continuous in μ too. Let the property* (4.1) *for $g(\mu)$ hold: $g_\mu^j W \supset [\bar{p}_\mu, \bar{p}_\mu + 1]$. Let $\bar{g}_\mu$ and $\bar{p}_\mu$ be the lifts of g_μ and p_μ, $\bar{g}_\mu(\bar{p}_\mu) = \bar{p}_\mu$. Then in any neighborhood W of $\bar{p}_\mu$ there exists a finite union of segments I_μ continuously depending on μ in the sense of the Hausdorff distance and a positive integer j such that*

$$g_\mu^j I_\mu = \mathbb{I}_\mu = [\bar{p}_\mu, \bar{p}_\mu + 1];$$

the endpoints of I_μ are mapped to those of $\mathbb{I}_\mu$.

PROOF. The map $g(\mu)$ has a transversal homoclinic orbit of a repelling fixed point p_μ as shown in the proof of Lemma 4.1. This orbit may be lifted to a sequence $\{y_{k,\mu}\}$:

$$y_{k,\mu} = \bar{p}_\mu + 1 \text{ for } k \geq 0, \ \bar{p}_\mu < \cdots < y_{k,\mu} < y_{k+1,\mu} < \cdots < y_{0,\mu} = \bar{p}_\mu + 1, k \geq 1,$$

$$y_{k,\mu} \to \bar{p}_\mu \text{ as } k \to \ -\infty, \ g_\mu(y_{k,\mu}) = y_{k+1,\mu}.$$

To find the desired union of segments I_μ consider a sequence of sets:

$$E(0, \mu) = \mathbb{I}_\mu, \ E(k, \mu) = g_\mu^{-1}(E(k-1, \mu)) \cap [\bar{p}_\mu, y_{-k,\mu}], \ k \geq 1.$$

For $k = j$ large enough, $E(j, \mu) \subset W$. This is the desired set I_μ. Let us prove that it has the properties required in Proposition 4.1.

First, it is a finite union of segments. Indeed, the map $g(\mu)$ is regular. Hence, it has a finite number of critical points. Therefore, the inverse image of any point under $g(\mu)$ is a finite set.

Second, $g_\mu(E(k, \mu)) = E(k-1, \mu)$ because $g_\mu([\bar{p}_\mu, y_{-k,\mu}]) \supset [\bar{p}_\mu, y_{-k+1,\mu}]$.

At last, g_μ brings the endpoints of the segments constituting $E(k, \mu)$ to those of $E(k-1, \mu)$. Indeed, any point of the interval $(\bar{p}_\mu, y_{-k,\mu})$ that is mapped by g_μ to the interior point of $E(k-1, \mu)$ is interior for $E(k, \mu)$. This follows from the continuity of g_μ. On the other hand, the endpoints of the segment $[\bar{p}_\mu, y_{-k,\mu}]$ are mapped by g_μ to those of $[\bar{p}_\mu, y_{-k+1,\mu}]$. Hence, g_μ brings the endpoints of $E(k,\mu)$ to those of $E(k-1,\mu)$. By induction, g_μ^j brings the endpoints of I_μ to those of $\mathbb{I}_\mu$.

The continuous dependence of $I_\mu = E(j,\mu)$ on μ follows from the construction. This proves Proposition 4.1.

REMARK. The number of segments of the set I_μ may change when the critical value of g_μ passes through a point of the homoclinic orbit $\{y_{k,\mu}\}$. Yet the dependence of I_μ on μ keeps to be continuous. Hence, the maximum value over I_μ of any function continuous both in $x \in W$ and μ depends on μ in a continuous way.

The second step is to prove (4.2)–(4.4). This will complete the proof of Lemma 4.2. It is sufficient to find positive integers n, l such that

a) The maximum value $\sigma_\mu = \max_{\mathbb{I}_\mu} \bar{f}_\mu^n$ is taken at an interior point of $\mathbb{I}_\mu$;

b) $\sigma_0 < p_0 + l$;

c) $\sigma_\delta > p_\delta + l$.

Indeed, let I_μ and j be the set and the integer found for $g_\mu = \bar{f}_\mu^q - p$ in Proposition 4.1. Then σ_μ from a) at the same time fits (4.2) with $N = n + j$ by Proposition 4.1. Moreover, σ_μ depends continuously on μ by the remark above.

Let us find n and l for which a)–c) hold. For δ small, $|\bar{p}_\mu - \bar{p}_0| < 1/2$ provided that $\mu \in [\delta, 0]$. Let us take such a small $\delta < 0$.

For any negative δ, we have $\bar{f}_\delta^+ \geqslant \bar{f}^+$, and somewhere the inequality is strict. Indeed, the graph of $\bar{f}_\delta^+$ is that of $\bar{f}^+$ shifted to the left.

By assumption 1 of Lemma 4.1, $r(\bar{f}^+)$ is irrational. By the monotonicity lemma (Lemma 5.3.2), we have $r((\bar{f}_\delta)^+) > r(\bar{f}^+)$. Hence,

$$r_+(\bar{f}_\delta) > r_+(\bar{f}). \tag{4.5}$$

Recall that

$$p/q \in r(\bar{f}), \qquad p/q \in r(\bar{f}_\delta). \tag{4.6}$$

Therefore, $r(\bar{f}_\delta)$ contains an interval. Let us now define n and l.

First, take a large multiple n of q, $n = aq$, such that three rational numbers with the following properties exist:

$$\frac{ap+2}{n} < r_+(\bar{f}) < \frac{m}{n}, \qquad \frac{m+2}{n} \in r(\bar{f}_\delta). \tag{4.7}$$

This choice is possible because of (4.6) and (4.5). Now take $l = m + 1$.

Let us prove a). Denote $h_\mu = f_\mu^n$ for brevity. We must show that for any $\mu \in [\delta, 0]$, the maximum value σ_μ of h_μ is taken inside the segment $\mathbb{I}_\mu$, not at its ends. This follows from the left inequality in (4.7). Indeed, σ_μ is monotonically decreasing. Hence any lower estimate for σ_0 holds for any σ_μ, $\mu \in [\delta, 0]$. We have $[p/q, r_+(f)] \subset r(\bar{f})$. Hence,

$$\frac{ap+2}{n} \in r(\bar{f}).$$

Consequently, there is an $x_0 \in \mathbb{I}_0$ such that $h_0(x_0) = x_0 + ap + 2 \geqslant \bar{p}_0 + ap + 2$. On the other hand, $h_0(\bar{p}_0) = \bar{p}_0 + ap$. Hence,

$$h_0(\bar{p}_0 + 1) = h_0(\bar{p}_0) + 1 = \bar{p}_0 + ap + 1 \leq h_0(x_0) - 1 \leqslant \sigma_0 - 1.$$

This proves a) because $\sigma_\mu \geq \sigma_0 > \bar{p}_\mu + ap + 1 = h_\mu(\bar{p}_\mu + 1)$ for $\mu \in [\delta, 0]$.

Statement b) follows from the second inequality in (4.7). Namely, for all $x \in \mathbb{I}_0$ we have

$$h_0(x) < x + m \leqslant \bar{p}_0 + m + 1;$$

hence, $\sigma_0 < \bar{p}_0 + l$. This proves b).

Statement c) follows from the last inclusion in (4.7). Namely, there exists an $x \in \mathbb{I}_\delta$ such that

$$h_\delta(x) = x + m + 2 \geqslant \bar{p}_\delta + m + 2 > \bar{p}_\delta + l.$$

Hence, $\sigma_\delta > \bar{p}_\delta + l$. This proves c) and Lemma 4.2.

4.3. Generic endomorphisms of the circle.

LEMMA 4.3. *There exists an open dense set A in the space of all endomorphisms of the circle in the C^2 topology such that for any $f \in A$ there exists a point $\varkappa \in [0, 1)$ for which $f_\varkappa$ satisfies the conditions of Lemma* 4.1.

This lemma is proved below. Now we shall use it.

PROOF OF LEMMA 1.2 AND THE APPENDIX TO IT. Let A be the same as in Lemma 4.3, and $f \in A$. By Lemma 4.3, the map $f_\varkappa$ satisfies the conditions of Lemmas 4.1, 4.2 for some $\varkappa \in [0, 1]$. The conclusions of these lemmas coincide with the conclusions of Lemma 1.2 and the appendix to it. Hence, the set A from Lemma 4.3 fits Lemma 1.2 and the appendix to it as well.

It remains to prove Lemma 4.3. The proof is based on the Block–Franke theorem.

PROOF OF LEMMA 4.3. 1. *The set A is dense in the space of all regular endomorphisms of the circle.*

Indeed, suppose g is an arbitrary regular C^2 endomorphism of the circle and $\bar{g}$ is its lift. Consider the map $g_\mu = g \circ R_\mu$. The upper rotation number $r(\mu) = r_+(g_\mu)$ takes an irrational value for some μ. Indeed, $r(\mu)$ is a continuous function on μ by Lemma 3.2. On the other hand,

$$\bar{g}_{\mu+1} = \bar{g} \circ (\mathrm{id} + 1) = \bar{g} + 1, \qquad r(\mu + 1) = r(\mu) + 1.$$

Hence, $r(\mu)$ is not a rational constant, and therefore takes irrational values.

Let $r(\varkappa) \notin \mathbb{Q}$. Then, by the Block–Franke theorem, the rotation set $r(f_\varkappa)$ contains rational points. Let $\bar{f}_\varkappa$ be a lift of $f_\varkappa$, $p/q \in r(\bar{f}_\varkappa)$.

PROPOSITION 4.2. *For the above map $f_\varkappa$, any periodic orbit with rotation number p/q has nonempty intersection with the set $f_\varkappa^+ \neq f_\varkappa$.*

PROOF. In the opposite case there would exist a short periodic orbit of $f_\varkappa$ with the rotation number p/q, which is also a periodic point of $f_\varkappa^+$. But the map $f_\varkappa^+$ is monotonic; its rotation set consists of only one point. It equals p/q. Hence, by Theorem 3.1, we would have $r_+(f_\varkappa) = p/q$. This contradicts the choice of $\varkappa$.

Now take all the periodic points of f with rotation number p/q. The map f may be slightly perturbed near the points of these orbits in the set $f_\varkappa^+ \neq f_\varkappa$. This will not change f^+, hence the upper rotation number will be preserved. The perturbation may be chosen to make all the p/q-periodic orbits of the perturbed map hyperbolic. This proves the density of A.

2. *The set A is open.* Let $g \in A$, and let $g_\varkappa$ satisfy assumptions 1 and 2 of Lemma 4.1. Then for all the maps h close to g and for all μ close to $\varkappa$, the map h_μ has periodic orbits with rotation number p/q, and all of them are hyperbolic. On the other hand, by Lemma 5.3.2, $r_+(g_\mu)$ strictly increases as μ passes through the value $\varkappa$. Hence, $r_+(h_\mu)$ varies when μ ranges near $\varkappa$. Therefore, $r_+(h_\mu)$ takes irrational values. Consequently, $h \in A$.

This completes the proof of Lemma 4.3, hence of Lemma 1.2, and hence of Theorem 1.1.

CHAPTER 7

Bifurcations of Homoclinic Trajectories of Hyperbolic Saddles

In this chapter we consider bifurcations of homoclinic loops of hyperbolic saddles in typical one-parameter families of vector fields. The simplest case, namely, the saddle in $\mathbb{R}^3$ with real eigenvalues, is studied in §1. Section 2 is devoted to the description of vector fields in $\mathbb{R}^3$ with a hyperbolic saddle having two complex conjugate eigenvalues in the left half-plane. The last condition may be satisfied by time reversal. When the saddle value is negative, i.e., stability wins, the homoclinic loop of the saddle generates a stable limit cycle. In the opposite case, the unperturbed vector field has a complicated invariant set, a countable number of Smale horseshoes. It cannot occur on the boundary of the Morse–Smale set, and formally is beyond the scope of this book. Yet we include the subject for completeness. The third section describes the bifurcation of a homoclinic loop of the saddle in the phase space of any dimension, provided that the unperturbed vector field belongs to the boundary of the Morse–Smale set.

§1. The homoclinic trajectory of a hyperbolic saddle with three real eigenvalues in $\mathbb{R}^3$

1.1. Generation of a periodic orbit. Consider a vector field with homoclinic trajectory of a hyperbolic saddle in the space $\mathbb{R}^3$. The singular point is in general position. The degeneration is the existence of the nontransversal intersection of the stable and unstable manifolds of the saddle. In other words, the degeneration is the existence of a homoclinic trajectory. The vector fields with such a homoclinic orbit occur in an unavoidable way in typical one-parameter families of vector fields. In this section we suppose that all eigenvalues of the saddle are real. The case of complex eigenvalues will be considered in the next section. Without loss of generality we can assume that two eigenvalues are negative and one is positive (otherwise we reverse time). Such a saddle has a two-dimensional stable manifold and a one-dimensional unstable manifold (see Figure 7.1). We say that the sum of the maximum negative eigenvalue and the positive one is the *saddle value.*

THEOREM 1.1. *Suppose that in a typical one-parameter family of smooth vector fields in $\mathbb{R}^3$ the zero value of the parameter corresponds to a "critical" vector field with homoclinic orbit γ of a saddle having two negative eigenvalues and a positive one. Then the vector fields corresponding to all sufficiently small values of the parameter on one side of zero have a hyperbolic periodic orbit which tends to the homoclinic orbit γ of the critical vector field as the parameter tends to zero. This periodic orbit is stable if the saddle value is negative, and has two-dimensional*

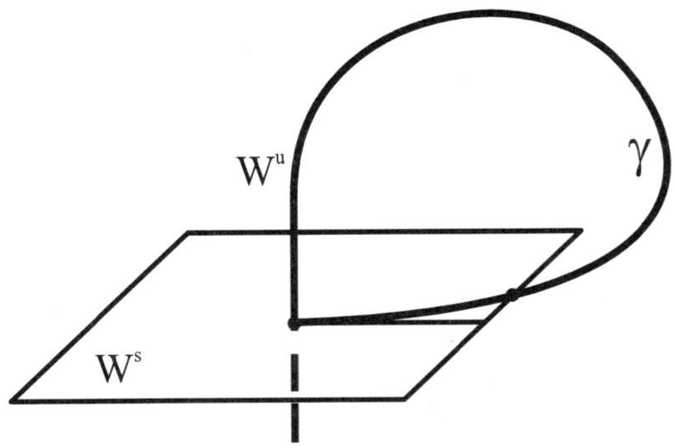

FIGURE 7.1. A hyperbolic saddle in the space, with real eigenvalues and a homoclinic loop.

stable and unstable manifolds if the saddle value is positive. The vector fields corresponding to all sufficiently small values of the parameter on the other side of zero have no periodic orbits in some neighborhood of γ.

1.2. Genericity assumptions. Before proving the theorem above, let us specify the genericity assumptions that the family of vector fields must satisfy.

1. *The eigenvalues of the hyperbolic saddle corresponding to the critical value of the parameter are nonresonant and pairwise disjoint.*

The above assumption implies that in some neighborhood of the saddle the vector field is smoothly equivalent to its linear part. Therefore, in the normalizing chart, its stable manifold is a plane and the unstable one is a line. On the stable plane, all phase curves except the singular point and two others tend to a singular point along the *leading stable direction* corresponding to the maximum negative eigenvalue as $t \longrightarrow +\infty$.

2. *The homoclinic trajectory tends to the saddle along the leading stable direction as* $t \longrightarrow +\infty$.

The next assumption makes use of the following remark. A saddle satisfying assumption 1 has an invariant plane W spanned by the eigenvectors associated to the two maximum eigenvalues. This invariant surface W may be prolonged along the homoclinic orbit. Now we can state the third assumption.

3. *The manifold* W *and the stable manifold of the saddle intersect transversally along the homoclinic orbit* (*see Figure* 7.2).

The next assumption is a property of the family itself.

4. *As the parameter passes through zero, the homoclinic orbit occurs and breaks in a transversal manner* (*a precise definition will be given below*).

In the proof of Theorem 1.1, only the assumptions 1–4 will be used.

1.3. Outline of the proof. In the same way as in the two-dimensional case in Chapter 3, we study the Poincaré map Δ of the homoclinic orbit by splitting it into two factors, the singular and the regular one: $\Delta = \Delta_\varepsilon^{\text{reg}} \circ \Delta_\varepsilon^{\text{sing}}$ (see Figure 7.3). To do this, in some neighborhood of the saddle, we will choose cross-sections Γ^+ and Γ^- close to the singular point and transversal to the homoclinic orbit. The

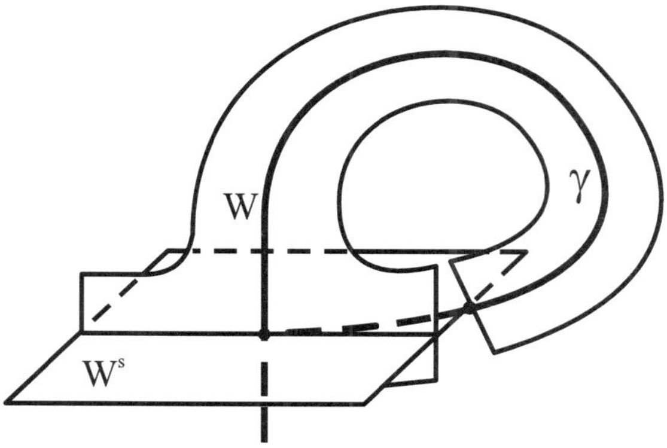

FIGURE 7.2. Transversality assumption for the homoclinic orbit of the saddle.

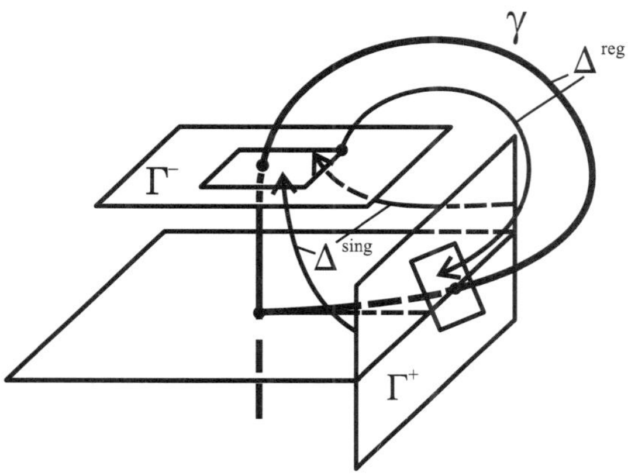

FIGURE 7.3. Singular and regular maps.

orbits of the vector field with initial points on Γ^+ enter the neighborhood and the orbits beginning on Γ^- leave it, which gives rise to the $+$ and $-$ in the notation.

The map $\Delta_\varepsilon^{\rm sing}$ is the map $\Gamma^+ \to \Gamma^-$ along the orbits passing near the saddle; it is defined on some subset of Γ^+ (see Figure 7.3). Its study constituted the principal technical difficulty in previous investigations. Now the theory of normal forms for local families gives explicit expression for $\Delta_\varepsilon^{\rm sing}$ in elementary functions. This makes the proofs much easier and the geometrical effects become almost transparent.

The map $\Delta_\varepsilon^{\rm reg}$ is smooth with respect to the phase variables, which are coordinates on Γ^-, and the parameters. In the heuristic proof it will be convenient to consider $\Delta_\varepsilon^{\rm reg}$ as simple as possible in order to describe the geometric effects. The hyperbolic fixed point theorem provides regular proofs of the effects discovered in this way.

1.4. The singular correspondence map. By the genericity assumption 1 and the theory of finitely smooth normal forms for local families, we can find a

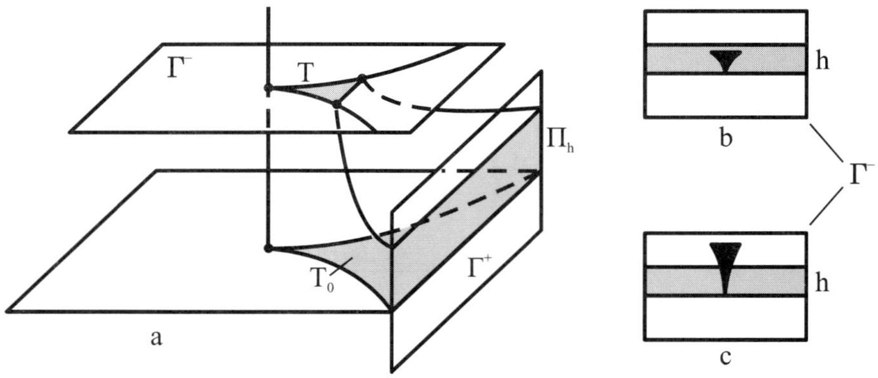

FIGURE 7.4. (a) Singular correspondence map for a saddle with real eigenvalues in the space. Images of Π_h in the cases (b) $\sigma < 0$ and (c) $\sigma < 0$.

chart (x, y, z) depending on the parameter near the hyperbolic saddle such that this family has the form

$$\begin{cases} \dot{x} = \lambda_1(\varepsilon)x, \\ \dot{y} = \lambda_2(\varepsilon)y, \\ \dot{z} = \mu(\varepsilon)z, \end{cases} \qquad \lambda_2(0) < \lambda_1(0) < 0 < \mu(0). \tag{1.1}$$

Let

$$\Gamma^+ = \{x = 1,\ |y| \leqslant 1,\ |z| \leqslant 1\}, \qquad \Gamma^- = \{z = 1,\ |x| \leqslant 1,\ |y| \leqslant 1\}.$$

The calculation of the singular correspondence map is elementary. Nevertheless, we give the geometric description of the result before the calculation.

Note that in system (1.1) the variables are separated, and (1.1) can be regarded as the product of two systems: one on the stable and the other on the unstable manifold. Hence the projection of any orbit along z-axis to the (x, y)-plane is the orbit of the system

$$\begin{cases} \dot{x} = \lambda_1(\varepsilon)x, \\ \dot{y} = \lambda_2(\varepsilon)y. \end{cases} \tag{1.2}$$

Thus any orbit starting from Γ^+ lies in a cylinder parallel to the z-axis. The base of this cylinder is the phase curve of system (1.2). The projections of all the phase curves starting at Γ^+ fill the curvilinear triangle T_0 formed by the orbits of the node (1.2) saturating the base of Γ^+ on $\{z = 0\}$ (see Figure 7.4).

Let Π_h denote the rectangle in Γ^+ with lower base on $\{z = 0\}$ and higher base of height h. In order to draw the image $T = \Delta^{\text{sing}}(\Pi_h)$, one should lift T_0 along z-axis to the plane Γ^-, and then cut the part of the lifted triangle by the image of the upper base of Π_h. The time in which the orbits starting at the upper base of Π_h run to the plane $\{z = 1\}$ is $t = -(\ln h)/\mu(\varepsilon)$. Thus the x coordinate of all the points on the base of T is

$$x = \exp \lambda_1(\varepsilon)t = h^{\beta(\varepsilon)}, \tag{1.3}$$

where $\beta(\varepsilon) = -\lambda_1(\varepsilon)/\mu(\varepsilon)$.

REMARK. The map $\Delta_\varepsilon^{\rm sing}$ is a strong contraction in the y-direction. The formula (1.3) shows the crucial difference between the two cases $\sigma > 0$ and $\sigma < 0$, where $\sigma = \lambda_1(0) + \mu(0)$ is the saddle value. In the case $\sigma > 0$, we have $\mu > |\lambda_1|$. "Instability wins":

$$0 < \beta(\varepsilon) < 1, \qquad h^{\beta(\varepsilon)} \gg h,$$

for h small. In the opposite case $\sigma < 0$, "stability wins":

$$\beta(\varepsilon) > 1, \qquad h^{\beta(\varepsilon)} \ll h,$$

for h small. In the first case the map $\Delta_\varepsilon^{\rm sing}$ is hyperbolic: strong contraction in the "horizontal" y-direction and strong expansion in the orthogonal direction, and the coefficient of contraction and expansion tends to 0 and ∞ respectively as $h \to 0$. In the second case the map is a strong contraction with coefficient tending to 0 as $h \to 0$.

The same calculation that led to (1.3) gives the formula for the singular correspondence map:

$$\Delta_\varepsilon^{\rm sing}(y, z) = (yz^\alpha, z^\beta), \qquad \alpha = \alpha(\varepsilon) = -\lambda_2(\varepsilon)/\mu(\varepsilon) > \beta(\varepsilon). \tag{1.4}$$

In the case $\sigma < 0$, we have $\beta > 1$, hence the map (1.4) is strongly contracting for small z.

1.5. Generation of the stable cycle. Below we prove Theorem 1.1 in the case $\sigma < 0$. Let the normalizing chart (x, y, z) and cross-sections Γ^+, Γ^- be the same as in 1.4. Let (y, z) and (y', x) be the coordinates on the cross-sections Γ^+ and Γ^- respectively defined as the restrictions of the normalizing coordinates to these cross-sections. Let γ be the homoclinic orbit of the saddle. The intersection point $\Gamma^- \cap \gamma$ is $O = (x = y = 0,\ z = 1)$. Denote by p the point $\Gamma^+ \cap \gamma$. The stable manifold intersects Γ^+ in a segment $H = \{z = 0,\ x = 1\}$. Hence, $z(p) = 0$.

By assumption 4, the image $O(\varepsilon) = \Delta_\varepsilon^{\rm reg}(O)$ transversally intersects the line $z = 0$ as ε passes through zero (this is the explicit statement of the assumption). Changing the parameter, we can suppose that

$$z(O(\varepsilon)) = \varepsilon. \tag{1.5}$$

In the definition of Π_h let us take $h = 2|\varepsilon|$. Below we will prove that for positive ε small enough the map $\Delta_\varepsilon = \Delta_\varepsilon^{\rm reg} \circ \Delta_\varepsilon^{\rm sing}$ sends Π_h inside itself, and is contracting (see Figure 7.5). If $\Delta_0(\Pi_h)$ belongs to the half-plane $z \leqslant 0$, we have an *orientable case*. If $z|\Delta_0(\Pi_h) < 0$, we have a *nonorientable case*.

After that we will prove that for small negative ε and some h not depending on ε, the map Δ_ε has no fixed points in Π_h. For such ε there are no periodic orbits in the neighborhood of the homoclinic orbit γ.

By (1.4), $\operatorname{diam} \Delta_\varepsilon^{\rm sing}\Pi_h \leqslant C\varepsilon^\beta$ for $h = 2\varepsilon$ and some positive C. On the other hand, the distance from $\Delta_\varepsilon(O)$ to the boundary of Π_h is equal to ε. Hence, for small $\varepsilon > 0$, we have $\Delta_\varepsilon\Pi_h \subset \Pi_h$. By (1.4),

$$D\Delta_\varepsilon^{\rm sing} = \begin{pmatrix} z^\alpha & \alpha y z^{\alpha-1} \\ 0 & \beta z^{\beta-1} \end{pmatrix} \tag{1.6}$$

for $\beta > 1$, $\|D\Delta_\varepsilon^{\rm sing}\| \to 0$ as $\varepsilon \to 0$. Hence, the map Δ_ε is contracting in Π_h, therefore it has a stable fixed point. This proves the existence of a stable periodic orbit for $\varepsilon > 0$.

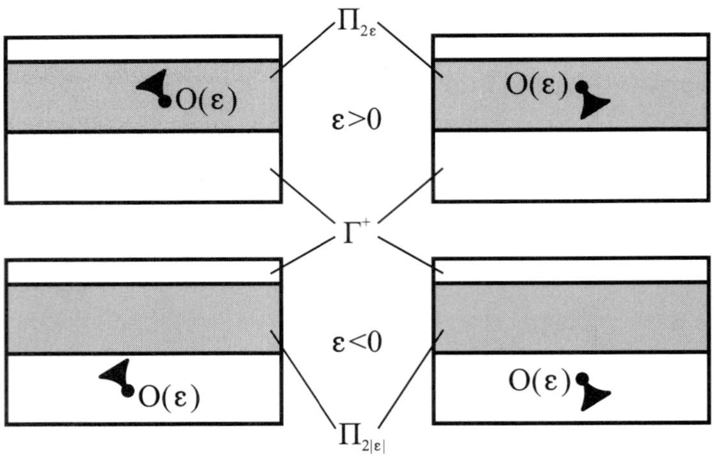

FIGURE 7.5. The Poincaré map of the homoclinic loop of a saddle with negative saddle value in the orientable (left) and nonorientable (right) cases.

In order to prove the absence of fixed points for $\varepsilon < 0$, we shall check that the map Δ_ε decreases the z-coordinate in Π_h for small negative ε and small h.

Indeed, let the Lipschitz constant of Δ_ε in Π_h be smaller than $1/2$. Take any $a = (y, z) \in \Pi_h$ and let $b = (y, 0)$. Note that $\Delta_\varepsilon(b) = \Delta_\varepsilon^{\text{reg}}(O) = (y(\varepsilon), \varepsilon)$. Then

$$z(\Delta_\varepsilon(a)) \leqslant z(\Delta_\varepsilon(b)) + z/2 = \varepsilon + z/2 < z,$$

because $\varepsilon < 0$. This proves Theorem 1.1 for $\sigma < 0$.

Now we pass to the case $\sigma > 0$.

1.6. Heuristic description. In order to describe the geometric effects, we will suppose that the map $\Delta_\varepsilon^{\text{reg}}$ is as simple as possible. The constructions used in the simplified case yield the so-called "principal family" which is in a sense "a topological normal form" for the bifurcation discussed. Suppose that $\Delta_\varepsilon^{\text{reg}}$ has one of the following forms:

$$\Delta_\varepsilon^{\text{reg}} = g_\varepsilon^{\pm}, \quad g_\varepsilon^{+} : (x, y, 1) \mapsto (1, y, x + \varepsilon), \quad g_\varepsilon^{-} : (x, y, 1) \mapsto (1, -y, -x + \varepsilon).$$

The first case is orientable, and the second case is nonorientable. The return map

$$\Delta_\varepsilon = g_\varepsilon^{\pm} \circ \Delta_\varepsilon^{\text{sing}}$$

is shown in Figure 7.6 for the case $\sigma > 0$.

In order to prove Theorem 1.1, we must study the fixed points of the map Δ_ε. Let us begin with the study of the model described above for the case $\sigma > 0$.

In the nonorientable case, the cusp point of the image of $\Delta_\varepsilon^{\text{sing}}$ should be located outside the preimage in order to make the hyperbolic fixed point principle applicable (see Figure 7.6). For this reason we take the domain of $\Delta_\varepsilon^{\text{sing}}$ to be Π_h with $h = |\varepsilon|/2$. Let V_ε be the segment $\Pi_h \cap \{y = 0\}$. Then, in the orientable case we have $\Delta_\varepsilon(V_\varepsilon) = [\varepsilon, \varepsilon + h^\beta]$. In the nonorientable case, $\Delta_\varepsilon(V_\varepsilon) = [\varepsilon, \varepsilon - h^\beta]$. In the first case, $\Delta_\varepsilon(V_\varepsilon) \supset V_\varepsilon$ for any $\varepsilon < 0$. In the second case, the same inclusion

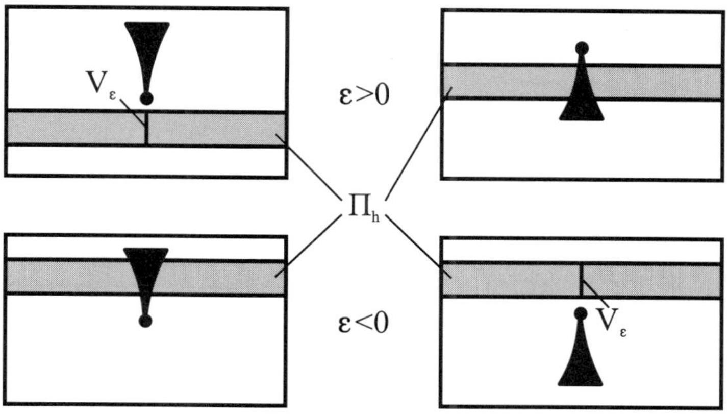

FIGURE 7.6. The Poincaré map of the homoclinic loop of a saddle with the positive saddle value in the orientable (left) and nonorientable (right) cases; heuristic description.

holds for $\varepsilon > 0$. In both cases, the map Δ_ε is strictly expanding on V_ε for small ε. Consequently, it has an unstable fixed point in V. This is the saddle fixed point for the map Δ_ε on Π_h.

This ends the heuristic investigation. We turn now to the rigorous proof of Theorem 1.1 for $\sigma > 0$.

1.7. Generation of a hyperbolic cycle: construction of domains and maps. The proof of Theorem 1.1 will be parallel to that of Theorem 4.3.1 on the generation of a hyperbolic cycle via the bifurcation of the homoclinic orbit of the saddlenode. First recall

DEFINITION 2.4.4. A (μ_h, μ_v)-*rectangle* D in $\mathbb{R}^n \oplus \mathbb{R}^m$ is a domain of the form $D = F(B^n \times B^m)$, where B^n and B^m are unit balls in $\mathbb{R}^n$ and $\mathbb{R}^m$ respectively, F is a homeomorphism $B^n \times B^m \to \mathbb{R}^n \oplus \mathbb{R}^m$ and the surfaces $F(B^n \times \{y\})$ and $F(\{x\} \times B^m)$ are μ_h-horizontal and μ_v-vertical respectively for any $y \in B^m$, $x \in B^n$. These surfaces are called *horizontal* and *vertical fibers* of D. The horizontal and vertical parts of the boundary of D are defined as the images under F of the horizontal and vertical parts of the boundary of $B^n \times B^m$ respectively:

$$\partial_h D = F(B^n \times \partial B^m), \qquad \partial_v D = F(\partial B^n \times B^m).$$

DEFINITION 2.4.5. A (μ_h, μ_v)-rectangle is called *vertically cylindric* (v-*cylindric*) in a standard rectangle B if $D \subset B$, $\partial_v D \in \partial_v B$. It is called *horizontally cylindric* (h-*cylindric*) if the same holds for h and v exchanged, i.e., $D \subset B$, $\partial_h D \in \partial_h B$.

HYPERBOLIC FIXED POINT THEOREM. *Let $B = D_h \times D_v$ be a standard rectangle. Let $D \subset B$ and $D' \subset B$ be v-cylindric and h-cylindric (μ_h, μ_v)-rectangles in B. Let $f\colon D \to D'$ be a map satisfying the (μ_h, μ_v) cone condition. Then f has a unique fixed point $\mathbf{O}$ in D*:

$$\mathbf{O} = \bigcap_{i=-\infty}^{\infty} f^i D.$$

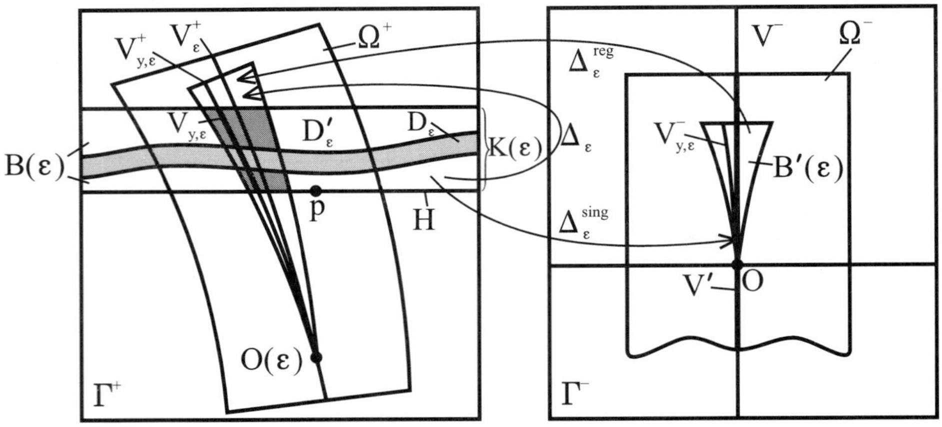

FIGURE 7.7. Domains and maps in the study of the homoclinic orbit of a hyperbolic saddle for the orientable case and $\varepsilon < 0$ in the space.

We shall construct the domains B, D, D' as in 4.3.6 except that the domain B will depend on ε. The necessity of this dependence was explained above in 1.6.

Now we pass to the explicit construction. Let the normalizing charts, the cross-sections, and the points O and p be the same as in 1.5. The stable manifold intersects Γ^+ in the segment $H = \{x = 1,\ z = 0\}$. Hence, $z(p) = 0$. The manifold W defined in 1.2 intersects Γ^- in a countable number of curves. One of them is $V^- = \{y' = 0\}$ (see Figure 7.7).

Suppose that

$$\Delta_\varepsilon^{\mathrm{reg}}(O) = (y(\varepsilon), \varepsilon) \tag{1.7}$$

which can be achieved by the reparametrization (1.5). The map $\Delta_\varepsilon^{\mathrm{reg}}$ sends some neighborhood $\Omega^- \subset \Gamma^-$ of O to some neighborhood $\Omega_\varepsilon^+ \subset \Gamma^+$ of p. By the transversality assumption,

$$\Delta_\varepsilon^{\mathrm{reg}} V^- \text{ is transversal to } H \text{ at } O \text{ for } \varepsilon = 0. \tag{1.8}$$

Choose the domain of $\Delta_\varepsilon^{\mathrm{reg}}$ in the form $\Omega^- = B_\rho^s \times B_\rho^u$, where

$$B_\rho^s = \{y \in \mathbb{R}^1,\ |y| \leqslant \rho\}, \qquad B_\rho^u = \{x \in \mathbb{R}^1,\ |x| \leqslant \rho\}.$$

Let $V' = \Omega^- \cap V^-$. Denote $V_\varepsilon^+ = \Delta_\varepsilon^{\mathrm{reg}} V'$. Take

$$B(\varepsilon) = B_y \times K(\varepsilon) \subset \Gamma^+, \quad B_y = \{y \in \mathbb{R}^1,\ |y| \leqslant 1\}, \quad K(\varepsilon) = \{0 \leqslant z \leqslant |\varepsilon|/2\}.$$

By the transversality assumption (1.8), V_ε^+ is the graph of a smooth map

$$v_\varepsilon\colon B_r^z = \{z \in \mathbb{R}^1,\ |z| \leqslant r\} \to B_y$$

for sufficiently small r and ε. Denote by π_z the projection $(y, z) \mapsto z$. For small ε, the curve

$$V_{y,\varepsilon}^- = \Delta_\varepsilon^{\mathrm{sing}}(\{y\} \times K(\varepsilon)), \qquad y \in B_y,$$

is close to $\Delta_\varepsilon^{\text{sing}}(\{0\} \times K(\varepsilon))$. Hence, $V_{y,\varepsilon}^+ = \Delta_\varepsilon^{\text{reg}} V_{y,\varepsilon}^-$ is a graph of a smooth map

$$v_{y,\varepsilon}^+ : \pi_z V_{y,\varepsilon}^+ \to B_y.$$

Let $\mathbf{L}$ be a common Lipschitz constant for all the maps $v_{y,\varepsilon}^+$.

The lower end of the arc $V_{y,\varepsilon}^+$ is the point $O(\varepsilon)$, with z-coordinate equal to ε (see (1.5) and Figure 7.7 for the orientable case). The length of the curve $V_{y,\varepsilon}^-$ for $y = 0$ is $(|\varepsilon|/2)^\beta$. Hence, there exists a constant c depending on $\mathbf{L}$ such that in the orientable case we have

$$[\varepsilon, \varepsilon + c|\varepsilon|^\beta] \subset \pi_z V_{y,\varepsilon}^+$$

for $\varepsilon < 0$ (see Figure 7.7). For the same c, in the nonorientable case we have $[\varepsilon, \varepsilon - c|\varepsilon|^\beta] \subset \pi_z V_{y,\varepsilon}^+$ for $\varepsilon > 0$. Consequently, $K(\varepsilon) \subset \pi_z V_{y,\varepsilon}^+$ for $\varepsilon < 0$ in the orientable case and $\varepsilon > 0$ in the nonorientable case. The intersections $V_{y,\varepsilon} = V_{y,\varepsilon}^+ \cap B(\varepsilon)$ are the graphs of smooth maps $v_{y,\varepsilon} : K(\varepsilon) \to B_y$. The boundary of the curve $V_{y,\varepsilon}$ belongs to the horizontal boundary of $B(\varepsilon)$:

$$\partial V_{y,\varepsilon} \subset \partial_h B(\varepsilon). \tag{1.9}$$

Let ε be so small that $B'(\varepsilon) = \Delta_\varepsilon^{\text{sing}} B(\varepsilon) \subset \Omega^-$. Then the map

$$\Delta_\varepsilon = \Delta_\varepsilon^{\text{reg}} \circ \Delta_\varepsilon^{\text{sing}} : B(\varepsilon) \to \Delta_\varepsilon B(\varepsilon)$$

is well defined. Consider the domains

$$D'_\varepsilon = B(\varepsilon) \cap \Delta_\varepsilon B(\varepsilon), \qquad D_\varepsilon = \Delta_\varepsilon^{-1} D'_\varepsilon.$$

These domains together with $B(\varepsilon)$ and the map Δ_ε will be the subject of the application of the hyperbolic fixed point theorem. Let us check that the assumptions of this theorem hold for $D_\varepsilon = D$, $D'_\varepsilon = D'$, $B(\varepsilon) = B$, and $\Delta_\varepsilon = f$ provided that $\varepsilon < 0$ in the orientable case and $\varepsilon > 0$ in the nonorientable case is sufficiently small.

1.8. Cone condition for the Poincaré map.

LEMMA 1.1. *The Poincaré map Δ_ε in D_ε satisfies the (μ_h, μ_v) cone condition for appropriate constants μ_h, μ_v, $\mu_h \mu_v < 1$ for $\varepsilon < 0$ in the orientable case and $\varepsilon > 0$ in the nonorientable case.*

PROOF. The cone condition of the lemma follows from Lemma 4.2.2 and transversality assumption (1.8). The arguments below are similar to those in 4.3.8. In more detail, let the Jacobian matrix of $\Delta_\varepsilon^{\text{reg}}$ in the chart x, y on Γ^- and y, z on Γ^+ be

$$H = \begin{pmatrix} a & b \\ c & d \end{pmatrix}.$$

The Jacobian matrix Σ of $\Delta_\varepsilon^{\text{sing}}$ is given by (1.6). The chain rule is applicable to Δ_ε in D_ε under the assumptions of Lemma 1.1. It is sufficient to check that for any $\mathbf{a} \in \Omega_\rho^-$ and any $\mathbf{b} \in B(\varepsilon)$ the matrices $H = D\Delta_\varepsilon^{\text{reg}}(\mathbf{a})$ and $\Sigma = D\Delta_\varepsilon^{\text{sing}}(\mathbf{b})$ satisfy the assumptions of Lemma 4.2.2. Recall this lemma for the planar case.

LEMMA 4.2.2. *Consider two invertible linear operators from $\mathbb{R}^2$ to itself:*

$$H = \begin{pmatrix} a & b \\ c & d \end{pmatrix}, \qquad \Sigma = \begin{pmatrix} 1 & B \\ 0 & 1 \end{pmatrix} \begin{pmatrix} \Lambda & 0 \\ 0 & M \end{pmatrix} \begin{pmatrix} 1 & 0 \\ D & 1 \end{pmatrix}. \tag{1.10}$$

Let

$$\|H\| \leqslant L, \quad \|H^{-1}\| \leqslant L, \quad |d^{-1}| \leqslant L. \tag{1.11}$$

Then for any μ_v greater than some constant depending on L only and for any μ_h, $0 < \mu_h < \mu_v^{-1}$, there exists a positive δ such that under the assumptions

$$|\Lambda| < \delta, \quad |M^{-1}| < \delta, \quad |B| < L, \quad |D| < \delta, \quad |cB| < \delta \tag{1.12}$$

the linear map $A = H\Sigma$ satisfies the (μ_h, μ_v) cone condition with $\lambda = 2$.

Let us now check conditions (1.11), (1.12) for the matrices $H = D\Delta_\varepsilon^{\text{reg}}(\mathbf{a})$ and $\Sigma = D\Delta_\varepsilon^{\text{sing}}(\mathbf{b})$.

Inequalities (1.11) follow from the regularity of $\Delta_\varepsilon^{\text{reg}}$ and the transversality assumption (1.8) for small ρ. Note that the matrix Σ given by (1.6) has the form (1.10) with

$$D = 0, \quad \Lambda = z^\alpha, \quad M = \beta z^{\beta-1}, \quad B = \frac{\alpha}{\beta} y z^{\alpha-\beta}.$$

Recall that $\beta < 1$, $\beta < \alpha$, $z = O(\varepsilon)$ in $B(\varepsilon)$. Hence, the inequalities (1.12) may be obtained for arbitrary δ by choosing ε sufficiently small. This proves Lemma 1.1.

1.9. Rectangular structure in the source and target of the Poincaré map. We shall define the rectangular structure on D_ε, D'_ε in the same way as in 4.3.7. The same notation as there will be used, but for slightly different geometrical objects.

LEMMA 1.2. *There exist positive constants μ_h, μ_v, $\mu_h\mu_v < 1$ such that $D_\varepsilon, D'_\varepsilon$ are (μ_h, μ_v)-rectangles horizontally and vertically cylindric in $B(\varepsilon)$ respectively.*

PROOF. Let us first investigate the domain D'_ε. Its horizontal fibers are parts of horizontal segments parallel to the y-axis. The vertical fibers are the curves $V_{y,\varepsilon}$, i.e., the graphs of the maps $v_{y,\varepsilon}$ (see Figure 7.7). As before, let $\mathbf{L}$ be a Lipschitz constant for all the maps $v_{y,\varepsilon}$. Take any $\mu_v > \mathbf{L}$ that fits the constructions of Lemma 1.1 (where μ_v had to be larger than some constant). Take any $\mu_h < \mu_v^{-1}$. We shall prove that D'_ε is a (μ_h, μ_v)-rectangle with these constants. Specify the map from Definition 2.4.4 as

$$F'_\varepsilon \colon B(\varepsilon) \to D'_\varepsilon, \qquad (y, z) \mapsto (v_{y,\varepsilon}(z), z).$$

The vertical and horizontal fibers of this chart will be μ_v-vertical and μ_h-horizontal surfaces respectively. Indeed, the vertical fibers are the graphs of the maps $v_{y,\varepsilon}$, and the horizontal ones belong to horizontal lines. The previous statement now follows from the choice of μ_v. The inclusion $\partial_h D'_\varepsilon \in \partial_h B(\varepsilon)$ follows from (1.9).

Now consider D_ε. The vertical fibers of the rectangular structure of D_ε will be segments of vertical lines. Namely, let

$$F_\varepsilon = \Delta_\varepsilon^{-1} \circ F'_\varepsilon \colon B(\varepsilon) \to D_\varepsilon.$$

Then, by the definitions of F'_ε and Δ_ε, we have $y \circ F_\varepsilon(y, z) = y$. Denote by $H'_{z,\varepsilon}$ the horizontal fiber of the chart F'_ε on D'_ε. It belongs to a horizontal line, hence it is μ_h-horizontal for any $\mu_h > 0$. By Lemma 1.1 the curve $H_{z,\varepsilon} = \Delta_\varepsilon^{-1} H'_{z,\varepsilon}$ is μ_h-horizontal. By definition of D'_ε, we have $\partial H_{z,\varepsilon} \in \partial_v B(\varepsilon)$. Hence, $\partial_v D_\varepsilon \in \partial_v B(\varepsilon)$. This completes the proof of Lemma 1.2 and, together with it, that of Theorem 1.1.

§2. The homoclinic trajectory of a hyperbolic saddle with two complex eigenvalues in $\mathbb{R}^3$

In this section we will consider the bifurcation of a three-dimensional vector field which has a homoclinic orbit of a saddle with two complex eigenvalues and a real one. We assume that the two complex eigenvalues have negative real part and the real one is positive. The other case may be obtained from this by time reversal. The sum of the real part of the complex eigenvalues and the real one is the *saddle value.* The bifurcation of the homoclinic trajectory for the cases of negative and positive saddle values produces quite different geometric effects. In the case of a negative saddle value, the effect is the same as in the real case. A vector field with positive saddle value has a "horseshoe" containing an infinite number of periodic trajectories in any neighborhood of the homoclinic orbit, which implies that it cannot belong to the boundary of the set of Morse–Smale systems.

2.1. Bifurcation of a homoclinic trajectory of a saddle with negative saddle value. The first part of the section is devoted to the following

THEOREM 2.1. *Suppose that in a typical one-parameter family of smooth vector fields in $\mathbb{R}^3$, the zero value of the parameter corresponds to a vector field with a homoclinic orbit γ of a saddle having a pair of complex conjugate eigenvalues with negative real part and a positive real eigenvalue. If the saddle value is negative, then the vector fields corresponding to all sufficiently small values of the parameter on one side of zero have a stable periodic orbit tending to the homoclinic orbit γ as the parameter tends to zero. The vector fields corresponding to all sufficiently small values of the parameter on the other side of zero have no periodic orbits in some neighborhood of γ.*

The genericity assumptions in the theorem are the following:

1. *The hyperbolic saddle corresponding to the critical value zero of the parameter is nonresonant.*

2. *As the parameter passes through zero, the homoclinic orbit occurs and breaks in a transversal manner.*

2.2. Heuristic proof. Before giving a formal proof of Theorem 2.1 in the next subsection, let us first describe the geometric effect. Hence we will suppose that the Poincaré map is as simple as possible. We assume that the vector field in the neighborhood of the singular point is a linear one:

$$\begin{cases} \dot{w} = (\lambda + i\omega)w, \\ \dot{z} = \mu z, \end{cases} \tag{2.1}$$

where $i = \sqrt{-1}$, $w = x + iy = r\exp(i\theta)$, $z \in \mathbb{R}$, $\lambda < 0 < \mu$, $\omega \neq 0$. Let

$$\Gamma^+ = \{r = 1\}, \qquad \Gamma^- = \{z = 1\}.$$

Without loss of generality, we assume that

$$\gamma \cap \Gamma^+ = p: \{(r,\theta,z) = (1,0,0)\}, \qquad \gamma \cap \Gamma^- = O: \{(w,z) = (0,1)\}.$$

Consider the rectangle $\Pi_h = \{(\theta, z) \in \Gamma^+ \mid 0 \leqslant z \leqslant h, |\theta| \leqslant 1\}$. Note that the variables in (2.1) are separated; in other words, (2.1) is the product of two systems: one on the stable and the other on the unstable manifold. Hence the projection of any orbit along z-axis to the (x, y) plane is the orbit of the system $\dot{w} = (\lambda + i\omega)w$.

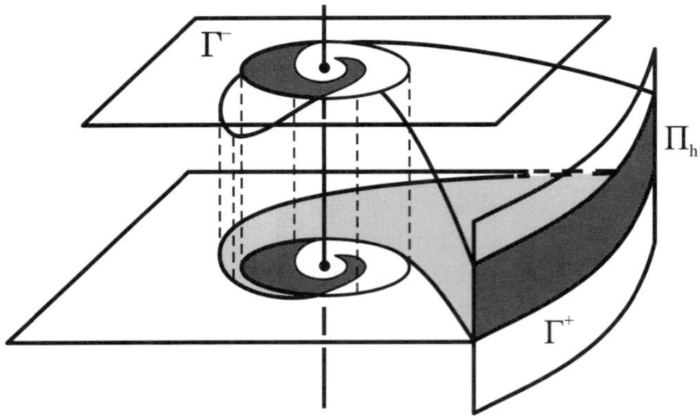

FIGURE 7.8. The singular correspondence map of a saddle with a focus on the stable manifold.

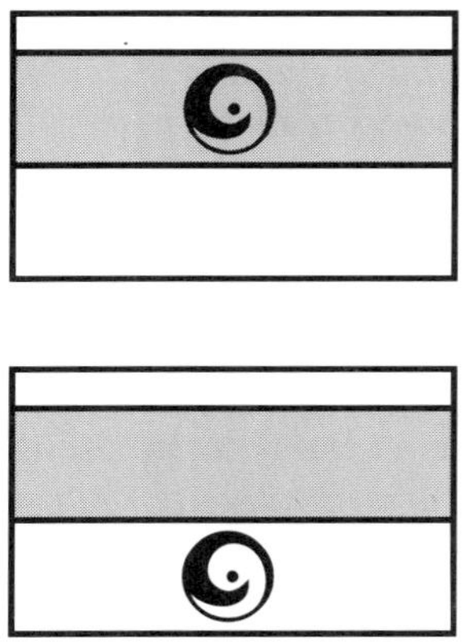

FIGURE 7.9. The Poincaré map of the homoclinic orbit of a saddle with a focus on the stable manifold and negative saddle value.

Thus all the orbits starting on Π_h lie in the cylinder parallel to the z-axis. The base of this cylinder is the "thick spiral" T_0 formed by the phase curves of the focus saturating the base of Π_h on $\{z = 0\}$. The map Δ^{sing} is the map from Π_h to Γ^- along the orbits of (2.1). In order to draw the image $T = \Delta^{\text{sing}}(\Pi_h)$, one must lift T_0 along the z-axis to the plane Γ^- and then cut the part of the lifted thick spiral by the image of the upper base of Π_h (see Figure 7.8).

The time in which the orbits starting at the points of the high base of Π_h run to the plane $\{z = 1\}$ is $t = -(\ln h)/\mu$. Thus, the image of the upper base of Π_h is an arc of the circle $|w| = H$, $H = h^\alpha$, $\alpha = -\lambda/\mu$. There are two cases:

1) $\sigma = \lambda + \mu < 0$, the assumption of Theorem 2.1. Then $\alpha > 1$, therefore $H = h^\alpha \ll h$.
2) $\sigma > 0$, the assumption of Theorem 2.2 below. Then $\alpha < 1$, therefore $H = h^\alpha \gg h$.

In this subsection we consider the case $\sigma < 0$. For simplicity, we assume that $\Delta_\varepsilon^{\text{reg}}$, which is defined as a map from a neighborhood of O on Γ^- to Γ^+ along the orbits, is a translation:

$$(\theta, z) = \Delta_\varepsilon^{\text{reg}}(x, y) = (x, y + \varepsilon).$$

Let $O(\varepsilon) = \Delta_\varepsilon^{\text{reg}}(O) = (0, \varepsilon)$. The image $\Delta_\varepsilon^{\text{reg}} \circ \Delta_\varepsilon^{\text{sing}}(\Pi_h)$ is a thick spiral with the "center" point $O(\varepsilon)$ and is contained in the circle C_H:

$$|(\theta, z) - O(\varepsilon)| = H = h^\alpha.$$

When $\varepsilon > 0$, we consider the rectangle $\Pi_{2\varepsilon}$. Since the image $\Delta_\varepsilon(\Pi_{2\varepsilon})$ is contained in the circle C_H, $H = (2\varepsilon)^\alpha$, the map Δ_ε is a contraction of $\Pi_{2\varepsilon}$ into itself. This implies that Δ_ε has a unique attracting fixed point. When $\varepsilon < 0$, the circle C_H lies below the line $z = h$ for any $h \ll 1$, which means that Δ_ε diminishes the z-coordinate and hence has no fixed points (see Figure 7.9).

2.3. Proof of Theorem 2.1: generation of the stable cycle. Let us describe the singular correspondence map. By assumption 1 and the theory of finitely smooth normal forms for local families, there exists a family of charts (x, y, z) in some neighborhood U of the hyperbolic saddle that brings the initial family to the form (2.1) with λ, ω, and μ depending on ε.

Let Γ^+, Γ^-, Π_h, Δ_ε be the same as in 2.2.

The singular correspondence map may be calculated by the method explained in the previous section. It yields $\Delta_\varepsilon^{\text{sing}}(\theta, z) = (z^{\alpha + i\beta} e^{i\theta})$, where $\alpha = -\lambda/\mu$, $\beta = -\omega/\mu$. In the real coordinates on Γ^- we have

$$\Delta_\varepsilon^{\text{sing}}(\theta, z) = z^\alpha(\cos\psi, \sin\psi), \tag{2.2}$$

where $\psi = \beta \log z + \theta$. Then

$$D\Delta_\varepsilon^{\text{sing}} = z^{\alpha - 1} \begin{pmatrix} \alpha\cos\psi - \beta\sin\psi & -z\sin\psi \\ \alpha\sin\psi + \beta\cos\psi & z\cos\psi \end{pmatrix}. \tag{2.3}$$

For $\sigma < 0$, we have $\alpha > 1$. Hence, the Poincaré map Δ_ε in the rectangle Π_h is strongly contracting for small h by (2.3). By (2.2), the diameter of the image $\Delta_\varepsilon(\Pi_h)$ tends to zero faster than h.

After this, Theorem 2.1 is proved by the same arguments word for word as the first part of Theorem 1.1.

2.4. Homoclinic trajectory of a saddle with positive saddle value: existence of an infinite set of horseshoes.

THEOREM 2.2. *Let X be a smooth vector field in $\mathbb{R}^3$ which has a homoclinic orbit of a saddle with one real positive eigenvalue and a pair of complex eigenvalues having negative real part. If the saddle value is positive, then X has a nontrivial hyperbolic invariant set containing a countable number of periodic orbits in any neighborhood of the homoclinic orbit.*

We prove the theorem under the additional assumption that the saddle under consideration is nonresonant. This allows us to apply formula (2.2) for the singular map.

Note that we deal with one vector field, not with a family. When the vector field from Theorem 2.2 bifurcates in a one-parameter family, any finite number of Smale horseshoes survives. More precisely, for any neighborhood of the union of

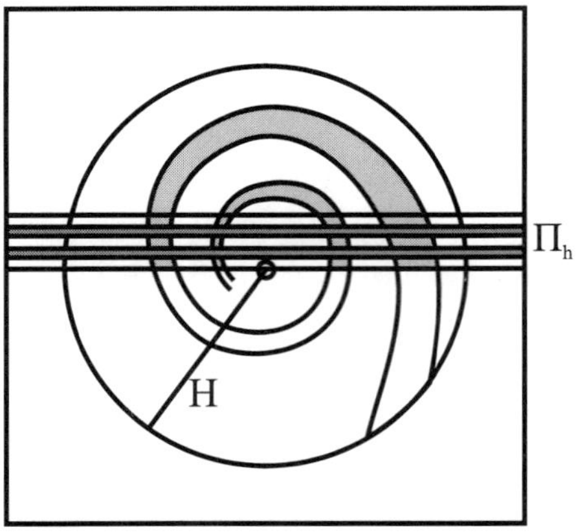

FIGURE 7.10. Smale horseshoes in the heuristic proof of Theorem 2.2.

the saddle and the homoclinic orbit and for any N there exists an interval of the parameter axis such that the corresponding vector field has at least N nontrivial hyperbolic invariant sets in the chosen neighborhood. This is an easy consequence of Theorem 2.2 and the structural stability of the Smale horseshoe; we omit the details of the proof.

HEURISTIC PROOF OF THEOREM 2.2. Let the cross-sections Γ^+, Γ^-, the maps $\Delta_\varepsilon^{\text{sing}}$, $\Delta_\varepsilon^{\text{reg}}$, the charts (θ, z) on Γ^+, w on Γ^- and the rectangle $\Pi_h \subset \Gamma^+$ be the same as in 2.2.

Denote by Δ^{sing}, Δ^{reg} the singular and regular correspondence maps $\Delta_\varepsilon^{\text{sing}}$, $\Delta_\varepsilon^{\text{reg}}$ respectively for $\varepsilon = 0$. By (2.2), the image of Π_h under Δ^{sing} is a "thick spiral" winding around zero and cut by a circle $|w| = H = h^\alpha$ (see Figure 7.10). Since the saddle value $\sigma = \lambda + \mu$ is positive, we have $\alpha = -\lambda/\mu < 1$. Hence, $H \gg h$ for small h.

Let the homoclinic orbit cut Γ^+ at $(\theta, z) = (\theta_0, 0)$ and Γ^- at $w = 0$.

To simplify the heuristic proof, suppose that the regular map is a shift:

$$\Delta^{\text{reg}}\colon w \mapsto (\theta, z) = (\operatorname{Re} w + \theta_0, \operatorname{Im} w).$$

The inverse images of the intersection of the thick spiral with the upper w-half-plane are curvilinear rectangles (see Figure 7.10). For any of these rectangles, the map $\Delta = \Delta^{\text{reg}} \circ \Delta^{\text{sing}}$ forms a Smale horseshoe, shown in the same figure. The countable number of revolutions of the thick spiral produces a countable number of horseshoes.

Now we pass to the formal proof of Theorem 2.2.

2.5. Construction of the domains and maps. Let Γ^+, Γ^-, Δ^{sing}, Δ^{reg}, (θ, z), w, $\Pi_h \subset \Gamma^+$, γ, θ_0 be the same as in 2.4, only Δ^{reg} is now no longer a shift. Let ψ_0 be the argument of the inverse image of the vector $\partial/\partial\theta = (1, 0) \in T_p\Gamma^+$ under $D\Delta^{\text{reg}}(0)$. Here $p = (\theta_0, 0) = \Gamma^+ \cap \gamma$. Note that the coordinate change

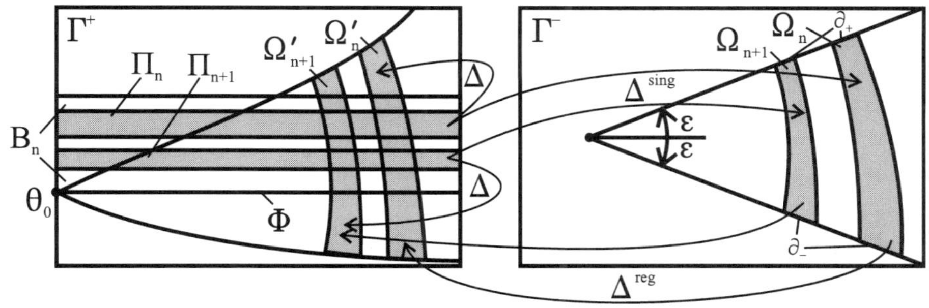

FIGURE 7.11. Domains and maps in Theorem 2.2.

$\theta' = \theta - \psi_0$, $w' = we^{-i\psi_0}$ preserves the form (2.2) of Δ^{sing}. In these coordinates

$$H_0 = D\Delta^{\text{reg}}(0) = \begin{pmatrix} a_0 & b_0 \\ 0 & d_0 \end{pmatrix}, \qquad a_0 d_0 \neq 0. \tag{2.4}$$

In what follows we drop the primes and write θ, w instead of θ', w'.

To simplify the formula for the map Δ^{sing}, consider a new coordinate z' instead of z by setting

$$z' = z\exp(\theta/\beta).$$

In this chart (we drop primes once more)

$$\Delta^{\text{sing}}(\theta, z) = z^{\alpha+i\beta}\exp\Big(-\frac{\alpha}{\beta}\theta\Big). \tag{2.5}$$

The benefit of this chart is that $\arg\Delta^{\text{sing}}$ now depends on z only:

$$\psi = \arg\Delta^{\text{sing}} = \beta\log z.$$

Take ε sufficiently small for what follows. Let S be a sector $|\arg w| < \varepsilon$, $I = [-\varepsilon, \varepsilon]$. Let $\beta < 0$. Then the "thick spiral" $\Delta^{\text{sing}}\Pi_h$ winds counterclockwise. The inclusion

$$\arg\Delta^{\text{sing}} \in 2\pi n + I, \qquad n \in \mathbb{N},$$

implies

$$z \in q^n I', \quad \text{where } q = \exp(2\pi/\beta) < 1,\ I' = \{\exp(\theta/\beta) \mid \theta \in I\}. \tag{2.6}$$

Denote by b the right end of I'. Take a small δ (to be specified later) and

$$\Phi = [\theta_0, \theta_0 + \delta], \quad K_n = [0, 2bq^n], \quad B_n = \Phi \times K_n$$

(see Figure 7.11). Let

$$\Pi_n = \{(\theta, z) \in B_n \mid z \in q^n I'\}, \qquad \Pi_{n+1} = \{(\theta, z) \in B_n \mid z \in q^{n+1} I'\}.$$

Further, let $\Omega_n = \Delta^{\text{sing}}\Pi_n$, $\Omega_{n+1} = \Delta^{\text{sing}}\Pi_{n+1}$. The map $\Delta = \Delta^{\text{reg}} \circ \Delta^{\text{sing}}$ is well defined in Π_n for large n. The restriction of Δ to $\Pi_n \cup \Pi_{n+1}$ already produces a Smale horseshoe (see Figure 7.11, where $\Omega'_k = \Delta^{\text{reg}}\Omega_k$, $k = n, n+1$).

But the application of the Smale horseshoe existence theorem requires some more constructions. First we shall check the cone condition for the map Δ, and before that recall the theorem.

SMALE HORSESHOE EXISTENCE THEOREM. *Let μ_h, μ_v with $\mu_h\mu_v < 1$ be two positive constants, and B a standard rectangle. Let $D_i \subset \mathbb{R}^n \oplus \mathbb{R}^m$, $i = 1, \dots, N$, $N > 1$, be N pairwise disjoint (μ_h, μ_v)-rectangles. Further let*

$$f\colon D = \bigcup_{i=1}^{N} D_i \to f(D) \in B$$

be a diffeomorphism of the domain D onto its image satisfying the following assumptions for $i, j = 1, \dots, N$:

1) *the map f satisfies the (μ_h, μ_v) cone condition in D;*
2) *the domains D_i are horizontally cylindric and $D_i' = f(D_i)$ are vertically cylindric in B.*

Then the set $\Lambda = \bigcap_{-\infty}^{\infty} f^k D$ is a hyperbolic invariant Cantor set. The restriction of f to Λ is topologically conjugated to the Bernoulli shift σ on the space Σ^N of sequences of the elements taking N different values $1, \dots, N$. This means that there exists a homeomorphism ϕ such that the diagram

$$\begin{array}{ccc} \Lambda & \xrightarrow{\ f\ } & \Lambda \\ {\scriptstyle\phi}\downarrow & & \downarrow{\scriptstyle\phi} \\ \Sigma^N & \xrightarrow{\ \sigma\ } & \Sigma^N \end{array}$$

is commutative.

2.6. Cone condition for the map Δ.

LEMMA 2.1. *There exist μ_h, μ_v, with $\mu_h\mu_v < 1$ such that the map Δ satisfies the (μ_h, μ_v) cone condition in $\Pi_n \cap \Pi_{n+1}$ for sufficiently large n.*

PROOF. It is convenient to use polar coordinates in Γ^-: $w = re^{i\psi}$. By (2.5), in these coordinates

$$\Delta^{\text{sing}}(\theta, z) = (r, \psi) = \Big(z^\alpha \exp\Big(-\frac{\alpha}{\beta}\theta\Big), \beta \log z\Big).$$

We shall check the cone condition related to the splitting $\mathbb{R}^2 = \mathbb{R}^1 \oplus \mathbb{R}^1$, the direct sum of the θ and z axes. Then

$$\Sigma = \frac{\partial \Delta^{\text{sing}}}{\partial(\theta, z)} = \begin{pmatrix} -(\alpha/\beta)r & \alpha r/z \\ 0 & \beta/z \end{pmatrix} = \begin{pmatrix} 1 & (\alpha/\beta)r \\ 0 & 1 \end{pmatrix} \begin{pmatrix} -(\alpha/\beta)r & 0 \\ 0 & \beta/z \end{pmatrix}.$$

On the other hand, $w = x + iy$, and in Ω_n we have

$$\frac{\partial \Delta^{\text{reg}}}{\partial(x, y)} = H_0 + O(r),$$

where H_0 is given by (2.4) and r is restricted to Ω_n. To calculate the derivative $A = D\Delta$, recall that

$$\frac{\partial(x \circ \Delta^{\text{sing}}, y \circ \Delta^{\text{sing}})}{\partial(\theta, z)} = \frac{\partial(x, y)}{\partial(r, \psi)} \circ \Delta^{\text{sing}} \cdot \Sigma.$$

We have

$$\frac{\partial(x, y)}{\partial(r, \psi)} = \begin{pmatrix} \cos\psi & -r\sin\psi \\ \sin\psi & r\cos\psi \end{pmatrix} = (E + O(\varepsilon)) \begin{pmatrix} 1 & 0 \\ 0 & r \end{pmatrix},$$

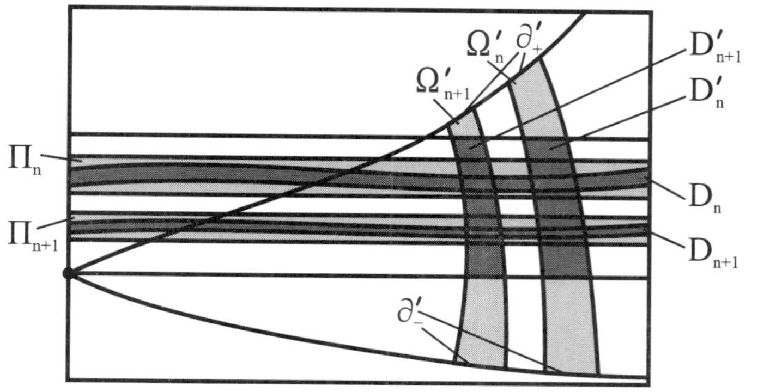

FIGURE 7.12. Diminished domains of the map Δ satisfying the conditions of the Smale horseshoe existence theorem.

because $|\psi| \leqslant \varepsilon$ in Ω_n, Ω_{n+1}. Hence,

$$A = (H_0 + O(r))(E + O(\varepsilon)) \begin{pmatrix} 1 & 0 \\ 0 & r \end{pmatrix} \begin{pmatrix} 1 & (\alpha/\beta)r \\ 0 & 1 \end{pmatrix} \begin{pmatrix} -(\alpha/\beta)r & 0 \\ 0 & \beta/z \end{pmatrix}.$$

In what follows, $*$ stands for some constants not depending on n. We have

$$\begin{pmatrix} 1 & 0 \\ 0 & r \end{pmatrix} \begin{pmatrix} 1 & *r \\ 0 & 1 \end{pmatrix} \begin{pmatrix} *r & 0 \\ 0 & \beta/z \end{pmatrix} = \begin{pmatrix} 1 & * \\ 0 & 1 \end{pmatrix} \begin{pmatrix} *r & 0 \\ 0 & *r/z \end{pmatrix}.$$

Let us take

$$H = (H_0 + O(r))(E + O(\varepsilon)) \begin{pmatrix} 1 & * \\ 0 & 1 \end{pmatrix}, \qquad \Sigma = \begin{pmatrix} *r & 0 \\ 0 & *r/z \end{pmatrix}.$$

The constants replaced by $*$ in Σ are nonzero. The matrix Σ in $\Pi_n \cup \Pi_{n+1}$ satisfies conditions (4.2.3) and (4.2.4) of Lemma 4.2.1 for sufficiently large n. Indeed, let z_n and r_n be the restrictions of z to Π_n and r to Ω_n, respectively. Then $r_n \to 0$ and $r_n/z_n = O(z_n^{\alpha-1}) \to \infty$ as $n \to \infty$ (see (2.5) and Figure 7.11).

The matrix H satisfies the conditions (4.2.1), (4.2.2) of Lemma 4.2.1 for n sufficiently large and ε sufficiently small by (2.4). This guarantees the conditions of Lemma 4.2.1, and thus proves Lemma 2.1.

In order to apply the Smale horseshoe existence theorem, we need to diminish the domains of the map Δ.

2.7. Rectangular structure on the modified domains of the map Δ. Recall that the Smale horseshoe existence theorem was stated in §2.5. The role of the standard rectangle B from this theorem will be played by B_n in our further construction. Take

$$D'_k = \Omega'_k \cap B_n, \quad D_k = \Delta^{-1} D'_k, \qquad k = n, n+1$$

(see Figure 7.12).

LEMMA 2.2. *The domains D'_n, D'_{n+1} are (μ_h, μ_v)-rectangles horizontally cylindric in B_n; the domains D_n, D_{n+1} are (μ_h, μ_v)-rectangles vertically cylindric in B_n; the constants μ_v and μ_h are the same as in Lemma 2.1.*

PROOF. Let $k = n$ or $n+1$; the domains Ω_k are foliated by the curves $V_{\theta,k}^-$,

$$V_{\theta,k}^- = \{z^{\alpha+i\beta}\exp(-(\alpha/\beta)\theta) \mid z \in q^k I'\}.$$

The endpoints of these curves belong to the rays $\arg w = \pm\varepsilon$, by the definition of I'. Their distance to zero is greater than $Cq^{\alpha k}$ for some $C > 0$ not depending on k. Denote the union of these points by ∂_+ and ∂_- respectively (see the right side of Figure 7.11). Then

$$\min_{\partial_+} \operatorname{Im} w > \varepsilon C q^{\alpha n}, \qquad \max_{\partial_-} \operatorname{Im} w < -\varepsilon C q^{\alpha n}.$$

Let $\partial'_+ = \Delta^{\mathrm{reg}}\partial_+$, $\partial'_- = \Delta^{\mathrm{reg}}\partial_-$. By (2.4), for large n, we have

$$\min_{\partial'_+} z > \max_{\Pi_n} z, \qquad \max_{\partial'_-} z < 0.$$

Denote by π_z the projection $(\theta, z) \mapsto z$. Let $V_{\theta,k}^+ = \Delta^{\mathrm{reg}} V_{\theta,k}^-$. The previous inequalities imply that $K_n \subset \pi_z V_{\theta,k}^+$. Here, as before, $K_n = [0, 2bq^n]$ and $\pi_z(y,z) = z$. Denote by $\sigma_{\theta,k}$, $k = n, n+1$, the vertical segment

$$\sigma_{\theta,n} = \{p \in \Pi_n \mid \theta(p) = \theta\}, \qquad \sigma_{\theta,n+1} = \{p \in \Pi_{n+1} \mid \theta(p) = \theta\}.$$

The segment $\sigma_{\theta,k}$ is μ_v-vertical for any positive μ_v. Note that $V_{\theta,k}^+ = \Delta\sigma_{\theta,k}$. Hence, by Lemma 2.1, the tangent lines to $V_{\theta,k}^+$ belong to the cones K^+ defined in the statement of the (μ_h, μ_v) cone condition. So, the intersection $V_{\theta,k} = V_{\theta,k}^+ \cap B_n$ is the graph of a smooth map $v_{\theta,k}\colon K_n \to \Phi$ with the Lipschitz constant μ_v from Lemma 2.1.

Now we can define the rectangular structure on D'_k:

$$F'_k\colon B_n \to D'_k, \qquad (\theta, z) \mapsto (v_{\theta,k}(z), z).$$

By Lemma 2.1, the images of the vertical segments of B_n are μ_v-vertical. The images of the horizontal ones are obviously horizontal. The inclusion requirement $\partial_h D'_k \subset \partial_h B_n$ follows from the construction.

Let us define the rectangular structure on D_k by setting $F_k = \Delta^{-1} \circ F'_k$. In order to investigate this map, put

$$H_z = \Phi \times \{z\}, \quad H_{z,k} = D'_k \cap H_z, \quad H_{z,k}^+ = \Delta^{-1} H_{z,k}.$$

By construction,

$$\partial H_{z,k}^+ \subset \partial_v B_n. \tag{2.7}$$

By Lemma 2.1, the images of the horizontal segments of B_n under F_k are μ_h-horizontal. The images of the vertical segments are purely vertical. By (2.7), $\partial_v D_k \subset \partial_v B_n$. This proves Lemma 2.2.

At the same time, the requirements of the Smale horseshoe existence theorem for the domains $D'_1 = D'_n$, $D'_2 = D'_{n+1}$, $D_1 = D_n$, $D_2 = D_{n+1}$, $B = B_n$ are checked. Lemma 2.1 allows the application of the theorem above to the map $\Delta = f$ restricted to $D = D_n \cup D_{n+1}$ for sufficiently large n. This proves Theorem 2.2.

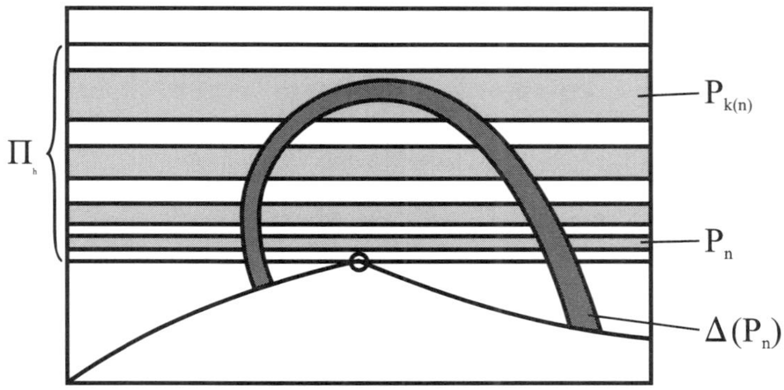

FIGURE 7.13. The combinatorics of intersections for the Poincaré map in Theorem 2.3.

2.8. Homoclinic trajectory of a saddle with complex eigenvalues: the continuous topological invariant.

THEOREM 2.3. *Suppose a smooth vector field in* $\mathbb{R}^3$ *has a homoclinic trajectory of a nonresonant hyperbolic saddle with eigenvalues* $\lambda \pm i\omega$, μ, $\lambda\mu < 0$, $\omega \neq 0$. *Then the ratio* λ/μ *is a topological invariant with respect to orbital topological equivalence.*

All notations in this subsection are the same as in 2.5. Let

$$P_n = \{p \in \Pi_h \mid \arg \Delta^{\text{sing}}(p) \in [-\varepsilon, \pi + \varepsilon] + 2\pi n\}.$$

Then $P_n = \{p \in \Pi_h \mid z(p) \in q^n I''\}$, where

$$I'' = \left[\exp \frac{\pi + \varepsilon}{\beta}, \exp\left(-\frac{\varepsilon}{\beta}\right)\right];$$

recall that $\beta < 0$. The following lemma is illustrated in Figure 7.13.

LEMMA 2.3. *For a fixed large* $n \in \mathbb{N}$, *set*

$$k(n) = \max\{k \in \mathbb{N} \mid P_k \cap \Delta P_n \neq \varnothing\}.$$

Then

$$\lim_{n\to\infty} \frac{k(n)}{n} = -\alpha,$$

where α *is the same as in* (2.5).

PROOF. The intersection $P_k \cap \Delta P_n$ is not empty if and only if

$$\min_{P_k} z \leqslant \max_{\Delta P_n} z.$$

By the definition of P_k, at any $\mathbf{a} \in P_k$ we have $z(\mathbf{a}) = q^{k+O(1)}$, where $O(1)$ is bounded with respect to k and $\mathbf{a}$. Relations (2.5) and (2.4) imply $\max_{\Delta P_n} z = q^{\alpha n + O(1)}$. By the definition of $k(n)$, $q^{k(n)} = q^{\alpha n + O(1)}$. This yields the lemma.

Now we prove Theorem 2.3. Let X_1, X_2 be two vector fields possessing a homoclinic orbit of the saddle with two complex eigenvalues in the left half-plane and one positive eigenvalue. Let X_1 and X_2 be topologically equivalent in the neighborhoods of their homoclinic orbits. Denote by H the conjugating homeomorphism. We shall prove that the sequences $k(n)$ defined in Lemma 2.3 for these two fields coincide up to a bounded correction. After that, Lemma 2.3 will imply Theorem 2.3. Let Γ_j^+ be two copies of the cross-section Γ^+ defined in 2.2. Let Δ_j be the Poincaré map defined on a subset of Γ_j^+ into Γ_j^+ in the same way as the map Δ at the end of 2.4 for the vector field X_j. Then the map H that brings the phase portrait of X_1 to that of X_2 near the homoclinic orbits, generates a topological conjugacy of Δ_1 and Δ_2, possibly, in diminished domains. Indeed, the image $H\Gamma_1^+$ is a (nonsmooth) surface, which is still transversal to the orbits of X_2 in the sense that it has a unique intersection with any phase curve of X_2 located in a sufficiently small neighborhood of $H\Gamma_1^+$. Let π be the projection of $H\Gamma_1^+$ to Γ_2^+ along the phase curves of X_2 passing near the arc of γ. Then the map $\pi \circ H \colon \Gamma_1^+ \to \Gamma_2^+$ conjugates Δ_1 and Δ_2. Hence, the sequences $k_1(n)$ and $k_2(n)$ defined for Δ_1 and Δ_2 have a bounded correction. This implies the theorem.

§3. Homoclinic orbits of hyperbolic saddles in high-dimensional spaces

In this section we will consider the bifurcation of vector fields in $\mathbb{R}^n$ that belong to the boundary of the Morse–Smale set and have a homoclinic orbit of a hyperbolic singular point. The result of this section is a generalization of Theorems 1.1 and 2.1 to higher dimension.

3.1. The leading direction and the saddle value. Consider the germ of a vector field $v(x) = Ax + \cdots$, $x \in (\mathbb{R}^n, 0)$ at the hyperbolic singular point O of saddle type, $\dim W^s = s > 0$, $\dim W^u = u > 0$, $u + s = n$. Let $\{\lambda_j, \mu_k\}$, $1 \leqslant j \leqslant s$, $1 \leqslant k \leqslant u$, be the eigenvalues of the operator A such that

$$\operatorname{Re}\lambda_s \leqslant \cdots \leqslant \operatorname{Re}\lambda_1 < 0 < \operatorname{Re}\mu_1 \leqslant \cdots \leqslant \operatorname{Re}\mu_u.$$

The sum $\sigma = \operatorname{Re}\lambda_1 + \operatorname{Re}\mu_1$ is called the *saddle value* of the singular point O. If $\operatorname{Re}\lambda_1 = \cdots = \operatorname{Re}\lambda_k > \operatorname{Re}\lambda_{k+1}$, then the invariant subspace of the operator A, corresponding to the eigenvalues $\lambda_1, \dots, \lambda_k$, is called the *leading stable direction* of the singular point O; the *leading unstable direction* is defined similarly. Almost all phase curves of the equation $\dot{x} = v(x)$ with initial values on the stable manifold W^s and unstable manifold W^u tend to O along the leading directions as $t \to \pm\infty$ respectively and the exceptional phase curves form submanifolds of W^s and W^u of dimension less than s and u. In the generic case, the leading direction is either one-dimensional (corresponding to a real eigenvalue) or two-dimensional (corresponding to a pair of complex conjugate eigenvalues). We shall say that the leading direction is *real* in the first case and *complex* in the second case.

If the saddle value σ is positive, then the leading stable direction is called *subordinate*. For $\sigma < 0$, the unstable leading direction is subordinate. Any of these directions may be either complex or real. In the theorem below this freedom is restricted.

3.2. Bifurcation of the homoclinic orbit of a saddle: creating a periodic orbit. The main result of this section is the following

THEOREM 3.1. *Suppose that in a typical one-parameter family of vector fields in $\mathbb{R}^n$ the zero value of the parameter corresponds to a vector field with a homoclinic orbit γ of the hyperbolic saddle O with real subordinate leading direction.*

Then the vector fields corresponding to all sufficiently small values of the parameter on one side of zero have a hyperbolic periodic orbit which tends to the homoclinic orbit γ of the critical vector field as the parameter tends to zero. The vector fields corresponding to all sufficiently small values of the parameter on the other side of zero have no periodic orbits in some neighborhood of γ.

If the subordinate leading direction is unstable, then the dimension of the stable manifold of the periodic orbit generated by γ is greater by one than the dimension of the stable manifold of the saddle. If the subordinate leading direction is stable, then the same is true for the unstable manifolds.

REMARK. We shall describe the genericity assumptions and prove the theorem in the case for which the subordinate leading direction is unstable. By the assumption of the theorem, this direction is real. The case of the subordinate *stable* leading direction is reduced to the previous one by time reversal. If both leading directions in the theorem are real, then we speak of the real case. If one of them is complex (it is not subordinate by the assumption of the theorem), then we are in the complex case.

The proof is based on the hyperbolic fixed point theorem (see 2.4.3). We recall this theorem below in 3.6.

The principal property of the vector field in the theorem is that the subordinate leading direction is real. If, on the contrary, it is complex, the vector field with a homoclinic loop cannot occur on the boundary of the Morse–Smale set. The properties of such a field are like those described in Theorem 2.2. In particular, the field has a countable number of periodic orbits, and so do all the nearby vector fields.

3.3. Genericity assumptions and corollaries. In this subsection we will describe the four genericity assumptions that the family of vector fields in Theorem 3.1 must satisfy. The first three concern the vector field corresponding to the critical value of the parameter, the last one is for the family itself.

1. *The saddle O is nonresonant.*

Hence, by Theorem 2.5.1, the local family of vector fields from the theorem may be finitely smoothly transformed to a linear one near the singular point.

2. *The leading directions are real one-dimensional or complex two-dimensional and the homoclinic orbit tends to O along the leading directions as $t \to \pm\infty$.*

In the real case the space $\mathbb{R}^n$ is split into the sum

$$\mathbb{R}^n = \mathbb{R}^{s-1} \oplus \mathbb{R} \oplus \mathbb{R} \oplus \mathbb{R}^{u-1}.$$

One-dimensional entries correspond to real leading directions. The corresponding splitting of the coordinates in $\mathbb{R}^n$ is the following:

$$\mathbf{x} = (y, x, z, v), \quad y \in \mathbb{R}^{s-1},\ x \in \mathbb{R},\ z \in \mathbb{R},\ v \in \mathbb{R}^{u-1}. \tag{3.1}$$

The linearization of the field of the family at the singular point in these coordinates is represented by a block diagonal matrix

$$\mathbf{A} = \operatorname{diag}(A, \lambda, \mu, B), \qquad \lambda < 0,\ \mu > 0, \tag{3.2}$$

where

$$\operatorname{Re}\operatorname{spec} A < \lambda < 0 < \mu < \operatorname{Re}\operatorname{spec} B, \qquad \lambda + \mu < 0. \tag{3.3}$$

The latter inequality holds because we assume that the subordinate leading direction is unstable.

In the complex case the space $\mathbb{R}^n$ is split into the sum

$$\mathbb{R}^n = \mathbb{R}^{s-2} \oplus \mathbb{C} \oplus \mathbb{R} \oplus \mathbb{R}^{u-1}.$$

The real one-dimensional entry corresponds to the unstable real leading direction. The complex one corresponds to the stable complex leading direction.

The corresponding splitting of the coordinates in $\mathbb{R}^n$ is the following:

$$\mathbf{x} = (y, w, z, v), \quad y \in \mathbb{R}^{s-2},\ w \in \mathbb{C},\ z \in \mathbb{R},\ v \in \mathbb{R}^{u-1}. \tag{3.4}$$

The linearization of the field of the family at the singular point in these coordinates is represented by a block diagonal matrix

$$\mathbf{A} = \operatorname{diag}(A, \varkappa, \mu, B), \qquad \varkappa \in \mathbb{C},\ \mu \in \mathbb{R}, \tag{3.5}$$

where

$$\varkappa = \lambda + i\omega, \qquad \lambda < 0,\ \mu > 0, \tag{3.6}$$

and (3.3) holds. Assumption 2 allows us to chose two cross-sections that will be definitely intersected by the homoclinic orbit. In the real case this will be cross-sections transversal to the axes of the stable leading direction x (the entrance cross-section Γ^+) and to the unstable leading direction z (the exit cross-section Γ^-). In the complex case the entrance cross-section Γ^+ will be transversal to the w plane at some point of this plane.

After an appropriate rescaling, Γ^- may be taken to be the product of unit balls of the y-space, x-space and v-space in the plane $z = 1$. In the real case, Γ^+ may be taken as a product of unit balls of the y-space, z-space and v-space in the hyperplane $x = 1$. In the complex case it may be taken to be the product of unit balls of the y-space, z-space, v-space and the arc $\{w = e^{i\theta} : |\theta| \leqslant 1\}$ of the circle $|w| = 1$.

3. To state assumption 3, denote by H and V the intersection of the stable and unstable manifolds of the saddle with the entrance and exit cross-sections respectively:

$$H = \Gamma^+ \cap W^s, \qquad V = \Gamma^- \cap W^u.$$

Let γ be the homoclinic curve of the saddle corresponding to the critical parameter value. Then the intersection of this curve with the entrance and exit cross-sections belong to H and V respectively:

$$p = \gamma \cap \Gamma^+ \in H, \qquad q = \gamma \cap \Gamma^- \in V.$$

A diffeomorphic *regular correspondence map* from some neighborhood of q to some neighborhood of p along the orbits of the vector fields of the family passing close to the arc of γ beginning at q and ending at p is well defined. Denote it by $\Delta_\varepsilon^{\mathrm{reg}}$.

The third genericity assumption asserts that

$$\Delta_\varepsilon^{\mathrm{reg}} V \text{ for } \varepsilon = 0 \textit{ is transversal to } H \oplus Oz \textit{ at } q \tag{3.7}$$

(see Figure 7.14).

That is all about the critical vector field.

4. The last assumption characterizes "separatrix splitting".

It asserts that the homoclinic orbit splits in a transversal manner as the parameter changes. The explicit statement will be given at the beginning of 3.6.

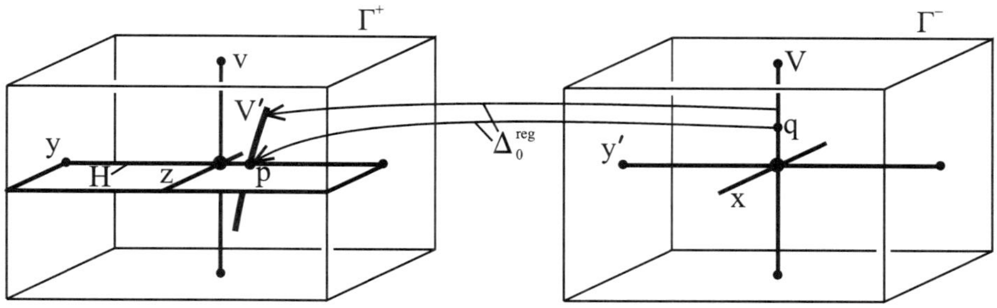

FIGURE 7.14. The transversality assumption.

3.4. Singular correspondence map in the real case. Let us now describe the singular correspondence map from the entrance to the exit cross-section along orbits passing near the saddle O and not leaving the neighborhood of the saddle where the normalizing chart is defined. To do this, we shall use the coordinates (3.1) and the matrix (3.2) of the normalized system

$$\dot{\mathbf{x}} = \mathbf{A}\mathbf{x}.$$

The time necessary for the orbit to pass from the point (3.1) or (3.4) to the plane $z = 1$ is

$$t(z) = -\frac{\log z}{\mu}. \tag{3.8}$$

The phase flow transformation for this time in the real case restricted to Γ^+ is

$$g^{t(z)}\colon (y, 1, z, v) \mapsto (y', x', z', v') = (z^{A'}y, z^{\alpha}, 1, z^{B'}v).$$

Here

$$A' = -\frac{A}{\mu}, \quad B' = -\frac{B}{\mu}, \quad \alpha = -\frac{\lambda}{\mu}. \tag{3.9}$$

By (3.3),

$$\operatorname{Re}\operatorname{spec} A' > 1, \quad \operatorname{Re}\operatorname{spec} B' < -1, \quad \alpha > 1. \tag{3.10}$$

Finally,

$$\Delta_\varepsilon^{\text{sing}}(y, z, v) = (z^{A'}y, z^{\alpha}, z^{B'}v). \tag{3.11}$$

The source and target of the singular correspondence map are shown in Figure 7.15.

3.5. Singular correspondence map in the complex case. The calculation of the map in the title follows the same lines as for the real case. Now we use the coordinates (3.4) and the matrix (3.5). The points on Γ^+ are $(y, e^{i\theta}, z, v)$. The time necessary for this point to reach Γ^- is given by (3.8). Take (y, θ, z, v) and (y', w, v) as coordinates on Γ^+ and Γ_- respectively. Then

$$\Delta_\varepsilon^{\text{sing}}(y, \theta, z, v) = (z^{A'}y, z^{\alpha}e^{i\theta}, z^{B'}v), \tag{3.12}$$

where $\alpha = -\varkappa/\mu$, while A' and B' are given by (3.9) and property (3.10) holds.

The source and target of the singular correspondence map in the complex case are shown in Figure 7.16.

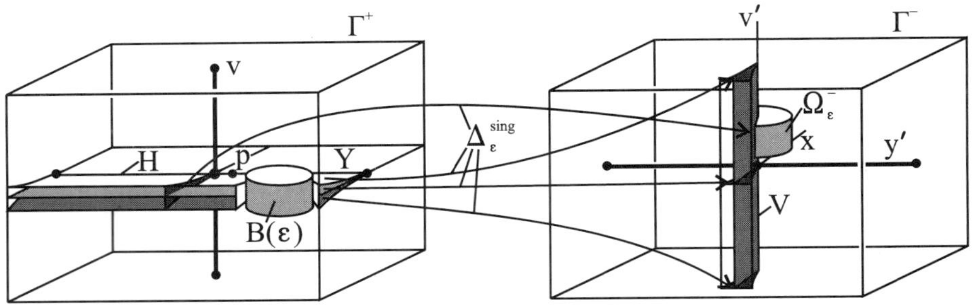

FIGURE 7.15. Singular correspondence map in the real case. The map from Figure 7.4b and the inverse to that from Figure 7.4c are seen in the lightly shadowed cross-sections. The shadowed cylinders show the domains $B(\varepsilon)$ (left) and $\Omega_\varepsilon^- \ni q$ (right) constructed below in 3.6.

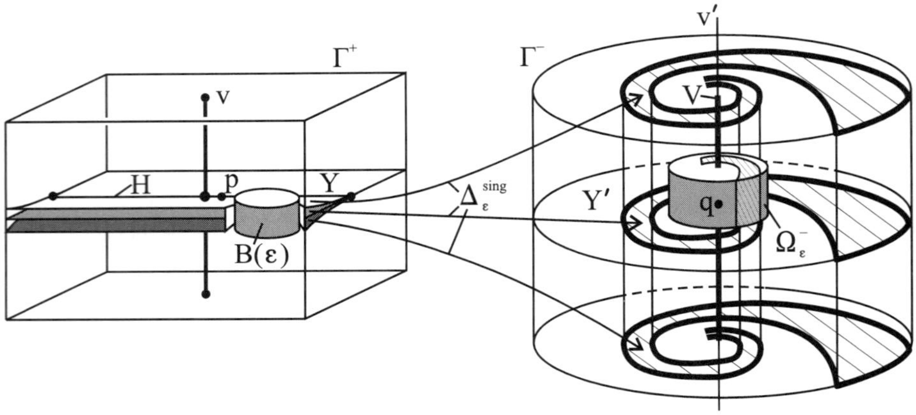

FIGURE 7.16. Singular correspondence map in the complex case. The map from Figure 7.8 is seen in the medium shadowed cross-section. The shadowed cylinders show the domains $B(\varepsilon)$ (left) and $\Omega_\varepsilon^- \ni q$ (right) constructed below in 3.6.

3.6. Source and target of the Poincaré map. Recall once more the following

HYPERBOLIC FIXED POINT THEOREM. *Let $B = D_h \times D_v$ be a standard rectangle. Let $D \subset B$ and $D' \subset B$ be v-cylindric and h-cylindric (μ_h, μ_v)-rectangles in B. Let $f\colon D \to D'$ be a map satisfying the (μ_h, μ_v) cone condition. Then f has a unique fixed point O in D:*

$$O = \bigcap_{i=-\infty}^{\infty} f^i D.$$

In what follows we will construct the domains $B(\varepsilon)$, D_ε, D'_ε that will be taken as B, D, D' in the previous theorem for any sufficiently small ε, provided that f in the theorem is Δ_ε. Let $(y, z, v) = (Y, v)$ or $(y, \theta, z, v) = (Y, v)$ be the chart on Γ^+ in the real and complex cases respectively. Let $(y', x, v') = (Y', v')$ or $(y', w, v') = (Y', v')$, $w \in \mathbb{C}$, be the chart on Γ^- in the real and complex cases respectively. Let $\mathbb{R}^s$ be the Y-space.

By the transversality assumption (3.7), a sufficiently small ball on V centered at q is transformed by Δ_0^{reg} to a surface V', which is the graph of a smooth map to $\mathbb{R}^s$ of some ball in the v-space in Γ^+. The same is true for any map and surface sufficiently close to Δ_0^{reg} and V.

Let

$$V_\varepsilon = \Delta_\varepsilon^{\mathrm{reg}} V, \quad B_r^{u-1} = \{v \in \mathbb{R}^{u-1} : |v| \leqslant r\}, \quad Q_\rho^{u-1} = \{v' \in \mathbb{R}^{u-1} : |v'| \leqslant \rho\}.$$

The radius r may be chosen so small that for all small ε the surfaces V_ε contain the graph of a smooth map $v_\varepsilon \colon B_r^{u-1} \to \mathbb{R}^s$ passing through the point p for $\varepsilon = 0$ and smoothly depending on ε. Moreover, the radius ρ and ε may be chosen so small that the surface $V_{Y',\varepsilon} = \Delta_\varepsilon^{\mathrm{reg}}(\{Y'\} \times Q_\rho^{u-1})$ is the graph of a smooth map $v_{Y',\varepsilon} \colon B_r^{u-1} \to \mathbb{R}^s$ for all Y' sufficiently small.

Denote by $p(\varepsilon)$ the point $(v_\varepsilon(0), 0) \in \Gamma^+$. Let $Z(\varepsilon) = z(p(\varepsilon))$. Now we can give the explicit statement of assumption 4 of the genericity condition (see 3.3). This statement is $Z'(0) \neq 0$. After that we can take $\varepsilon' = Z(\varepsilon)$ as the new parameter. All maps that are smooth in ε are smooth in ε' as well. In what follows, we drop the superscript and write ε instead of ε'.

Let

$$K(\varepsilon) = \{Y \in \mathbb{R}^s : |Y - Y(p(\varepsilon))| \leqslant \varepsilon/2\}.$$

Let $\mathbf{L}$ be a common Lipschitz constant of all the maps $v_{Y',\varepsilon}$. Here and below, $\varepsilon \in [0, \varepsilon_0]$, where ε_0 is sufficiently small for what follows. Let

$$B_\varepsilon^{u-1} = \{v \in \mathbb{R}^{u-1} : |v - v(p(\varepsilon))| \leqslant \varepsilon/(3\mathbf{L})\}.$$

Recall that $v(p(\varepsilon)) = 0$. The choice of the radius of this ball is motivated by the following inclusion of images: $v_{Y',\varepsilon}(B_\varepsilon^{u-1}) \subset K(\varepsilon)$ for $Y' = 0$, $\varepsilon = 0$, hence, for Y' and ε small enough. Further, let

$$B(\varepsilon) = K(\varepsilon) \times B_\varepsilon^{u-1}.$$

This is the counterpart of the domain B from the previous theorem, for sufficiently small ε (see Figures 7.15, 7.16).

Now we pass to the construction of the domains D_ε, D'_ε. This will be done in several steps, as in 4.3.6. Let us choose $\rho(\varepsilon)$ in such a way that for any Y' satisfying $|Y'| \leqslant \rho(\varepsilon)$ and any small ε the previous inclusion of images holds. Let

$$B^s(\varepsilon) = \{Y' \in \mathbb{R}^s : |Y'| \leqslant \rho(\varepsilon)\} \quad \text{and} \quad \Omega_\varepsilon^- = B^s(\varepsilon) \times Q_\rho^{u-1}.$$

Denote $\Omega_\varepsilon^+ = \Delta_\varepsilon^{\mathrm{reg}} \Omega_\varepsilon^-$ (see Figure 7.17). Set

$$W_\varepsilon^+ = B(\varepsilon) \cap \Omega_\varepsilon^+, \qquad W_\varepsilon^- = (\Delta_\varepsilon^{\mathrm{reg}})^{-1} W_\varepsilon^+.$$

Further, set

$$D_\varepsilon^- = W_\varepsilon^- \cap \Delta_\varepsilon^{\mathrm{sing}} B(\varepsilon), \quad D'_\varepsilon = \Delta_\varepsilon^{\mathrm{reg}} D_\varepsilon^-, \quad D_\varepsilon = (\Delta_\varepsilon^{\mathrm{sing}})^{-1} D_\varepsilon^-$$

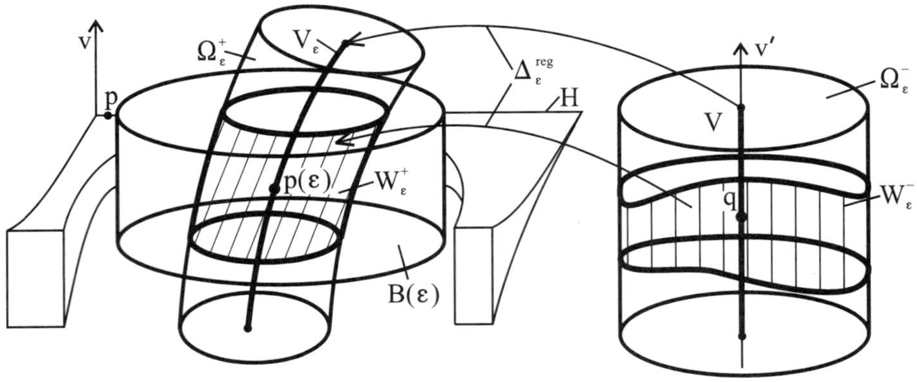

FIGURE 7.17. Domains $B(\varepsilon)$ and $W_\varepsilon^\pm$ in the real and complex cases.

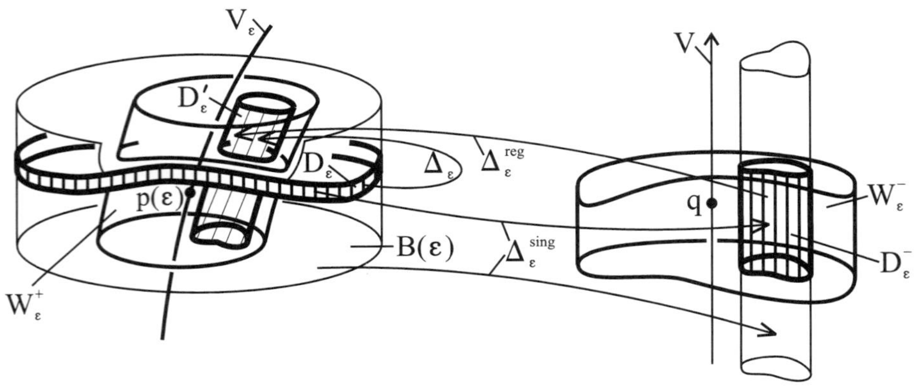

FIGURE 7.18. Domains D_ε^-, D'_ε, and D_ε^+ in the real and complex cases.

(see Figure 7.18). The map Δ_ε takes D_ε onto D'_ε. These are the map and the domains to which the hyperbolic fixed point theorem will be applied.

Let us check its assumptions.

3.7. Cone condition for the Poincaré map.

LEMMA 3.1. *There exist positive constants μ_h, μ_v, with $\mu_h\mu_v < 1$ such that Δ_ε satisfies the (μ_h, μ_v) cone condition in D_ε.*

PROOF. We shall prove that for any $a \in D_\varepsilon$ and any $b \in D'_\varepsilon$ (and even for $b \in W_\varepsilon^-$), the matrices

$$\Sigma = D\Delta_\varepsilon^{\mathrm{sing}}(a), \qquad H = D\Delta_\varepsilon^{\mathrm{reg}}(b)$$

satisfy the conditions of Lemma 4.2.2. This means that for any σ there exists an ε such that the assumptions of the lemma for the maps H and Σ above hold.

1. The real case. By (3.11),

$$\Sigma = \begin{pmatrix} \Lambda & 0 \\ D' & M \end{pmatrix}, \quad \text{where } \Lambda = \begin{pmatrix} z^{A'} & A' z^{A'-E} y \\ 0 & \alpha z^{\alpha-1} \end{pmatrix} \tag{3.13}$$

and

$$M = z^{B'}, \qquad D' = (0 \quad B' z^{B'-E} v). \tag{3.14}$$

Hence,

$$\Sigma = \begin{pmatrix} \Lambda & 0 \\ 0 & M \end{pmatrix} \begin{pmatrix} 1 & 0 \\ D & 1 \end{pmatrix}, \quad \text{where } D = B' z^{-1} v.$$

By (3.10), we have $\|\Lambda\| \to 0$, $\|M^{-1}\| \to 0$ as $\varepsilon \to 0$ everywhere in $B(\varepsilon)$. The estimate for D is more delicate. It is proved in D_ε, not in $B(\varepsilon)$, in the same way as in 5.7.8. Namely, since $\Delta_\varepsilon^{\text{sing}} D_\varepsilon \subset \Omega^-$, there exists a C such that

$$|z^{B'} v(a)| \leqslant C \text{ for any } a \in D_\varepsilon.$$

Hence, for such a, we have

$$|z^{-1} v(a)| \leqslant |z^{-B'-E}(z^{B'} v)(a)| \leqslant C \|z^{-B'-E}\|.$$

The last norm tends to zero in $B(\varepsilon)$ by (3.10). This gives the properties of Σ from Lemma 4.2.2. The desired properties of H now follow from the regularity of $\Delta_\varepsilon^{\text{reg}}$ and the transversality assumption (3.7).

This completes the verification of the cone condition in the real case.

2. The complex case. In this case, by (3.12), Σ has the form (3.13) (left) with M and D' given by (3.14) and

$$\Lambda = \begin{pmatrix} z^{A'} & 0 & A' z^{A'-E} y \\ 0 & w_1 & w_2 \end{pmatrix},$$

where $w_1 = i z^\alpha e^{i\theta}$, $w_2 = \alpha z^{\alpha-1} e^{i\theta}$. Here the string $(w_1\ w_2)$ stands for the operator $\mathbb{R}^2 \to \mathbb{R}^2$ with the matrix

$$(w_1 w_2) = \begin{pmatrix} \operatorname{Re} w_1 & \operatorname{Re} w_2 \\ \operatorname{Im} w_1 & \operatorname{Im} w_2 \end{pmatrix}.$$

Once more, in $B(\varepsilon)$ we have $\|\Lambda\| \to 0$ as $\varepsilon \to 0$. All the rest of the proof is the same as in the real case.

This proves Lemma 3.1.

3.8. Rectangular structure and boundary conditions.

LEMMA 3.2. *There exist μ_h, μ_v with $\mu_h \mu_v < 1$ such that the domains D_ε, D'_ε are (μ_h, μ_v)-rectangles for sufficiently small ε. Moreover, D_ε is vertically and D'_ε is horizontally cylindric in $B(\varepsilon)$.*

PROOF. We begin with some general remarks. Consider the triple (B, F, D) from the definition of the rectangular structure. Let $B = B_h \times B_v$ and y and z be the charts on B_h and B_v respectively. We refer to the set $F(B_h \times \{z\}) = H(z, F)$ as the *horizontal fiber* and the set $F(\{y\} \times B_v) = V(y, F)$ as the *vertical fiber* of D in the chart (B, F, D).

Now we construct the rectangular structure on D'_ε. Note that the map $\Delta_\varepsilon^{\text{sing}}$ preserves the vertical foliation, i.e., $Y' \circ \Delta_\varepsilon^{\text{sing}}$ depends on Y only. Let

$$\mathcal{S}_\varepsilon(Y) = Y' \circ \Delta_\varepsilon^{\text{sing}}.$$

The explicit formulas for $\mathcal{S}(Y)$ are different in the real and complex cases. They may be easily derived from (3.11) and (3.12). We need only the fact that $\mathcal{S}_\varepsilon(Y) \to 0$ as $\varepsilon \to 0$ uniformly in $Y \in K(\varepsilon)$.

For brevity we omit the subscript ε below and write $\mathcal{S}(Y)$ instead of $\mathcal{S}_\varepsilon(Y)$. Let

$$F'_\varepsilon\colon B(\varepsilon) \to D'_\varepsilon, \ (Y, v) \mapsto (v_{\mathcal{S}(Y),\varepsilon}(v), v).$$

The vertical fiber of this chart is the graph $V_{\mathcal{S}(Y),\varepsilon}$ of a map $v_{\mathcal{S}(Y),\varepsilon}$ restricted to $B^s(\varepsilon)$. The Lipschitz constant of this map is $\mathbf{L}$. Suppose that in Lemma 3.1 we have $\mu_v > \mathbf{L}$ and $\mu_h \in (0, \mu_v^{-1})$.

The surface $V_{\mathcal{S}(Y),\varepsilon}$ is μ_v-vertical in $B(\varepsilon)$. Consider the horizontal surface

$$H(v, F'_\varepsilon) \subset K(\varepsilon) \times \{v\}$$

of the chart $(B(\varepsilon), F'_\varepsilon, D'_\varepsilon)$. It is μ_h-horizontal for any $\mu_h > 0$. This completes the study of D'_ε.

Let us now pass to D_ε. Define $F_\varepsilon = \Delta_\varepsilon^{-1} F'_\varepsilon$. By the definition of D_ε and D'_ε, the map F_ε is onto. Note that $Y \circ F_\varepsilon = Y$. Indeed,

$$\Delta_\varepsilon^{-1} = (\Delta_\varepsilon^{\text{sing}})^{-1} \circ (\Delta_\varepsilon^{\text{reg}})^{-1}.$$

By the definition of the fiber $V_{\mathcal{S}(Y),\varepsilon}$,

$$(\Delta_\varepsilon^{\text{reg}})^{-1}(V_{\mathcal{S}(Y),\varepsilon}) \subset \{\mathcal{S}(Y)\} \times Q_\rho^{u-1} \cap D_\varepsilon^-.$$

Hence,

$$\Delta_\varepsilon^{-1}(V_{\mathcal{S}(Y),\varepsilon}) \subset \{Y\} \times B_\varepsilon^{u-1}.$$

Therefore, the fiber $V(Y, F_\varepsilon)$ belongs to the vertical plane $Y = \text{const}$ and is μ_v-vertical for any μ_v. Now consider the fiber $H(v, F_\varepsilon) = \Delta_\varepsilon^{-1} H(v, F'_\varepsilon)$. By the definitions of D'_ε and D_ε^-, we have

$$\partial(\Delta_\varepsilon^{\text{reg}})^{-1} H(v, F'_\varepsilon) \subset \Delta_\varepsilon^{\text{sing}} \partial_v B(\varepsilon).$$

Hence,

$$\partial H(v, F_\varepsilon) \subset \partial_v B(\varepsilon), \qquad \partial_v D_\varepsilon \subset \partial_v B(\varepsilon).$$

Now we check that $H(v, F_\varepsilon)$ is horizontal for sufficiently small ε. The map Δ_ε satisfies the (μ_h, μ_v) cone condition by Lemma 3.1. The surface $H(v, F'_\varepsilon)$ belongs to the plane $v = \text{const}$. Consequently, it is μ_h-horizontal for any positive μ_h. Hence, any tangent plane to $H(v, F_\varepsilon)$ belongs to some cone of the family K^-. Consequently, the projection $\pi\colon (Y, v) \mapsto Y$ is a local diffeomorphism on $H(v, F_\varepsilon)$. Therefore, the map of spheres

$$\pi \circ F_\varepsilon\colon (\partial K(\varepsilon) \times \{v\}) \to \partial K(\varepsilon)$$

is regular. Hence, it is a diffeomorphism. A local diffeomorphism of balls diffeomorphic on their boundaries is a global diffeomorphism. Hence, $H(v, F_\varepsilon)$ is a μ_h-horizontal surface for sufficiently small ε.

This proves Lemma 3.2.

Thus, all the conditions of the hyperbolic fixed point theorem for the map $\Delta_\varepsilon \colon D_\varepsilon \to D'_\varepsilon$ are checked. The application of this theorem provides a fixed point for the Poincaré map Δ_ε with s-dimensional stable manifold. Hence, the dimension of the stable manifold of the corresponding periodic orbit is $s+1$. Theorem 3.1 is proved.

CHAPTER 8

Elements of Hyperbolic Theory

In this chapter we will prove the theorems stated in §4 of Chapter 2 and §6 of Chapter 5 and used to study the homoclinic bifurcations in Chapters 4, 5, and 7.

We begin in §1 by reviewing some basic topics in hyperbolic theory such as hyperbolic sets, and the structurally stable and invariant manifold theorem for hyperbolic sets. The hyperbolic sets are generated by bifurcations of homoclinic orbits of singular points and cycles considered in Chapters 4, 6, and 7. The results of §1 describe the properties of dynamical systems with invariant sets of this kind. We also introduce the cone condition, which can be used to check the hyperbolicity of a map. In §2 we discuss some results in symbolic dynamics, which play a crucial role in the description of complicated dynamical behavior.

In §3 we give a proof of the hyperbolic fixed point theorem stated in 4.3 of Chapter 2. In §4, using the techniques of the previous section, we prove the Smale horseshoe existence theorem of 4.4 in Chapter 2. In §5 we will prove the generalized Smale horseshoe existence theorem stated in §6 of Chapter 5.

§1. Hyperbolic sets and their properties

1.1. Definition of hyperbolic invariant sets of a map. In Chapter 1 we gave the definition of a hyperbolic singular point and cycle. Now we generalize the concept of hyperbolicity to more complicated invariant sets.

Let us begin with the definition of a hyperbolic invariant set of a map.

DEFINITION 1.1. Let $f\colon \mathbb{R}^n \to \mathbb{R}^n$ be a diffeomorphism and let $\Lambda \subset \mathbb{R}^n$ be a closed set invariant under the action of f, i.e., $f\Lambda = \Lambda$. We say that Λ is *hyperbolic for f* if there exists a continuous direct sum decomposition $T_\Lambda\mathbb{R}^n = E^s_\Lambda \oplus E^u_\Lambda$ with the following property:

There exist constants $C > 0$, $0 < \lambda < 1$ such that for any $p \in \Lambda$, $n \geqslant 0$,

1) $Df(p)E^s_p = E^s_{f(p)}$, $Df(p)E^u_p = E^u_{f(p)}$;
2) if $u \in E^s_p$, then we have $\|Df^n(p)u\| \leqslant C\lambda^n\|u\|$; if $v \in E^u_p$, then we have $\|Df^{-n}(p)v\| \leqslant C\lambda^n\|v\|$.

REMARK. The continuity of the splitting $\mathbb{R}^n = E^s_p \oplus E^u_p$ means that as p varies in Λ, one can find continuously varying bases for E^s_p and E^u_p. In general, the splitting may be nonsmooth with respect to the point p.

EXAMPLE 1. Let $f\colon \mathbb{R}^2 \to \mathbb{R}^2$ be the linear map $(x,y) \mapsto (x/2, 2y)$. Then $\Lambda = (0,0)$ is a hyperbolic invariant set of f. In fact, Λ is a hyperbolic fixed point of f.

EXAMPLE 2. Let $f\colon D \to \mathbb{R}^2$ be the horseshoe map as described in §3 of Chapter 2. Then the set $\Lambda = \bigcap_{n=-\infty}^{\infty} f^n(D)$ is a hyperbolic invariant set. In fact for any point $p \in \Lambda$ we have

$$E_p^s = \{(dx, dy) \in T_p\mathbb{R}^2 \mid dy = 0\}, \qquad E_p^u = \{(dx, dy) \in T_p\mathbb{R}^2 \mid dx = 0\}.$$

We can choose $C = 1$ and $\lambda = 1/5$ for the constants in Definition 1.1.

1.2. Stable and unstable manifolds of a hyperbolic set. Just as for a hyperbolic fixed point, we can also define stable and unstable manifolds for a hyperbolic invariant set. Let Λ be a hyperbolic invariant set of f and $p \in \Lambda$. Define the local stable and unstable manifolds of p as follows:

$$W_\varepsilon^s(p) = \{q \in U_\varepsilon(p) : |f^n(p) - f^n(q)| \to 0 \text{ as } n \to +\infty\},$$
$$W_\varepsilon^u(p) = \{q \in U_\varepsilon(p) : |f^n(p) - f^n(q)| \to 0 \text{ as } n \to -\infty\},$$

where $U_\varepsilon(p)$ is the ε-neighborhood of p.

We now state, without proof, two theorems about hyperbolic invariant sets. The first is stated for completeness, the second was used in §5.8.

THEOREM 1.1 (Invariant Manifold Theorem) [**HPS**]. *Let Λ be a hyperbolic invariant set of a C^r ($r \geqslant 1$) diffeomorphism f. Then for $\varepsilon > 0$ sufficiently small and for each point $p \in \Lambda$ the following holds:*

1. *For sufficiently small ε, $W_\varepsilon^s(p)$ and $W_\varepsilon^u(p)$ are C^r manifolds with tangent planes E_p^s and E_p^u at p respectively.*
2. *There are constants $C > 0$, $0 < \lambda < 1$ such that for any $n \geqslant 0$, if $q \in W_\varepsilon^s(p)$, then we have $|f^n(p) - f^n(q)| \leqslant C\lambda^n|p - q|$; if $q \in W_\varepsilon^u(p)$, then we have $|f^{-n}(p) - f^{-n}(q)| \leqslant C\lambda^n|p - q|$.*
3. *The following invariance conditions hold:*

$$f(W_\varepsilon^s(p)) \cap U_\varepsilon(f(p)) \subset W_\varepsilon^s(f(p)),$$
$$f^{-1}(W_\varepsilon^u(p)) \cap U_\varepsilon(f^{-1}(p)) \subset W_\varepsilon^u(f^{-1}(p)).$$

For any point $p \in \Lambda$, the global stable and unstable manifolds of p are defined as follows:

$$W^s(p) = \bigcup_{n=0}^{\infty} f^{-n}(W_\varepsilon^s(f^n(p))), \qquad W^u(p) = \bigcup_{n=0}^{\infty} f^n(W_\varepsilon^u(f^{-n}(p))).$$

1.3. The structural stability of a hyperbolic set.

THEOREM 1.2 (Structural Stability Theorem) [**Ni, An**]. *Let $\Lambda \subset \mathbb{R}^n$ be a hyperbolic invariant set for f. Then for any $\varepsilon > 0$, there exists a $\delta > 0$ such that for any g with C^1-distance between f and g smaller than δ, in the ε-neighborhood of Λ there exists a hyperbolic invariant set Λ_g of g and a homeomorphism $\Phi\colon \Lambda \to \Lambda_g$ with the following properties:*

1) $\|\Phi - \mathrm{id}\| < \varepsilon$;
2) *the following diagram is commutative:*

$$\begin{array}{ccc} \Lambda & \xrightarrow{\;f\;} & \Lambda \\ {\scriptstyle\Phi}\downarrow & & \downarrow{\scriptstyle\Phi} \\ \Lambda_g & \xrightarrow{\;g\;} & \Lambda_g \end{array}$$

1.4. Cone condition for a hyperbolic set. In practice, verifying the conditions of Definition 1.1 for an invariant set of a nonlinear map is quite difficult. Therefore we introduce below an equivalent definition of hyperbolicity, which is easier to check.

DEFINITION 1.2. Let $D \subset \mathbb{R}^n$ be a closed set, and $f\colon D \to \mathbb{R}^n$ a diffeomorphism. We say that f satisfies the *cone condition* on D if there exists a continuous direct sum splitting $T_p\mathbb{R}^n = T_p^- \oplus T_p^+$ on the region $D \cup f(D)$, two real-valued continuous functions $C^+ \geqslant C^- : D \cup f(D) \to \mathbb{R}_+$ and a constant $\lambda > 1$ such that the two families of cones

$$\begin{aligned} K_p^+ &= \{(\xi_p^-, \xi_p^+) \in T_p^- \oplus T_p^+ : |\xi_p^+| \geqslant C^+(p)|\xi_p^-|\}, \\ K_p^- &= \{(\xi_p^-, \xi_p^+) \in T_p^- \oplus T_p^+ : |\xi_p^+| \leqslant C^-(p)|\xi_p^-|\} \end{aligned}$$

possess the following properties:
1) $df(K_p^+) \subset K_{f(p)}^+$ for $p \in D$;
2) $df^{-1}(K_p^-) \subset K_{f^{-1}(p)}^-$ for $p \in f(D)$;
3) $|(df\xi_p)^+| \geqslant \lambda|\xi_p^+|$ for $p \in D$, $\xi_p \in K_p^+$;
4) $|(d(f^{-1})\xi_p)^-| \geqslant \lambda|\xi_p^-|$ for $p \in f(D)$, $\xi_p \in K_p^-$.

REMARK. In 4.2 of Chapter 2, we gave a special version of the cone condition defined above: the so-called (μ_h, μ_v) cone condition, i.e., in the definition above the splitting $T_p^\pm$ and the function $C(p)$ are independent of the point p. That version was sufficient for studying the bifurcations in the previous chapters and easier to check.

THEOREM 1.3 [**NP**]. *Let $\Lambda \subset \mathbb{R}^n$ be a closed set and let $f\colon \Lambda \to \Lambda$ be a diffeomorphism. Then Λ is a hyperbolic set for f if and only if there exists an integer $n > 0$ such that f^n satisfies the cone condition on Λ.*

§2. Introduction to symbolic dynamics

As we have seen in 3.7 and 3.8 of Chapter 2, symbolic dynamics plays a key role in the description of periodic, homoclinic, and heteroclinic orbits of the horseshoe map. In this section we will describe some general aspects of symbolic dynamics in more detail.

2.1. The space of symbolic sequences and its structure. Let

$$\begin{aligned} &S = \{1, \dots, N\}, \qquad N \geqslant 2, \\ &\omega = \dots \omega_{-n} \dots \omega_0 \dots \omega_n \dots, \qquad \omega_j \in S. \end{aligned} \tag{2.1}$$

DEFINITION 2.1. Denote by Σ^N the space of all the sequences ω with the metric

$$\rho(\omega, \omega') = \sum_{-\infty}^{\infty} \frac{|\omega_n - \omega_n'|}{2^{|n|}}.$$

For any $\omega \in \Sigma^N$, denote by $\omega(n)$ the finite subsequence

$$\omega(n) = \omega_{-n} \dots \omega_0 \dots \omega_n.$$

REMARK. It follows from Definition 2.1 that

$$\rho(\omega,\omega') < \frac{1}{2^n} \Longrightarrow \omega(n) = \omega'(n), \qquad \omega(n) = \omega'(n) \Longrightarrow \rho(\omega,\omega') < \frac{C}{2^n}$$

for some $C > 0$ depending on N only.

Before stating the next proposition about the structure of Σ^N, let us recall the following definitions. A set is called *perfect* if it is closed and has no isolated points. It is *totally disconnected* if the connected component of any point contains that point only. The classical example having both properties is the Cantor perfect set. Another example is given by the following

PROPOSITION 2.1. *The space* Σ^N *equipped with the metric* (2.2) *is* 1) *compact*; 2) *totally disconnected*; 3) *perfect.*

REMARK. The three properties of Σ^N stated in the proposition are just the definition of a general Cantor set.

We shall not use the above proposition but give the proof for completeness.

PROOF OF PROPOSITION 2.1. 1. Let K be an arbitrary infinite subset of Σ^N. Since the set of symbols S is finite, we see that there exists an infinite subset K_0 of K such that all the sequences from K_0 contain the same symbol ω_0.

Now we construct by induction a nested sequence of infinite subsets $K_n \subset K$:

$$K_n = \{\omega \in K_{n-1} \mid \omega(n) \text{ is the same for all } \omega \in K_n\}.$$

Such an infinite set exists for any n. Indeed, by induction, the set K_{n-1} is infinite. On the other hand, the set S^{2n+1} of subsequences $\omega(n)$, $\omega \in \Sigma^N$, is finite. Now take a sequence of sequences $\omega^n \in K$. Any finite segment from this sequence is stabilized, i.e., $\omega^m(n) = \omega^n(n)$ for $m > n$. The sequence ω defined by $\omega(n) = \omega^n(n)$ is the limit of ω^n.

2. For any two different points $\omega', \omega'' \in \Sigma^N$ with $\omega'(n) \neq \omega''(n)$, the neighborhoods

$$\rho(\omega',\omega) < \frac{1}{2^{n+1}}, \qquad \rho(\omega'',\omega) < \frac{1}{2^{n+1}}$$

do not intersect. On the other hand, each of these neighborhoods is open and closed in Σ^N. Hence, ω' and ω'' belong to different connected components of Σ^N.

3. Any compact set is closed.

2.2. The Bernoulli shift.

DEFINITION 2.2. The map

$$\sigma\colon \Sigma^N \to \Sigma^N, \quad \omega \mapsto \omega', \quad (\omega')_{n-1} = \omega_n$$

is called the *Bernoulli shift*.

REMARK. Definition 2.1 implies that the Bernoulli shift is a homeomorphism of Σ^N onto itself.

For any $\omega \in \Sigma^N$ denote by O_ω the orbit of ω under σ. The main result of this section is the following

PROPOSITION 2.2. *The Bernoulli shift* $\sigma\colon \Sigma^N \to \Sigma^N$ *has the following properties*:

1. *The set of periodic orbits is dense in* Σ^N.
2. *For any pair of periodic points* ω', ω'' *there exists a dense set of points* ω *whose orbit is heteroclinic to* ω' *and* ω'', *that is,*

$$\operatorname{dist}(\sigma^n\omega, O_{\omega'}) \to 0 \text{ as } n \to +\infty, \quad \operatorname{dist}(\sigma^n\omega, O_{\omega''}) \to 0 \text{ as } n \to -\infty.$$

3. *There exists a dense orbit of* σ.

PROOF. For any finite set a of symbols from S denote by (a) a bi-infinite periodic sequence with the subsequence a periodically repeated. By $(a)^+$ and $(a)^-$ denote the sequences infinite to the right, respectively, to the left, with the above periodicity property.

1. Take any point $\omega \in \Sigma^N$. Then the points $(\omega(n))$ are $(2n+1)$-periodic and tend to ω as $n \to \infty$.

2. Let $\omega' = (a)$, $\omega'' = (b)$, and let ω be an arbitrary point. Then the sequence

$$\omega_n = (b)^-\omega(n)(a)^+$$

tends to ω as $n \to \infty$. On the other hand, for any n the orbit of ω_n is heteroclinic to ω' and ω''.

3. Let us enumerate all the finite sequences of elements of S and put them in a row as one bi-infinite sequence ω^0. Then for any ω and any n there exists a k such that we have $(\sigma^k\omega^0)(n) = \omega(n)$. This implies the density of the orbit of ω^0 under the Bernoulli shift.

§3. Hyperbolic fixed point theorem

In this section we will prove the generalized version of the hyperbolic fixed point theorem stated in 4.2 of Chapter 2. In what follows we always assume that μ_h and μ_v are two positive constants with $\mu_h\mu_v < 1$, and n, m are two natural numbers.

3.1. Statement of the theorem. For the definition of the (μ_h, μ_v) cone condition, see Definition 4.2 of Chapter 2. Let us recall some of the contents of §2.4.

DEFINITION 2.4.1. Consider the splitting $\mathbb{R}^{n+m} = \mathbb{R}^n \oplus \mathbb{R}^m$. Let D_h and D_v be compact connected manifolds with boundary in $\mathbb{R}^n$ and $\mathbb{R}^m$, diffeomorphic to unit balls in these spaces respectively. A *standard rectangle* B is the Cartesian product of these domains, $B = D_h \times D_v$. Moreover,

$\partial_h B = D_h \times \partial D_v$ is called the *horizontal part* of the boundary of B,
$\partial_v B = \partial D_h \times D_v$ is called the *vertical part* of the boundary of B.

DEFINITION 2.4.3. Consider a standard rectangle $B = D_h \times D_v$ and two positive constants μ_h, μ_v with $\mu_h\mu_v < 1$. Call the graph H of a Lipschitz map $G \to D_v$ with the constant μ_h a μ_h-*horizontal surface.* Here G is a domain (with boundary) in D_h. The surface H is *horizontal* in B if $G = D_h$ in the previous definition. A μ_v-*vertical surface* (in B) is defined in the same way by taking D_v, μ_v instead of D_h, μ_h and vice versa.

REMARK. Recall that the Lipschitz map φ with constant L is defined by the inequality

$$|\varphi(x) - \varphi(y)| \leqslant L \operatorname{dist}(x, y)$$

for any x, y in the domain of f. If this domain is convex, then the distance $\operatorname{dist}(x, y)$ is the same as in the ambient Euclidean space. If the domain is only connected, then the distance is intrinsic, that is, induced by the Riemannian metric of the ambient Euclidean space. It is equal to the infimum of the length of curves that join x to y in the domain of f. In what follows we consider domains for which this minimum exists.

DEFINITION 3.1. Let B be a standard rectangle. Let H and V be the graphs of Lipschitz maps

$$h\colon D_h \to D_v, \qquad v\colon D_v \to D_h$$

with Lipschitz constants no greater than μ_h and μ_v respectively. Then H and V are called *horizontal* and *vertical surfaces in* B respectively (see Figure 8.2).

EXAMPLE. The graph of a C^1 map $f\colon D_h \to D_v$ is a μ_h-horizontal surface in $B = D_h \times D_v$ if and only if its tangent plane at each point belongs to the cone of the family $\{K^+\}$ at this point.

DEFINITION 2.4.4. A (μ_h, μ_v)-*rectangle* D in $\mathbb{R}^n \oplus \mathbb{R}^m$ is a domain of the form $D = F(B^n \times B^m)$, where B^n and B^m are the unit balls in $\mathbb{R}^n$ and $\mathbb{R}^m$ respectively, F is a homeomorphism $B^n \times B^m \to \mathbb{R}^n \oplus \mathbb{R}^m$ and the surfaces $F(B^n \times \{y\})$ and $F(\{x\} \times B^m)$ are μ_h-horizontal and μ_v-vertical respectively for any $y \in B^m$, $x \in B^n$. The *horizontal* and *vertical parts* of the boundary of D are defined as the images under F of the horizontal and vertical parts of the boundary of $B^n \times B^m$ respectively:

$$\partial_h D = F(B^n \times \partial B^m), \qquad \partial_v D = F(\partial B^n \times B^m).$$

DEFINITION 2.4.5. A (μ_h, μ_v)-rectangle is called *vertically cylindric* (*v-cylindric*) in a standard rectangle B if $D \subset B$ and $\partial_v D \subset \partial_v B$. It is called *horizontally cylindric* (*h-cylindric*) if the same holds for h and v interchanged, i.e., $D \subset B$ and $\partial_h D \subset \partial_h B$.

THEOREM 3.1 (Hyperbolic Fixed Point Theorem). *Let $B = D_h \times D_v$ be a standard rectangle. Let $D \subset B$ and $D' \subset B$ be v-cylindric and h-cylindric (μ_h, μ_v)-rectangles in B. Let $f\colon D \to D'$ be a map satisfying the (μ_h, μ_v) cone condition. Then f has a unique fixed point O in D*:

$$O = \bigcap_{i=-\infty}^{\infty} f^i D.$$

Here $f^i D$ is the image of the set of all those points in D where the iterate $f^i D$ is well defined. (See Figure 8.1.)

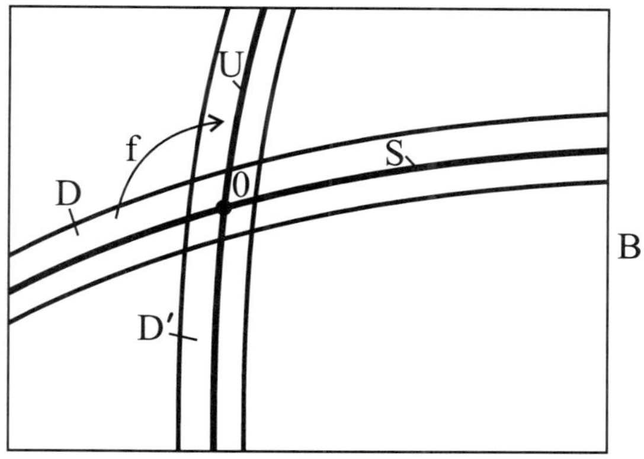

FIGURE 8.1. Source, target and fixed point in Theorem 3.1.

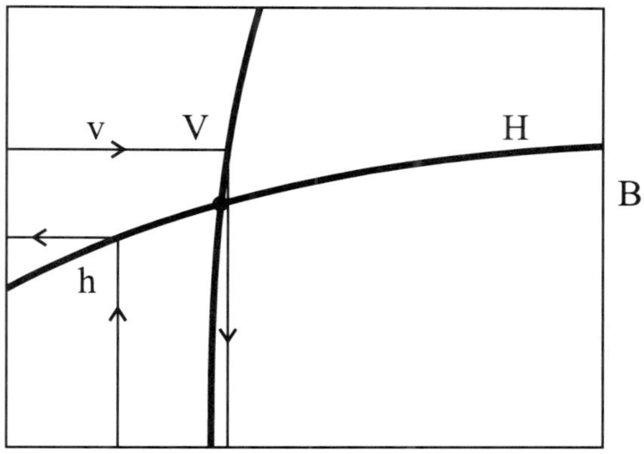

FIGURE 8.2. Intersection of horizontal and vertical surfaces.

3.2. Several intersection lemmas.

LEMMA 3.1. *Let $B = D_h \times D_v$. Suppose H and V are horizontal and vertical surfaces in B. Then H and V have exactly one intersection point.*

PROOF. A point

$$p = (p_h, p_v), \qquad p_h \in D_h,\ p_v \in D_v,$$

belongs to the intersection $H \cap V$ if and only if p_h and p_v are fixed points of the maps $v \circ h\colon D_h \to D_h$ and $h \circ v\colon D_v \to D_v$, respectively. Both maps are Lipschitz with Lipschitz constant $\mu_h\mu_v < 1$, hence contracting. The existence and uniqueness of a fixed point for any of these maps proves the existence and uniqueness of the intersection point $H \cap V$. (See Figure 8.2.)

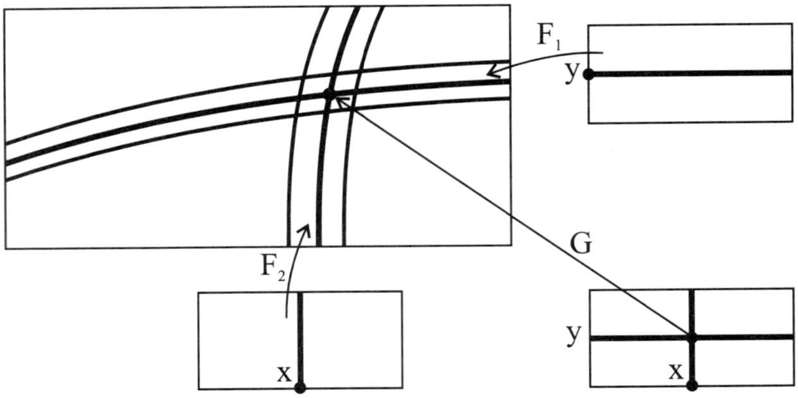

FIGURE 8.3. Constructing a new (μ_h, μ_v)-rectangle in Lemma 3.2.

LEMMA 3.2. *Let B be a standard rectangle. Let D_1 and D_2 be vertically and horizontally cylindric (μ_h, μ_v)-rectangles in B respectively. Then the intersection $D_1 \cap D_2$ is a (μ_h, μ_v)-rectangle with*

$$\partial_h(D_1 \cap D_2) \subset \partial_h D_1, \qquad \partial_v(D_1 \cap D_2) \subset \partial_v D_2.$$

PROOF. Let $D_1 = F_1(B^n \times B^m)$, $D_2 = F_2(B^n \times B^m)$; see Definition 2.4.4. We will construct a homeomorphism

$$G\colon B^n \times B^m \to D_1 \cap D_2$$

with the properties required by this definition (see Figure 8.3). Let

$$H_y = F_1(B^m \times \{y\}), \qquad V_x = F_2(\{x\} \times B^n).$$

By Lemma 3.1, there exists a unique point $G(x, y) = V_x \cap H_y$. The map G thus defined is continuous and injective. Moreover, G is onto, because any point in D_1 belongs to some V_x, and any point in D_2 belongs to some H_y.

PROPOSITION 3.1. *Let V be a μ_v-vertical surface. Let f be a map satisfying the (μ_h, μ_v) cone condition in a neighborhood of V, with the image of V in a standard rectangle B, and $f(\partial V) \subset \partial_h B$. Then the surface $f(V)$ is vertical in B. The same is true for horizontal surfaces and f replaced by f^{-1}.*

PROOF. Let V be C^1 smooth. Then $f(V)$ is C^1 smooth and locally μ_v-Lipschitz by the cone condition. Let π be the projection $B \to B^m$ along B^n. Then the map $\pi\colon f(V) \to B^m$ is a local diffeomorphism. It sends the boundary to the boundary because $f(\partial V) \subset \partial_h B$. By definition, V is the graph of a map $\varphi\colon G \to \mathbb{R}^n$. The map $\pi \circ f\colon V \to D_h$ is a local homeomorphism of balls that sends the boundary to the boundary. Such a map is a global homeomorphism. Therefore,

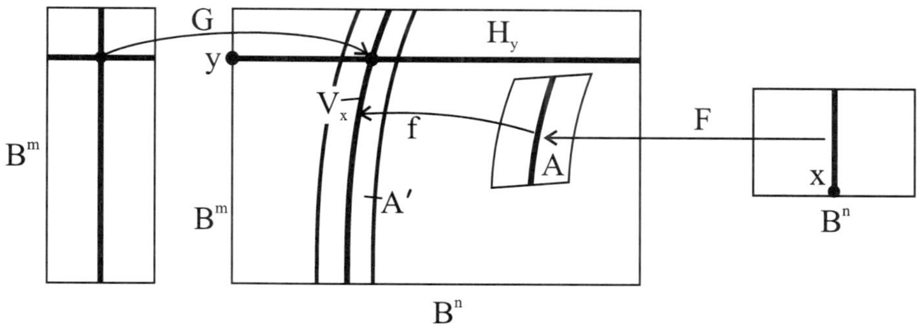

FIGURE 8.4. Constructing a new (μ_h, μ_v)-rectangle in Lemma 3.3.

the surface $f(V)$ is the graph of a μ_v-Lipschitz map $D_v \to D_h$, hence, it is vertical in B.

The case of a nonsmooth Lipschitz surface V may be reduced to the previous one by smoothing.

The second statement is proved in the same way as the first.

LEMMA 3.3. 1. *Let B be a standard rectangle, $A \subset B$ a (μ_h, μ_v)-rectangle, and $f\colon A \to B$ a map satisfying the (μ_h, μ_v) cone condition and the inclusion $f(\partial_h A) \subset \partial_h B$. Then $A' = f(A)$ is a (μ_h, μ_v)-rectangle horizontally cylindric in B.*

2. *Let B, A, and f be the same as before with f^{-1} well defined in A and such that $f^{-1}(\partial_v A) \subset \partial_v B$. Then $A'' = f^{-1}(A)$ is a (μ_h, μ_v)-rectangle vertically cylindric in B.*

PROOF. We shall prove the first statement only. The second is proved in the same way, word for word.

The proof is similar to the previous one. Let $A = F(B^n \times B^m)$ (see Definition 2.4.4). We will construct a homeomorphism $G\colon B^n \times B^m \to f(A)$ with the properties required by Definition 2.4.4 (see Figure 8.4).

Let

$$V_x = f \circ F(\{x\} \times B^m), \qquad H_y = B^n \times \{y\}.$$

Then, by Proposition 3.1, V_x is a vertical surface. Obviously, H_y is a horizontal surface. By Lemma 3.1, there exists a unique point $G(x, y) = V_x \cap H_y$. The map G thus defined is continuous. Moreover, G is onto, because any point in $f(A)$ belongs to some V_x, and at the same time to some H_y. Hence G, being a continuous one-to-one map of compact sets, is a homeomorphism.

3.3. Width of a (μ_h, μ_v)-rectangle. Below we measure distances between points on horizontal and vertical surfaces along *Lipschitz curves*. The latter are defined as Lipschitz maps $[0, 1] \to \mathbb{R}^N$. These maps are almost everywhere differentiable, and the length of a Lipschitz curve is given by the same formula as for a smooth curve.

DEFINITION 3.2. Let H be a horizontal surface. The intrinsic distance between two points a and b in H is the infimum of the lengths of Lipschitz curves that connect a and b and belong to H. The *diameter* of H is the supremum of the intrinsic distances between two points in H (it may be infinite for surfaces with a bad boundary). It is denoted by $\operatorname{diam}_h(H)$. The diameter of a vertical surface is defined in the same way.

DEFINITION 3.3. Let D be a (μ_h, μ_v)-rectangle. The *horizontal width of* D, $w_h(D)$, is the supremum of the diameters of horizontal surfaces that belong to D and whose boundary belongs to $\partial_v D$ (it may be infinite as well). The *vertical width of* V, $w_v(D)$, is defined in the same way replacing "horizontal" by "vertical".

REMARK. In what follows we shall deal with rectangles for which the vertical and horizontal width is well defined. The same holds for the vertical and horizontal surfaces and their diameters.

EXAMPLE. Consider a standard rectangle $B = D_h \times D_v$. The diameter of D_h, as well as of D_v, is finite, because D_h and D_v are diffeomorphic images of the ball. Consider a (μ_h, μ_v)-rectangle A vertically cylindric in B. Any horizontal surface H that belongs to A with $\partial H \subset \partial_v A$ has the same property with respect to B. Then

$$w_h(A) \leqslant w_h(B). \tag{3.1}$$

A similar estimate holds for any (μ_h, μ_v)-rectangle A' horizontally cylindric in B:

$$w_v(A') \leqslant w_v(B). \tag{3.2}$$

On the other hand, the diameter of any horizontal surface in B is no greater than $(1 + \mu_h) \operatorname{diam} D_h$. Hence

$$w_h(B) \leqslant (1 + \mu_h) \operatorname{diam} D_h, \qquad w_v(B) \leqslant (1 + \mu_v) \operatorname{diam} D_v.$$

LEMMA 3.4 (Width Lemma). 1. *Let B, A, f, and A' be the same as in Lemma 3.3. Then*

$$w_v(A) \leqslant \frac{(1 + \mu_v)}{\lambda} w_v(B).$$

2. *Let B, A, f, and A'' be the same as in Lemma 3.3. Then*

$$w_h(A) \leqslant \frac{(1 + \mu_h)}{\lambda} w_h(B).$$

PROOF. Once more, we shall prove the first statement only. The second is proved in the same way.

Let V be an arbitrary vertical surface in A with $\partial V \subset \partial_h A$. Let $V' = f(V)$. Then $\partial V' \subset \partial_h B$, by assumption. Hence, by Proposition 3.1, V' is vertical in B (see Figure 8.5).

PROPOSITION 3.2. *Let γ' be an arbitrary Lipschitz curve in V', $\gamma = f^{-1}(\gamma')$. Then*

$$|\gamma'| \geqslant \frac{\lambda}{1 + \mu_v} |\gamma|.$$

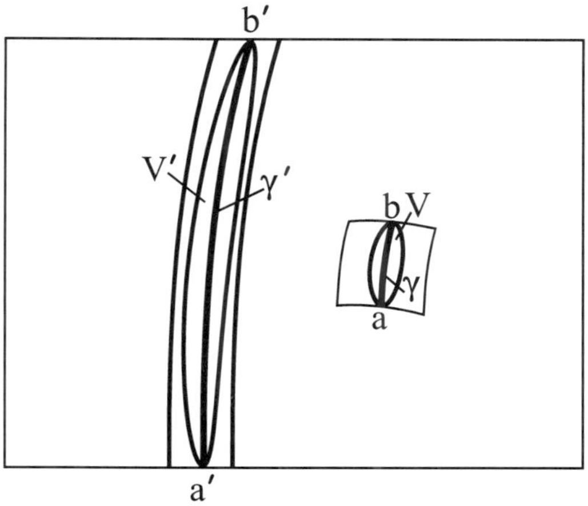

FIGURE 8.5. Horizontal surfaces and the width lemma.

PROOF. A Lipschitz map of a segment is almost everywhere differentiable. The length of a Lipschitz curve is given by the same formula as for a smooth curve. Let $\xi(t) = \dot{\gamma}(t)$, $\eta(t) = \dot{\gamma}'(t)$, $\xi = (\xi^-, \xi^+)$, $\eta = (\eta^-, \eta^+)$. The cone condition for f implies $|\eta^+(t)| \geqslant \lambda|\xi^+(t)|$. By the definition of the cone K_p^+, we have $|\xi(t)| \leqslant (1+\mu_v)|\xi^+(t)|$. On the other hand

$$|\gamma| = \int_0^1 |\xi(t)|\, dt, \qquad |\gamma'| = \int_0^1 |\eta(t)|\, dt.$$

Hence,

$$|\gamma'| \geqslant \int_0^1 |\eta^+(t)|\, dt \geqslant \lambda \int_0^1 |\xi^+(t)|\, dt \geqslant \frac{\lambda}{1+\mu_v}\, |\gamma|.$$

This proves Proposition 3.2.

Now Lemma 3.4 follows from Definitions 3.2 and 3.3. Proposition 3.2 implies

$$\operatorname{diam} V' \geqslant \frac{\lambda}{1+\mu_h} \operatorname{diam} V.$$

Therefore,

$$w_v(A) \leqslant \frac{1+\mu_v}{\lambda}\, w_v(A').$$

Together with (3.1), this implies the estimate in the lemma.

LEMMA 3.5. *Let B be a standard rectangle. Then a nested sequence of (μ_h, μ_v)-rectangles vertically cylindric in B with vertical width tending to zero has a non-empty intersection. This intersection is a horizontal surface in B. The same holds if "vertical" is changed to "horizontal" and vice versa.*

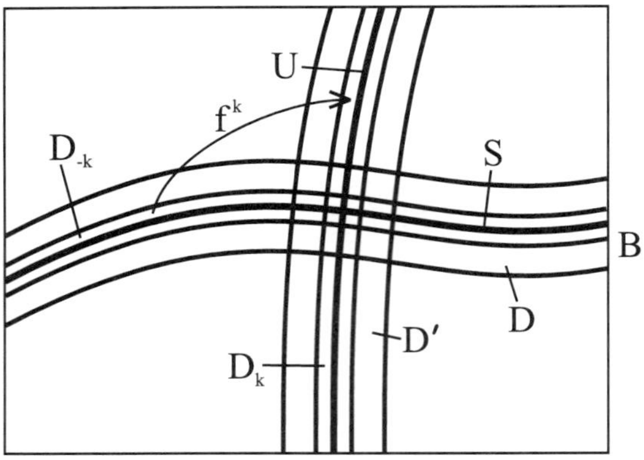

FIGURE 8.6. Proof of the hyperbolic fixed point theorem.

PROOF. Let D_k be the (μ_h, μ_v)-rectangles from the lemma (see Definition 3.1). Let $F_k\colon B^n \times B^m \to \mathbb{R}^n \oplus \mathbb{R}^m$ be the corresponding maps. Let $0 \in B^m$ and denote $V_k = F_k(B^n \times \{0\})$. Then, by Definition 2.4.4 (see 2.1), V_k is a μ_h-horizontal surface in B. By Definition 2.4.3, it is the graph of a Lipschitz map $\varphi_k\colon D_h \to D_v$ with Lipschitz constant no greater than μ_h. By assumption, the rectangles D_k are nested, and $w_v(D_k) \to 0$. Therefore, the maps φ_k form a Cauchy sequence in the C norm.

Indeed, the intersection of the disk $\{y\} \times D_v$ with D_k is a μ_v-vertical surface for any μ_v. The supremum of the diameters of these surfaces with respect to $x \in D_h$ is no greater than the vertical width of D_k. On the other hand, it is no smaller than the distance in C between the maps φ_i, φ_j with $i > k$, $j > k$.

Hence, there exists a limit φ of the sequence φ_k, which is, once more, a Lipschitz map. Its graph V is a μ_h-horizontal surface in B. It forms the desired intersection.

3.4. Proof of Theorem 3.1. Let us repeat the reasoning of 2.4.3, now in a rigorous way, making use of the above preparations. Let

$$\begin{aligned} D_1 &= D', & D_k &= f(D_{k-1} \cap D), & k &\geqslant 2,\\ D_0 &= D, & D_{-k} &= f^{-1}(D_{1-k} \cap D'), & k &\geqslant 1 \end{aligned}$$

(see Figure 8.6).

Let us prove by induction on k that the D_k's are (μ_h, μ_v)-rectangles horizontally cylindric in B for $k > 0$ and vertically cylindric in B for $k < 0$.

Base of induction, $k = 0$, $k = 1$: the domains $D_0 = D$, $D_1 = D'$ are the desired (μ_h, μ_v)-rectangles by assumption.

Induction step from $k - 1$ *to* k. By Lemma 3.2, and the induction assumption, $D_{k-1} \cap D$ is a (μ_h, μ_v)-rectangle. By Lemma 3.3, D_k is the desired (μ_h, μ_v)-rectangle.

The induction step from $1 - k$ to $-k$ is carried out in the same way.

Now note that D_{-k+1} is the set of all points for which the iterate f^k is well defined. The same is true for D_k and f^{-k}. This implies that $f^k(D_{-k+1}) = D_k$. Note that the maps f^k and f^{-k} satisfy the (μ_h, μ_v) cone condition. On the other hand, $\partial_h D_k \subset \partial_h B$, $\partial_v D_{-k} \subset \partial_v B$. Consequently, the domains $D_{-k+1} = A$ and $D_k = A'$ satisfy the assumptions of Lemma 3.4 with f, λ replaced by f^k, λ^k. Hence, by Lemma 3.4,

$$w_h(D_k) \leqslant \frac{1+\mu_h}{\lambda^k} w_h(B).$$

Therefore, the rectangles D_k satisfy the condition of Lemma 3.5. By this lemma, the surface $U = \bigcap D_k$ is vertical. The definition of U implies that $f(U \cap D) = U$. This gives meaning to the statement that U is f-invariant.

In the same way, the surface $S = \bigcap D_{-k}$ is f^{-1}-invariant. This surface is horizontal.

By Lemma 3.1, there exists a unique point $O = S \cap U$. This is the desired fixed point of the map f.

The existence of the fixed point is proved. Now let us prove its hyperbolicity.

Consider the linear operator $df(O)$. Since f satisfies the (μ_h, μ_v) cone condition, the operator $df(O)$ satisfies the same condition. It is easy to prove that the linear operator that satisfies the cone condition with any constants is hyperbolic.

The proof of the theorem is complete.

REMARK. The sets S and U defined above in the proof of Theorem 3.1 are in fact the stable and unstable manifolds of O respectively, and are therefore smooth.

§4. Sufficient conditions for the existence of a Smale horseshoe

In this section we will prove the Smale horseshoe existence theorem stated in 4.2 of Chapter 2.

4.1. Statement of the theorem. Recall the statement from 2.4.4.

THEOREM 4.1 (Smale Horseshoe Existence Theorem). *Let μ_h, μ_v with $\mu_h\mu_v < 1$ be two positive constants, and B a standard rectangle. Let*

$$D_i \subset \mathbb{R}^n \oplus \mathbb{R}^m, \qquad i = 1, \dots, N,\ N > 1,$$

be N pairwise disjoint (μ_h, μ_v)-rectangles. Let $f\colon D = \bigcup_{i=1}^N D_i \to f(D) \subset B$ be a diffeomorphism of the domain D onto its image satisfying the following assumptions for $i = 1, \dots, N$:

1) *the map f satisfies the (μ_h, μ_v) cone condition in D;*
2) *the domains D_i are vertically cylindric and $D_i' = f(D_i)$ are horizontally cylindric in B.*

Then the set

$$\Lambda = \bigcap_{-\infty}^{\infty} f^k D$$

is a hyperbolic invariant Cantor set. The restriction of f to Λ is topologically conjugated to the Bernoulli shift σ on the space Σ^N of sequences of elements taking N different values $1, \dots, N$. This means that there exists a homeomorphism Φ such that the diagram

$$\begin{array}{ccc} \Lambda & \xrightarrow{\ f\ } & \Lambda \\ {\scriptstyle\Phi}\big\downarrow & & \big\downarrow{\scriptstyle\Phi} \\ \Sigma^N & \xrightarrow{\ \sigma\ } & \Sigma^N \end{array} \tag{4.1}$$

is commutative.

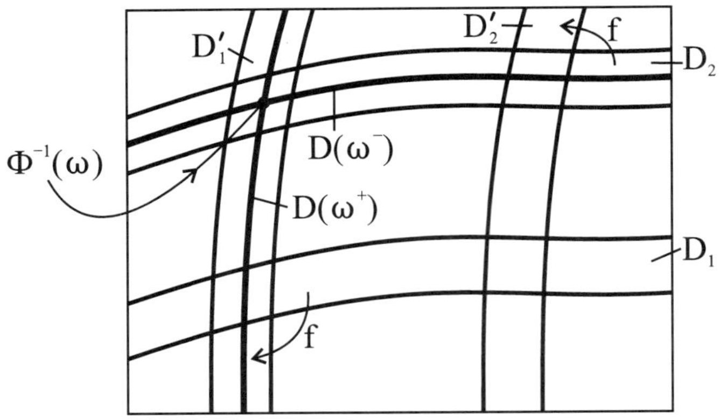

FIGURE 8.7. Smale horseshoe existence theorem.

REMARK. The hyperbolicity of Λ follows from the cone condition for f and Theorem 1.3.

COROLLARY. *The map f satisfying the assumptions of Theorem* 4.1 *has an invariant Cantor set Λ such that*

1) *the set Λ contains periodic points of arbitrary long periods and periodic points are dense in Λ;*
2) *the union of homoclinic and heteroclinic orbits of periodic points is dense in Λ;*
3) *the set Λ contains a dense orbit.*

This corollary follows immediately from the commutativity of diagram (4.1) and Proposition 2.2.

4.2. Beginning of the proof of the Smale horseshoe existence theorem. (See Figure 8.7.) We must find a homeomorphism $\Phi\colon \Lambda \to \Sigma^N$ such that the diagram (4.1) becomes commutative. This map is very natural. It is just the correspondence: point $\to$ fate. The following four statements prove the theorem.

1. Φ is well defined.
2. Φ is onto and injective.
3. Φ is continuous.
4. Diagram (4.1) with Φ defined above is commutative.

The statements 1, 3, 4 are simple. Namely

1. By definition, Λ is the set of all points for which the complete f-orbit exists. For such points the fate, which is an element of Σ^N, is well defined.

3. By continuity of f, a small change of a point preserves a long finite past and future fate unchanged. By the definition of the metrics in Σ^N, this implies a small change of the fate in Σ^N.

4. The fate of the image of a point is the fate of the point itself shifted by 1:

$$\omega_{n-1}(f(x)) = \omega_n(x).$$

Hence, the diagram (4.1) is commutative.

The only nontrivial statement is 2. It is equivalent to the following

LEMMA 4.1 (Main Lemma). *For any sequence from Σ^N there exists exactly one point having this sequence as a fate.*

This lemma will be proved following the same lines as for Theorem 3.1. The proof is given in the next two subsections. Once more, as in §2, the "linear" arguments will be replaced by "hyperbolic" ones.

4.3. Points with prescribed finite future and past fate. Suppose that for a point in the domain of the map f from Theorem 4.1, all the iterates of f are well defined. Recall that the fate of a point x is the sequence

$$\omega(x) = \{\omega_n(x)\}, \quad \omega_n(x) = j \quad \text{if and only if} \quad f^n(x) \in D_j. \tag{4.2}$$

Consider the finite sequence of symbols $\omega_j \in \{1, \dots, N\}$:

$$\omega_n^+ = \omega_0 \dots \omega_{n-1}.$$

It will be called the *future fate of length* n. Consider the set

$$D(\omega_n^+) = \{x \in D \mid \omega_j(x) = \omega_j,\, j = 0, \dots, n-1\}.$$

It consists of points with the given future fate of length n equal to ω_n^+. In the same way the past fate of length n: $\omega_n^- = \omega_{-n} \dots \omega_{-1}$ and the set $D(\omega_n^-)$ are defined. We will describe these sets in Lemma 4.2 below.

LEMMA 4.2. 1. *The set $D(\omega_n^+)$ is a vertically cylindric (μ_h, μ_v)-rectangle in B with exponentially small vertical width*:

$$w_v(D(\omega_n^+)) \leqslant \frac{1 + \mu_h}{\lambda^n} w_v(B). \tag{4.3}$$

Moreover,

$$\partial_h f^{n+1} D(\omega_n^+) \subset \partial_h B. \tag{4.4}$$

2. *The same is true for $D(\omega_n^-)$ with "vertical" replaced by "horizontal", and v and h exchanged.*

PROOF. Statement 2 is an immediate consequence of statement 1.

Statement 1 is proved by induction on n. The base of induction, the case $n = 1$, is elementary. Namely, $D(\omega_0^+) = D_j$ for $\omega_0 = j$. Moreover, $D_j = f^{-1} D_j'$. The sets D_j, D_j' satisfy the assumptions of Lemma 3.4 as A and A'. This proves (4.3) for $n = 1$. In this case (4.4) is evident.

Now we pass to the induction step (see Figure 8.8).

Suppose that Lemma 4.2 holds with n replaced by $n - 1$. Let

$$\omega_n^+ = \omega_{n-1}^+ j, \qquad j \in \{1, \dots, N\}.$$

Further, put

$$X_{n-1} = D(\omega_{n-1}^+), \qquad Y_{n-1} = f^n(X_{n-1}).$$

By the induction assumption and Lemma 3.3, it follows that X_{n-1} and Y_{n-1} are (μ_h, μ_v)-rectangles respectively vertically and horizontally cylindric in B (see Figure 8.8).

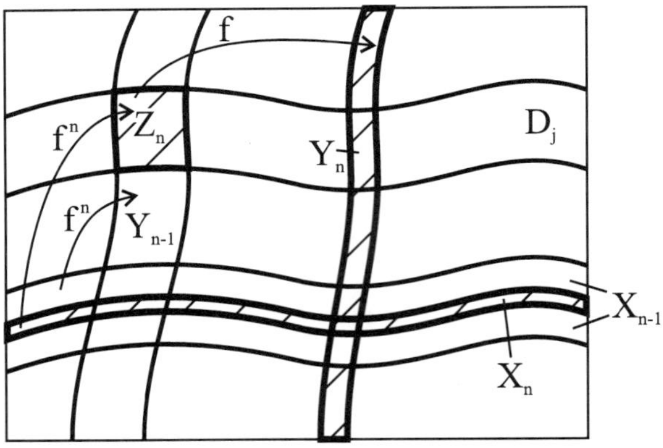

FIGURE 8.8. Points with prescribed finite future fate.

Let $Z_n = Y_{n-1} \cap D_j$. By Lemma 3.2, Z_n is a (μ_h, μ_v)-rectangle with $\partial_h Z_n \subset \partial_h D_j$, $\partial_v Z_n \subset \partial_v Y_{n-1}$. By definition

$$X_n = D(\omega_n^+) = \{x \in X_{n-1} \mid f^n(x) \in D_j\}.$$

Hence, $X_n = f^{-n} Z_n$. Note that

$$f^{-n}(\partial_v Z_n) \subset f^{-n}(\partial_v Y_{n-1}) \subset \partial_v X_{n-1} \subset \partial_v B.$$

By statement 2 of Lemma 3.3 with Z_n, f^{-n}, and X^n replacing A, f^{-1}, and A'', respectively, we conclude that X_n is a (μ_h, μ_v)-rectangle.

On the other hand, $f(\partial_h D_j) = \partial_h D_j' \subset \partial_h B$ (see Figure 8.8). Therefore,

$$\partial_h f(Z_n) = \partial_h f^{n+1}(X_n) \subset \partial_h B.$$

This proves (4.4). By statement 1 of Lemma 3.3, $f(Z_n)$ is a (μ_h, μ_v)-rectangle horizontally cylindric in B. Lemma 3.4 applied to f replaced by f^{n+1} and $A = X_n$, $A' = f(Z_n)$ implies (4.3) and proves Lemma 4.2.

4.4. Points with prescribed fate and proof of the Smale horseshoe existence theorem (Theorem 4.1). Given a sequence

$$\omega = \{\dots, \omega_{-1}, \omega_0, \omega_1, \dots\} \in \Sigma^N,$$

we define the following finite and semi-infinite subsequences:

$$\omega_n^+ = \{\omega_0, \omega_1, \dots, \omega_{n-1}\}, \qquad \omega_n^- = \{\omega_{-n}, \dots, \omega_{-1}\},$$
$$\omega^+ = \{\omega_0, \omega_1, \dots\}, \qquad \omega^- = \{\dots, \omega_{-2}, \omega_{-1}\}.$$

The sets

$$D(\omega^+) = \bigcap_0^\infty D(\omega_n^+), \qquad D(\omega^-) = \bigcap_1^\infty D(\omega_n^-)$$

consist of all points x with future fate ω^+ and with past fate ω^- respectively.

By Lemma 3.5, we see that $D(\omega^+)$ is a μ_h-horizontal surface with $\partial D(\omega^+) \subset \partial_v B$ (see Figure 8.7). In the same way, we can prove that $D(\omega^-)$ is a μ_v-vertical surface with $\partial D(\omega^-) \subset \partial_h B$. Now, by Lemma 3.1, $D(\omega^+)$ and $D(\omega^-)$ have a unique intersection point. This point has the future fate ω^+ and the past fate ω^-, and is uniquely determined by this property. Hence, its fate is ω. This proves Lemma 4.1, and therefore, Theorem 4.1.

§5. The generalized Smale horseshoe

In this section the most sophisticated result of hyperbolic theory that we need, Theorem 5.6.1, is proved. Below we repeat the statement of the theorem. The necessary definitions are contained in 5.6.1, 5.6.2.

5.1. Generalized Smale horseshoe existence theorem. As above, μ_h, μ_v are two positive constants with product smaller than 1. Below we recall the generalized (μ_h, μ_v) cone condition.

We split the tangent space at any point p of $\mathbb{R}^n = \mathbb{R}^s \oplus \mathbb{R}^c \oplus \mathbb{R}^u$ into the sum $T_p\mathbb{R}^s \oplus T_p\mathbb{R}^c \oplus T_p\mathbb{R}^u$. Tangent vectors will be denoted by ξ, η, and so on, and split to the components, i.e., $\xi = (\xi^s, \xi^c, \xi^u)$ according to the previous splitting of the tangent space. Denote for brevity

$$\xi^{cs} = (\xi^c, \xi^s), \quad \xi^{cu} = (\xi^c, \xi^u), \quad |\xi^{cs}|^2 = |\xi^c|^2 + |\xi^s|^2, \quad |\xi^{cu}|^2 = |\xi^c|^2 + |\xi^u|^2.$$

Now we define the (μ_h, μ_v) cones family as a tuple:

$$K_p^s = \{\xi \in T_p\mathbb{R}^n : |\xi^u| \leqslant \mu_h |\xi^{cs}|\}, \qquad K_p^u = \{\xi \in T_p\mathbb{R}^n : |\xi^s| \leqslant \mu_v |\xi^{cu}|\}.$$

Denote by U_p and S_p respectively the complement of K_p^s and K_p^u. For any map f and any point p from the domain of $f, \xi \in T_p\mathbb{R}^n$, denote by η the vector $df(p)\xi$. Recall that η is split to the components (η^s, η^c, η^u) according to the initial splitting of $\mathbb{R}^n$.

DEFINITION 5.6.1. Let $D \subset \mathbb{R}^s \oplus \mathbb{R}^c \oplus \mathbb{R}^u$ be a closed region, and suppose that $f \colon D \to \mathbb{R}^s \oplus \mathbb{R}^c \oplus \mathbb{R}^u$ is a diffeomorphism. We say that f satisfies the *generalized* (μ_h, μ_v) *cone condition* on D if there exists a constant $\lambda > 1$ such that for any $p \in D$

1) $df(K_p^u) \subset K_{f(p)}^u$;
2) $df^{-1}(K_{f(p)}^s) \subset K_p^s$;
3) for any $\xi \in U_p$, $|\eta^u| \geqslant \lambda |\xi^u|$;
4) for any $\eta \in S_{f(p)}$, $|\xi^s| \geqslant \lambda |\eta^s|$.

In the theorem below, the image $f^l D$ is the image of the set of all points in D at which the iterate f^l is well defined.

THEOREM 5.6.1 (Generalized Smale Horseshoe Existence Theorem). *Let the set $\mathcal{B} \subset \mathbb{R}^s \times T^c \times \mathbb{R}^u$ be a standard solid and $S = \{1, \dots, N\}$. Let D_i be vertically cylindric and D_i' be horizontally cylindric in $\mathcal{B}$, $i \in S$. Let*

$$\mathbf{D} = \bigcup_{i=1}^N D_i, \qquad \mathbf{D}' = \bigcup_{i=1}^N D_i',$$

and let $f\colon \mathbf{D} \to \mathbf{D}'$ *be a map with the following properties:*

1) f *satisfies the generalized* (μ_h, μ_v) *cone condition;*
2) $f(D_i) = D_i'$, $f(\partial_h D_i) = \partial_h D_i'$, $f^{-1}(\partial_v D_i') = \partial_v D_i$.

Then there exists a map Φ *of*

$$\Lambda = \bigcap_{l=-\infty}^{\infty} f^l(\mathbf{D})$$

onto Σ^N *such that the diagram*

$$\begin{array}{ccc} \Lambda & \xrightarrow{\ f\ } & \Lambda \\ {\scriptstyle\Phi}\downarrow & & \downarrow{\scriptstyle\Phi} \\ \Sigma^N & \xrightarrow{\ \sigma\ } & \Sigma^N \end{array} \tag{5.1}$$

is commutative.

Moreover, there exists a homeomorphism $\Lambda \to T^c \times \Sigma^N$. *The fiber of the map* Φ *is a Lipschitz torus, that is, the graph of a Lipschitz map* $T^c \to \mathbb{R}^s \times \mathbb{R}^u$.

5.2. Intersection of generalized vertical and horizontal surfaces.

LEMMA 5.1. *Let* $\mathcal{B}$ *be a standard solid, and* V *and* H *generalized* μ_v*-vertical and* μ_h*-horizontal surfaces in* $\mathcal{B}$ *respectively (see Definition 5.6.2). Then the intersection* $H \cap V$ *is the graph of a Lipschitz map* $T^c \to B^s \times B^u$ *with Lipschitz constant depending on* μ_h *and* μ_v *only.*

PROOF. For any $\varphi \in T^c$, the set $B_\varphi = B^s \times \{\varphi\} \times B^u$ is a standard rectangle (see Definition 2.4.1 in 3.1). The intersections $B_\varphi \cap H$ and $B_\varphi \cap V$ are μ_h-horizontal and μ_v-vertical in B_φ respectively in the sense of Definition 2.4.3. Denote by h_φ and v_φ the corresponding maps.

By Lemma 3.1, the surfaces $B_\varphi \cap H$ and $B_\varphi \cap V$ have a unique intersection point $\tau(\varphi) \in B_\varphi$. Therefore, the intersection $V \cap H$ is the graph of a map $\tau\colon T^c \to B^s \times B^u$.

Let us now prove that this map is Lipschitz with the constant mentioned in the lemma.

For this we recall the proof of Lemma 3.1.

The projections y and z on B^s and B^u of the point $\tau(\varphi)$ are the fixed points of the maps $f_\varphi = v_\varphi \circ h_\varphi$ and $g_\varphi = h_\varphi \circ v_\varphi$. These maps are contracting with coefficient $q = \mu_v \mu_h < 1$. For any $y \in B_s$ the Lipschitz constant L_y of the map $\varphi \mapsto f_\varphi(y)$ may be estimated uniformly in y. Indeed:

$$\begin{aligned} |f_{\varphi+\delta}(y) - f_\varphi(y)| &\leqslant |v_{\varphi+\delta} \circ h_{\varphi+\delta}(y) - v_\varphi \circ h_{\varphi+\delta}(y)| + |v_\varphi \circ h_{\varphi+\delta}(y) - v_\varphi \circ h_\varphi(y)| \\ &\leqslant (\mu_v + \mu_v \mu_h)|\delta|. \end{aligned}$$

Now let us recall a well-known fixed point property of a parameter-depending contracting map.

PROPOSITION 5.1. *Consider a family of* q*-contracting maps* $f_\varepsilon\colon M \to M$, *depending on* ε, *of a closed set* M *in a Euclidean space. Suppose* f *depends on* ε *in a Lipschitz way with constant* L. *Then* $\tau(\varepsilon)$, *the fixed point of* f_ε, *depends on* ε *in a Lipschitz way with constant* $L/(1-q)$.

PROOF OF PROPOSITION 5.1. It is well known that the distance from a fixed point a of a q-contracting map $f\colon M \to M$ to an arbitrary point b is estimated as

$$|a-b| \leqslant \frac{|b-f(b)|}{1-q}.$$

Let $a(\varepsilon)$ be the fixed point of f_ε. Then

$$|a(\varepsilon+h)-a(\varepsilon)| \leqslant \frac{|a(\varepsilon)-f_{\varepsilon+h}a(\varepsilon)|}{1-q}.$$

On the other hand,

$$|a(\varepsilon)-f_{\varepsilon+h}a(\varepsilon)| = |f_\varepsilon a(\varepsilon)-f_{\varepsilon+h}a(\varepsilon)| \leqslant Lh.$$

Therefore, $|a(\varepsilon+h)-a(\varepsilon)| \leqslant Lh/(1-q)$. This proves the proposition.

To complete the proof of Lemma 5.1, note that the fixed point $p(\varphi)$ of the map (f_φ, g_φ) depends on φ in a Lipschitz way with constant $L \leqslant (L_1^2+L_2^2)^{1/2}/(1-q)$. Here $L_1 = \mu_v + q$, $L_2 = \mu_h + q$ are Lipschitz constants in φ for f_φ, g_φ respectively.

5.3. Intersections of (μ_h, μ_v) annular solids. The material of this subsection is parallel to that of 3.2.

LEMMA 5.2. *Let $\mathcal{B}$ be a standard solid. Let D_1 and D_2 be vertically and horizontally cylindric (μ_h, μ_v) annular solids in $\mathcal{B}$ respectively. Then the intersection $D_1 \cap D_2$ is a (μ_h, μ_v) annular solid with*

$$\partial_h(D_1\cap D_2) \subset \partial_h D_1, \qquad \partial_v(D_1\cap D_2) \subset \partial_v D_2.$$

PROOF. Let $D_1 = F_1(\mathcal{B})$ and $D_2 = F_2(\mathcal{B})$ (see Definition 5.6.3). We will construct a homeomorphism $G\colon \mathcal{B} \to D_1 \cap D_2$ satisfying all the assumptions of this definition imposed on F. The construction is illustrated in Figure 8.3, but the angular variable is not shown in the figure. Let

$$H_z = F_1(B^s \times T^c \times \{z\}), \qquad V_y = F_2(\{y\} \times T^c \times B^u).$$

By Lemma 5.1, the intersection $H_z \cap V_y$ is the graph of a map $\tau_{yz}\colon T^c \to B^s \times B^u$. The map G we are looking for has the form $G(y,\varphi,z) = \tau_{yz}(\varphi)$. This is an injective map. Indeed, for different pairs (y,z), (y',z') either the surfaces V_y, $V_{y'}$, or H_z, $H_{z'}$ are pairwise disjoint. Hence, $\operatorname{Im}\tau_{yz} \cap \operatorname{Im}\tau_{y'z'} = \varnothing$. On the other hand, the map G preserves the φ coordinate. Hence, it is injective.

Moreover, it is onto, because any point of the intersection $D_1 \cap D_2$ belongs to some H_z and some V_y.

On the other hand, $G(y,\cdot,\cdot) \subset V_y$, $G(\cdot,\cdot,z) \subset H_z$ by construction.

Hence, all the requirements of Definition 5.6.3 are fulfilled.

The boundary requirement of the lemma is evident for G. This proves the lemma.

Before proving the counterpart of Lemma 3.3, let us study the images of the generalized horizontal and vertical surfaces under maps satisfying the cone condition. For brevity denote $M^{sc} = B^s \times T^c$, $M^{cu} = T^c \times B^u$.

PROPOSITION 5.2. 1. *Consider a map* $v\colon M^{cu} \to \mathcal{B}$ *such that* $V = v(M^{cu})$ *is a vertical surface. Let the diffeomorphism* f *be defined in some closed region* U *that contains* V. *Suppose that* f *satisfies the generalized* (μ_h, μ_v) *cone condition in* U, *and* $f(\partial V) \subset \partial_h \mathcal{B}$. *Moreover, let* $(f \circ v)_*\colon \pi_1(M^{cu}) \to \pi_1(\mathcal{B})$ *be an isomorphism. Then* $f(V)$ *is a vertical surface in* $\mathcal{B}$.

2. *Consider a map* $h\colon M^{sc} \to \mathcal{B}$ *such that* $H = h(M^{sc})$ *is a horizontal surface. Let the diffeomorphism* g *be defined in some closed region* U *that contains* H. *Suppose that the map* $f = g^{-1}\colon g(U) \to U$ *satisfies the generalized* (μ_h, μ_v) *cone condition in* $g(U)$, *and* $g(\partial H) \subset \partial_v \mathcal{B}$. *Moreover, let* $(g \circ h)_*\colon \pi_1(M^{sc}) \to \pi_1(\mathcal{B})$ *be an isomorphism. Then* $g(H)$ *is a horizontal surface in* $\mathcal{B}$.

PROOF. The second statement is proved word for word in the same way as the first. Only the vertical and horizontal surfaces and the maps f and g will change places. We will prove the first statement.

Suppose first that the surface V in the proposition is C^1 smooth. Then any tangent vector $\xi \in T_pV$ for any $p \in V$ satisfies the inequality $|\xi^s| \leqslant \mu_v |\xi^{cu}|$. Hence, $\xi \in K_p^u$ (see Figure 5.8). By assumption 1 of the generalized (μ_h, μ_v) cone condition, we have $\eta = df_p\xi \in K^u_{f(p)}$. Hence,

$$|\eta^s| \leqslant \mu_v |\eta^{cu}|.$$

This implies that the tangent plane $T_{f(p)}f(V)$ belongs to the cone $K^u_{f(p)}$. Consequently, the surface $f(V)$ is locally a graph of a Lipschitz map from M^{cu} to B^s.

Now let us prove that $f(V)$ is globally a graph. Consider the projection $\pi_{cu}\colon \mathcal{B} \to M^{cu}$ along B^s. The restriction $\pi = \pi_{cu}|_{f(V)}$ is a local diffeomorphism. It maps $f(\partial V)$ to $\partial(T^c \times B^u)$. We shall prove that the map $k = \pi \circ f \circ v\colon M^{cu} \to M^{cu}$ is a homeomorphism. Indeed, the map k is a local homeomorphism of a manifold with boundary to itself. It sends the boundary to the boundary. Moreover, $\mathcal{B}$ may be retracted to $M^{cu} \times \{0\}$ along B^s. Hence, the assumption about $(f \circ v)_*$ implies that $k_*\colon \pi_1(M^{cu}) \to \pi_1(M^{cu})$ is an isomorphism.

PROPOSITION 5.3. *Let* k *be a map of a manifold with boundary to itself. Suppose that* k *is locally homeomorphic, takes the boundary to the boundary and induces an isomorphism of the fundamental group of the manifold onto itself. Then* k *is a global homeomorphism of the manifold to itself.*

PROOF. Let us first prove that the map k is onto. Denote the manifold in the proposition by M. Let y be an arbitrary point in $\operatorname{int} M$. We want to prove that $y \in \operatorname{Im} k$. Take $x \in \operatorname{Im} k \cap \operatorname{int} M$, and a path γ that joins x and y in $\operatorname{int} M$. The lift $\widetilde{\gamma}$ of the path γ is well defined because k is a local homeomorphism that takes the boundary to itself. The endpoint of $\widetilde{\gamma}$ is the preimage of y. The map k is onto up to the boundary by continuity.

Let us now prove that k is one-to-one. Assume the contrary. Let $x, y \in \operatorname{int} M$ and $k(x) = k(y) = z$. Once more, let γ be a path that joins x to y in $\operatorname{int} M$. Then $k(\gamma) = \gamma'$ is a loop. Since k_* is an isomorphism of $\pi_1(M)$ to itself, there exists a class $[\gamma]$ of loops in M mapped into $[\gamma']$. Let $\Gamma \subset [\gamma]$, $\gamma'' \in [\gamma']$, $k(\Gamma) = \gamma''$. Then $k(\gamma) = \gamma'$, $k(\Gamma) = \gamma''$ and the loops γ' and γ'' are homotopic in $\operatorname{int} M$. Since the map $k\colon \operatorname{int} M \to \operatorname{int} M$ is a local homeomorphism onto, the homotopy from γ'' to γ' may be covered beginning with Γ. This contradicts the assumption that γ is not a closed curve. This proves Proposition 5.3.

Proposition 5.3 implies that the surface $f(V)$ is a graph of a Lipschitz map inverse to π. Hence, $f(V)$ is vertical in $\mathcal{B}$.

The case when V is only Lipschitz may be easily reduced to the previous one by smoothing. Namely, any Lipschitz map may be approximated uniformly on compact sets by a C^1 map with the same Lipschitz constant. This proves the proposition.

Now we can prove the counterpart of Lemma 3.3.

LEMMA 5.3. *Let $\mathcal{B}$ be a standard solid, $A \subset \mathcal{B}$ a (μ_h, μ_v) annular solid, and $F\colon \mathcal{B} \to A$ the corresponding map from Definition* 5.6.3.

1. *Let $f\colon A \to A' \subset \mathcal{B}$ be a map that satisfies the generalized (μ_h, μ_v) cone condition and $\partial_h f(A) \subset \partial_h \mathcal{B}$. Let the embedding $(f \circ F)_*\colon \pi_1(\mathcal{B}) \to \pi_1(\mathcal{B})$ be isomorphic. Then $A' = f(A)$ is a (μ_h, μ_v) annular solid.*
2. *Suppose that $g\colon A \to A'' \subset \mathcal{B}$ is a map such that $f^{-1}\colon A'' \to A$ satisfies the generalized (μ_h, μ_v) cone condition, $g(\partial_v A) \subset \partial_v \mathcal{B}$, and the embedding $(g \circ F)_*\colon \pi_1(\mathcal{B}) \to \pi_1(\mathcal{B})$ is an isomorphism. Then $A'' = gA$ is a (μ_h, μ_v) annular solid.*

PROOF. We shall prove only the first statement. The second is proved in the same way. The proof follows the same lines as that of Lemma 3.3. Let $A = F(\mathcal{B})$ (see Definition 5.6.3). We will construct a homeomorphism $G\colon \mathcal{B} \to f(A)$ with the properties from Definition 5.6.3. Let

$$V_y = f \circ F(\{y\} \times T^c \times B^u), \qquad H_z = B^s \times T^c \times \{z\}.$$

The assumptions of the lemma allow us to apply Proposition 5.2 to the maps $v = F(\{y\} \times M^{cu})$ and f. Hence, V_y is a generalized μ_v-vertical surface in $\mathcal{B}$. By Lemma 5.1, there exists a map $\tau_{yz}\colon T^c \to B^s \times B^u$ such that the graph of this map is the intersection $V_y \cap H_z$. Now let $G(\varphi, y, z) = \tau_{yz}(\varphi)$. The map G thus defined is continuous. Moreover, it is onto. Indeed, any point of any surface V_y belongs to some surface H_z because $V_y \subset \mathcal{B}$. Finally, a continuous one-to-one map of one compact set to another is homeomorphic.

5.4. Width of (μ_h, μ_v) annular solids and the width lemma. The width of (μ_h, μ_v) annular solids is defined in the same way as for (μ_h, μ_v)-rectangles.

DEFINITION 5.1. A *strongly μ_v-vertical* C^1 smooth surface in $\mathbb{R}^s \times T^c \times \mathbb{R}^u$ is the graph of a C^1 map of a domain in $\mathbb{R}^u$ to $T^c \times \mathbb{R}^s$ whose tangent planes T_p belong to the cones U_p. *Strongly μ_h-horizontal* C^1 smooth surfaces in $\mathbb{R}^s \times T^c \times \mathbb{R}^u$ are defined in the same way, only the cones U_p are replaced by S_p.

The next definitions repeat those of 3.3 word for word.

DEFINITION 5.2. The distance between two points on the same strongly vertical (horizontal) surface is the infimum of the lengths of curves that connect these points and belong to the surface.

DEFINITION 5.3. The *diameter* of a strongly vertical (horizontal) surface is the supremum of the distances between points of these surfaces.

DEFINITION 5.4. The *vertical width* w_v (*horizontal width* w_h) of a (μ_h, μ_v) annular solid is the supremum of diameters of strongly vertical (horizontal) surfaces that belong to this solid.

Now we can state the partially hyperbolic version of the width lemma (Lemma 3.4).

LEMMA 5.4. 1. *Let $\mathcal{B}$, A, and f be the same as in Lemma 5.3, and let $A' = f(A)$ be horizontally cylindric in $\mathcal{B}$. Then*

$$w_v(A) \leqslant \frac{(1+\mu_v)}{\lambda}\, w_v(\mathcal{B}).$$

2. *Let $\mathcal{B}$, A, and f be the same as in Lemma 5.3 with $f^{-1}\colon A \to \mathcal{B}$ well defined and $A'' = f^{-1}(A)$ vertically cylindric in $\mathcal{B}$. Then*

$$w_h(A) \leqslant \frac{(1+\mu_h)}{\lambda}\, w_h(\mathcal{B}).$$

PROOF. The proof is word for word the same as for Lemma 3.4, only Proposition 3.1 is replaced by Proposition 5.2.

LEMMA 5.5. *Let $\mathcal{B}$ be a standard solid. Then a nested sequence of vertically cylindric (μ_h, μ_v) annular solids with vertical width tending to zero has a nonempty intersection. This intersection is a generalized μ_h-horizontal surface in $\mathcal{B}$. The same holds if "vertical" is exchanged with "horizontal".*

PROOF. The proof is the same as for Lemma 3.5. Namely, let D_k be the (μ_h, μ_v) annular solids from the lemma (see Definition 5.6.3). Let $F_k\colon \mathcal{B} \to \mathcal{B}$ be the corresponding maps. Take the point $0 \in B^u$ and set $H_k = F_k(T^c \times B^s \times \{0\})$. Then, by Definition 5.6.3, H_k is a generalized μ_h-horizontal surface in $\mathcal{B}$. By Definition 5.6.2, it is the graph of a Lipschitz map $\varphi_k\colon T^c \times B^s \to B^u$ with Lipschitz constant no greater than μ_h. By assumption, the solids D_k are nested and $w_v(D_k) \to 0$. Therefore, the maps φ_k form a Cauchy sequence, and the limit φ is, once more, a Lipschitz map. Its graph V is a generalized μ_h-horizontal surface in B. It forms the desired intersection.

5.5. Symbolic dynamics on the invariant set. This subsection completes the proof of Theorem 5.6.1. The fate $\omega(x)$ is defined as in 4.3.

LEMMA 5.6 (Main Lemma). *Under the conditions of Theorem 5.6.1, for any sequence $\omega \in \Sigma^N$, the set of all points with fate ω constitutes a unique Lipschitz torus T^c_ω, the graph of a Lipschitz map $\tau_\omega\colon T^c \to B^s \times B^u$.*

PROOF. The proof follows the same lines as for Lemma 4.1. Consider a finite sequence of symbols $\omega_j \in \{1, \dots, N\}$:

$$\omega_n^+ = \omega_0 \dots \omega_{n-1}.$$

It will be called the *future fate of length* n. Consider the set

$$D(\omega_n^+) = \{x \in \mathbf{D} \mid \omega_j(x) = \omega_j,\ j = 0, \dots, n-1\}.$$

It consists of all points with future fate of length n equal to ω_n^+.

We will describe this set in the following

LEMMA 5.7. 1. *The set $D(\omega_n^+)$ is a vertically cylindric (μ_h, μ_v) annular solid in $\mathcal{B}$ with exponentially small vertical width:*

$$w_v(D(\omega_n^+)) \leqslant \frac{1+\mu_h}{\lambda^n} w_v(\mathcal{B}). \tag{5.2}$$

Moreover,

$$\partial_h f^{n+1} D(\omega_n^+) \subset \partial_h \mathcal{B}. \tag{5.3}$$

2. *The same is true for $D(\omega_n^-)$ with "vertical" replaced by "horizontal", and v and h exchanged.*

PROOF. The proof is almost the same as for Lemma 4.2. Let X_{n-1} and Y_{n-1} be defined as in 4.3:

$$\omega_n^+ = \omega_{n-1}^+ j, \quad X_{n-1} = D(\omega_{n-1}^+), \quad Y_{n-1} = f^n(X_{n-1}).$$

Suppose that X_{n-1} is a generalized (μ_h, μ_v) annular solid horizontally cylindric in $\mathcal{B}$ and Y_{n-1} is vertically cylindric in $\mathcal{B}$ (induction hypothesis). The domain $Z_n = Y_{n-1} \cap D_j$ is an annular solid and the corresponding parametrization preserves φ, by Lemma 5.2.

Let $F_j \colon \mathcal{B} \to D_j$ and $F_j' \colon \mathcal{B} \to D_j'$ be the parameterizations corresponding to D_j and D_j' according to the definition. The map $f_j = f|_{D_j} \colon D_j \to D_j'$ induces an isomorphism of fundamental groups, $(f_j \circ F_j)_* \colon \pi_1(\mathcal{B}) \to \pi_1(\mathcal{B})$. Indeed, the embedding F_j induces a similar isomorphism, because $\mathcal{B}$ may be retracted onto any fiber $M^{sc} \times \{z\}$, and $\pi_{sc} \circ F_j \colon M^{sc} \times \{z\} \to M^{sc}$ is a homeomorphism; here π_{sc} is the projection $\mathcal{B} \to M^{sc}$ along B^u. The same is true for F_j'. Finally, f_j is a diffeomorphism. Hence, f_j induces the isomorphism $\pi_1(D_j) \to \pi_1(D_j')$.

Now Lemma 5.3 may be applied to Z_n and f (statement 1) and to Z_n and f^{-n} (statement 2). The first application was motivated in the previous paragraph; the second is motivated in the same way. Let $X_n = f^{-n} Z_n$. Then $X_n = D(\omega_n^+)$. On the other hand, by Lemma 5.3, X_n is a generalized (μ_h, μ_v) annular solid vertically cylindric in $\mathcal{B}$. Moreover, $f(Z_n) = f^{n+1}(X_n)$ is a generalized (μ_h, μ_v) annular solid horizontally cylindric in $\mathcal{B}$. By Lemma 5.4 applied to $A = X_n$, $A' = f(Z_n)$ we infer that (5.2) holds. This proves Lemma 5.7.

Let us now complete the proof of the main lemma. The proof repeats the reasoning in 4.4. Only the references to Lemmas 3.6, 3.1 should be replaced by references to Lemmas 5.6, 5.1.

Finally, $D(\omega^+)$ and $D(\omega^-)$ are a generalized μ_h-horizontal surface and a generalized μ_h-horizontal surface, respectively. By Lemma 5.1, their intersection is a Lipschitz torus, namely, the graph of a map which we denote by τ_ω:

$$\tau_\omega \colon T^c \to B^s \times B^u.$$

This proves the main lemma (Lemma 5.6).

Together with it, the first statement of Theorem 5.1 is proved: the commutative diagram (5.1) is constructed.

Moreover, we have described the fibers of the map Φ, namely $\Phi^{-1}(\omega) = \operatorname{graph} \tau_\omega$.

The surfaces $D(\omega^+)$ and $D(\omega^-)$ depend continuously on ω^+ and ω^- in the metric

$$\rho(\omega^+, \alpha^+) = \sum_0^\infty \frac{\delta(\omega_j, \alpha_j)}{2^j}.$$

Here

$$\omega^+ = \omega_0, \dots, \omega_n, \dots, \qquad \alpha^+ = \alpha_0, \dots, \alpha_n, \dots, \qquad \delta(\omega_j, \alpha_j) = \begin{cases} 0 & \text{if } \omega_j = \alpha_j, \\ 1 & \text{if } \omega_j \neq \alpha_j. \end{cases}$$

The same goes for $\rho(\omega^-, \alpha^-)$. This follows from Lemma 5.6 and Proposition 5.3.

Now define the map

$$T^c \times \Sigma^N \to \Lambda, \qquad (\varphi, \omega) \mapsto \tau_\omega(\varphi).$$

The inverse map is

$$\Lambda \to T^c \times \Sigma^N, \qquad p \mapsto (\varphi(p), \omega(p)).$$

Hence, both maps are one-to-one, and the first is continuous by the above reasoning. Consequently, both maps are homeomorphisms, because Λ and $T^c \times \Sigma^N$ are compact sets.

This proves Theorem 5.6.1.

CHAPTER 9

Normal Forms for Local Families: Hyperbolic Case

The results of §5 in Chapter 2, which constitute the foundation of all the previous exposition, are proved in this and the next chapters. The method of proof allows us to adjoin numerous extensions to the quoted results. They are important for the study of nonlocal bifurcations in multiparameter families. We include these extensions, although they are not used in this book.

§1. Main results and their reduction to the Belitskii–Samovol theorem

In this section the main results of Chapters 9 and 10 are stated and some reductions are discussed.

1.1. General description. Nice finitely smooth normal forms for local families of vector fields and maps may be obtained for unfoldings of generic and slightly degenerate germs. In the case of more complicated degeneracies, functional moduli occur. For higher degeneracies even topological and formal normal forms are not obtained. The generic germs here are *hyperbolic nonresonant*; their normal forms are linear (see Theorems 1.1 and 1.2 below). Degeneracies of codimension one constitute the following list.

1. The fixed point is hyperbolic resonant; all the resonance relations are consequences of a single one. Both cases of vector fields and maps are considered.
2. One eigenvalue of the linearization of the vector field at its singular point is zero, while the others form a nonresonant tuple.
3. One multiplier of the map at the fixed point is equal to ± 1, while the others form a multiplicatively nonresonant and hyperbolic tuple.
4. A pair of eigenvalues of the singular point of the vector field is purely imaginary, while the others form a nonresonant tuple.
5. A pair of multipliers of the fixed point of the map lies on the unit circle, while the others form a multiplicatively nonresonant and hyperbolic tuple.

Unfoldings of germs in items 1 and 2 have polynomial integrable normal forms, or are linear suspensions over integrable systems. Explicit formulas are given below. The C^1 classification in cases 3 and 4 has functional moduli. On the contrary, case 5 seems to have neither reasonable topological, nor formal normal forms.

1.2. Deformations of hyperbolic germs. Now we pass to formal statements. Theorems 1.1 and 1.2 are just the same as 2.5.1 and 2.5.2 in §5 of Chapter 2.

THEOREM 1.1. *If the eigenvalues of the germ of a vector field at a fixed point form a nonresonant tuple, then the local family obtained by perturbation of this germ is finitely smooth equivalent to a linear one*: $\dot{x} = A(\varepsilon)x$.

THEOREM 1.2. *If the eigenvalues of the linearization of the germ of a diffeomorphism at a fixed point form a multiplicatively nonresonant tuple, then the local family obtained by perturbation of this germ is finitely smoothly equivalent to a linear one:* $x \mapsto A(\varepsilon)x$.

The statement of Theorem 1.3 requires the following definition.

DEFINITION 1.1. A tuple λ of n complex numbers is called *strongly one-resonant* if all the resonance relations for the numbers of the tuple are consequences of a single relation of the form

$$(r, \lambda) = 0, \qquad r \in \mathbb{Z}_+^n, \ |r| \geqslant 2,$$

where $\mathbb{Z}_+$ is the set of nonnegative integers. *A multiplicatively one-resonant tuple* is defined in the same way. All the multiplicative resonance relations in this case are generated by the single relation

$$\lambda^r = 1, \qquad r \in \mathbb{Z}_+^n, \ |r| \geqslant 2.$$

The monomial $u(x) = x^r$ is called *resonant* in both cases.

THEOREM 1.3. *Let $v(x, \varepsilon)$ be a local smooth family of vector fields, an unfolding of a germ $v(x, 0)$ having a hyperbolic one-resonant tuple of eigenvalues at the singular point zero. Then this family is finitely smoothly equivalent to a local family yielding the differential equation*

$$\dot{x} = Xg(u(x), \varepsilon), \qquad X = \operatorname{diag}(x_1, \dots, x_n), \ x \in (\mathbb{R}^n, 0), \tag{1.1}$$

where $g(u, \varepsilon)$ is a vector polynomial in the scalar variable $u \in \mathbb{R}^1$ with coefficients smoothly depending on ε and $u(x) = x^r$ is the resonant monomial. In more detail, for any natural N there exists a polynomial $g_N(u, \varepsilon)$ such that the initial family is C^N smoothly equivalent to the normalized family (1.1) with $g = g_N$ in some neighborhood of zero depending on N in the space of phase variables and parameters.

Suppose, in addition, that the germ $v_0 = v(\,\cdot\,, 0)$ has no formal first integral. This means that there is no formal power series such that its formal derivative along the formal Taylor expansion of v_0 is identically zero. Then g_N in the previous theorem may be taken independently of N. In this case the normal form depends on the initial family only, and not on the rate of smoothness of the normalizing transformation. The set of germs that violate this assumption is of infinite codimension. Therefore the above restriction is very mild.

For families of diffeomorphisms a similar statement also holds.

THEOREM 1.4. *Any deformation of a hyperbolic strongly one-resonant germ of a diffeomorphism is finitely smoothly equivalent to a local family of the form*

$$x \mapsto Xg(u(x), \varepsilon), \qquad X = \operatorname{diag}(x_1, \dots, x_n), \ x \in (\mathbb{R}^n, 0), \ u(x) = x^r.$$

The vector polynomial g in the normal form above has the same properties as in Theorem 1.3.

1.3. Deformations of multidimensional saddlenodes. Let us recall that a *multidimensional saddlenode* is a germ of a vector field at a singular point having one zero eigenvalue of the linearization while the other eigenvalues form a hyperbolic tuple.

THEOREM 1.5. *Let a germ of a multidimensional saddlenode satisfy the Łojasiewicz condition. This means that the modulus of the vector field decreases no faster than some power of the distance from the singular point. Moreover, suppose that the real parts of the nonzero eigenvalues form a nonresonant tuple.*

Then any smooth deformation of this germ is finitely smooth equivalent to a local family yielding the following system of differential equations:

$$\begin{aligned} \dot{x} &= \sum_{i=0}^{\mu-2} a_i(\varepsilon)x^i \pm x^\mu + a_{2\mu-1}(\varepsilon)x^{2\mu-1}, && x \in (\mathbb{R}^n, 0), \\ \dot{y} &= A(x,\varepsilon)y, && y \in (\mathbb{R}^{n-1}, 0),\ \varepsilon \in (\mathbb{R}^p, 0). \end{aligned}$$

Here $\mu \geqslant 2$ is an integer equal to the multiplicity of the singular point of the germ under deformation. The coefficients a_i are finitely smooth functions of the parameter ε, with $a_i(0) = 0$ for $0 \geqslant i \geqslant \mu - 2$, and the matrix $A(x,\varepsilon)$ is finitely smooth in x and ε.

The following statement gives the principal tool for the study of bifurcations of homoclinic orbits of saddlenodes carried out in Chapter 4. In fact, this is Theorem 2.5.3.

COROLLARY 1.1. *A generic one-parameter local family of vector fields with exactly one eigenvalue of linearization transversally passing through zero may be reduced by a finitely smooth change of coordinates and parameter to the form*

$$\begin{aligned} \dot{x} &= (x^2 + \varepsilon)(1 + a(\varepsilon)x)^{-1}, && x \in (\mathbb{R}^1, 0), \\ \dot{y} &= A(x,\varepsilon)y, \\ \dot{z} &= B(x,\varepsilon)z. \end{aligned}$$

The matrices $A(0,0)$ and $B(0,0)$ have eigenvalues in the open left and right half-planes respectively.

The next statement plays an important role in the study of nonlocal bifurcations in multiparameter families of *planar* vector fields. This study is not part of the main topic of this book, but we will prove the corollary as a free byproduct of our general study.

COROLLARY 1.2. *If in Theorem 1.5 the phase space is two-dimensional, then by a finitely smooth change of coordinates and multiplication by a finitely smooth function the family may be transformed to a polynomial one*:

$$\dot{x} = P(x,\varepsilon), \quad \dot{y} = \pm y, \qquad x, y \in \mathbb{R}^1,\ P(x,0) = x^\mu = \cdots,\ \varepsilon \in (\mathbb{R}^p, 0).$$

The variables in this family are split and it may be integrated in elementary functions.

Theorem 1.5, together with its corollaries and all the preliminary results, including the Takens theorem on smooth saddle suspensions, is proved in Chapter 10.

1.4. Belitskii–Samovol theorem. The theorem of the present subsection deals with local families as well as with single equations. A local family depending on the parameter ε may be transformed to a single equation by adding the extension $\dot{\varepsilon} = 0$. The equation thus obtained is called *family-like* and has an additional structure: the projection to the parameter space has the fibers $\varepsilon = \text{const}$ tangent to the vector field of the equation. The same is true for the family of maps.

THEOREM 1.6 (Belitskii–Samovol). *For any natural k and any tuple $\lambda \in \mathbb{C}^n$ there exists an integer $N = N(k, \lambda)$ such that the following holds. Suppose two germs of vector fields or diffeomorphisms at a fixed point with the spectrum of linearization equal to λ have a common center manifold, and their jets of order N coincide at all the points of this manifold. Then these germs are C^k equivalent.*

If, moreover, the equation or the map are family-like, then a conjugating coordinate change preserves the fibers $\varepsilon = \text{const}$.

Theorems 1.1–1.4 are direct consequences of the latter. Moreover, this theorem is one of the principal tools in the proof of the Takens theorem quoted above.

EXAMPLE 1: STERNBERG THEOREM.

THEOREM 1.7. *For any hyperbolic tuple $\lambda \in \mathbb{C}^n$ and any k there exists a number $N = N(k, \lambda)$ such that if the tuple λ satisfies the resonant relations of order greater than N only, then any germ of a smooth vector field with the tuple of eigenvalues λ is C^k equivalent to its linear part.*

A similar statement holds for germs of maps, provided that resonances are replaced by multiplicative resonances.

This is an immediate consequence of the previous theorem and formal normal form theory. In fact, it is well known since Poincaré that the germs of nonresonant vector fields and diffeomorphisms are formally equivalent to their linear parts. Hence, for any N, a polynomial coordinate change exists that transforms the initial field or map to a new one whose difference from its linear part is of order $o(|x|^N)$.

On the other hand, the fixed point in the theorem is hyperbolic. Hence, the center manifold of the germ under study is reduced to the fixed point only. Application of the Belitskii–Samovol theorem to this germ immediately yields the Sternberg theorem.

EXAMPLE 2: CHEN THEOREM.

THEOREM 1.8. *Suppose that two germs of smooth vector fields or maps at the hyperbolic fixed point are formally equivalent. Then they are C^k equivalent for any k.*

The proof is just the same as the previous one.

The reductions of Theorems 1.1–1.4 to the Belitskii–Samovol theorem are performed in the same way; they are similar, and we discuss only the first.

1.5. Normal forms for unfoldings of nonresonant germs. Here Theorems 2.5.1 and 2.5.2 are proved. They were called Theorems 1.1 and 1.2 at the beginning of the section.

PROOF OF THEOREM 1.1. The perturbation of the given germ is $\dot{x} = v(x, \varepsilon)$. The corresponding family-like equation is

$$\dot{x} = v(x, \varepsilon), \qquad \dot{\varepsilon} = 0.$$

The unperturbed germ is hyperbolic and hence nondegenerate. Therefore, singular points of the last equation form a germ of a smooth surface of the same dimension as the parameter space. This manifold is exactly the central one for the equation under consideration. A smooth coordinate change transforms it into the plane $x = 0$. The germ corresponding to the zero parameter value is nonresonant by assumption. Therefore, the base of the parameter may be chosen so small that the germs of the family will satisfy resonance relations of high order only. Hence, there exists a coordinate change, polynomial in x and smooth in ε, that brings the initial family to another one, whose difference with its linearization at the singular point has a high order of decreasing. The application of the Belitskii–Samovol theorem allows us to kill this discrepancy and to transform the family to its linear normal form.

The proof of Theorem 1.2 is completely the same as that of Theorem 1.1.

The remainder of the chapter contains the proof of the Belitskii–Samovol theorem. The proof is based on the homotopy method presented in the next section. The method is closely related to Frobenius theory.

§2. Introduction to Frobenius theory and the homotopy method

Any smooth line field has integral curves passing through any point of the phase space. Is a similar statement true if the lines are replaced by planes? In general, the answer is "no". Frobenius theory gives a criterion for the situation when the answer is "yes". This theory will be the subject of the present section. At the same time we will get a powerful tool, the homotopy method, that plays a crucial role in the proof of the Belitskii–Samovol theorem.

2.1. Integral surfaces of fields of planes.

DEFINITION 2.1. A field of l-dimensional planes, an *l-distribution*, is the correspondence that assigns to any point of some domain of Euclidean space an l-dimensional plane passing through this point.

EXAMPLE. Consider a tuple of l vector fields with an extra condition: at every point the vectors of these fields are linearly independent. We shall say that these vector fields are everywhere linearly independent. Then the planes spanned by these fields constitute a family of l-planes. By definition, this family is generated by these fields. The rate of smoothness of this family of planes is by definition that of the given vector fields.

DEFINITION 2.2. An *integral surface* of a field of planes is a manifold whose tangent plane at every point coincides with the plane of the field attached to this point. A field of planes is *integrable* if every point of the domain of the field belongs to some integral surface of the field.

In this section, except for the very end, the planes will be two-dimensional. In order to fix ideas, we begin with the following

WRONG THEOREM. *Any C^1 smooth field of planes generated by two everywhere linearly independent vector fields is integrable.*

PROOF. Denote the given vector fields by v and w. Let g_v^t and g_w^s the corresponding phase flows. Take an arbitrary point p in the phase space of these fields and a curve

$$\gamma = \{g_v^t p \mid t \in (\mathbb{R}, 0)\}. \tag{2.1}$$

A phase curve γ_q of the vector field w,

$$\gamma_q = \{g_w^s q \mid s \in (\mathbb{R}, 0)\}, \tag{2.2}$$

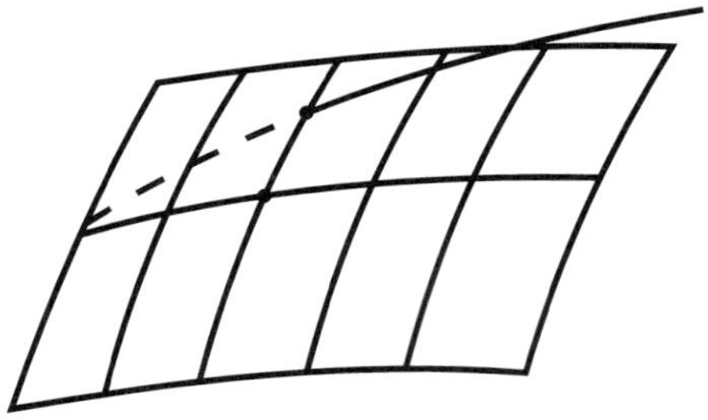

FIGURE 9.1. Uniqueness and occasional nonexistence of an integral surface.

passes through every point $q \in \gamma$ (see Figure 9.1). The curve γ_q depends smoothly on q. The required surface is

$$\Gamma = \bigcup_{q \in \gamma} \gamma_q . \tag{2.3}$$

This proof does not work for the following reason. The vector field V is in general not tangent to Γ outside the curve γ (see Figure 9.1). In fact, the following true statement lies in the background of the wrong proof above.

UNIQUENESS THEOREM. *If the field of planes is integrable, then every integral surface has the form* (2.3) *with* γ *and* γ_q *defined in* (2.1), (2.2).

PROOF. Since the field V is tangent to the integral surface, this surface contains γ. Since the field W is tangent to the integral surface, this surface contains γ_q.

Formula (2.3) may be written in the following form:

$$\Gamma_p = \{g_W^s \circ g_V^t p \mid (t, s) \in (\mathbb{R}^2, 0)\}. \tag{2.4}$$

PROPOSITION 2.1. *If two everywhere linearly independent smooth vector fields* V *and* W *have commuting phase flows, then they generate an integrable field of planes.*

PROOF. The surface Γ given by (2.4) contains, together with each of its points, the phase curves of the vector fields V and W passing through this point.

Now we pass to the study of vector fields with commuting phase flows. Such fields themselves are called *commuting.*

2.2. Commutator of vector fields and commuting phase flows. In this subsection a necessary and sufficient condition for vector fields to commute is given. By phase flows we mean local phase flows.

Let v, w be two smooth vector fields and g_v^t, g_w^t the corresponding phase flows. Define the family of vector fields

$$w_t = (g_v^t)_* w. \tag{2.5}$$

The following definition is the central one in this subsection.

DEFINITION 2.3. The *commutator* $[v, w]$ of vector fields v, w is

$$[v, w] = \frac{d}{dt} w_t \bigg|_{t=0}. \tag{2.6}$$

THEOREM 2.1. *The phase flows of two vector fields commute if and only if the commutator of these fields identically equals zero.*

PROOF. 1. Suppose the phase flows commute. This implies that the map g_v^t takes phase curves of the field w to phase curves of the same field preserving the time parametrization. This implies that the field of velocity vectors of these curves is also preserved. Hence

$$w_t \stackrel{t}{\equiv} w, \tag{2.7}$$
$$[v, w] \equiv 0. \tag{2.8}$$

2. Now suppose that (2.8) holds. We shall prove that this implies (2.7). If (2.7) holds, then the phase flow g_v^t preserves the field w and therefore commutes with the phase flow of this field.

In order to deduce (2.7) from (2.8), we must prove that

$$\frac{d}{dt} w_t \equiv 0.$$

Using the group property of phase flows, we have

$$\frac{d}{dt} w_t \bigg|_{t=a} = \frac{d}{dt} (g_v^t)_* w \bigg|_{t=a} = \frac{d}{dh} (g_v^{a+h})_* w \bigg|_{h=0} = \frac{d}{dh} (g_v^a)_* w_h \bigg|_{h=0}.$$

Now recall that the derivation of the parameter-depending vector field with respect to the parameter commutes with the transformation of this field by a diffeomorphism. In other words, the derivation with respect to the parameter is an invariant operation. Hence, for $t = a + h$

$$\frac{d}{dt} w_t = (g_v^a)_* \frac{d}{dh} w_h \bigg|_{h=0} = (g_v^a)_* [v, w] = 0.$$

This proves the theorem.

This theorem has two applications. The first is the integrability criterion for fields of planes. The second is the homotopy method, presented in the next subsection.

2.3. Homotopy method. The homotopy method provides a sufficient condition for two vector fields to be smoothly equivalent. Let v and w be two smooth vector fields defined in the same phase space $\Omega \subset \mathbb{R}^n$. Denote $R = w - v$, $v_s = v + sR$. Then $v_0 = v$, $v_1 = w$. Consider a vector field in $\Omega \times \mathbb{R}$

$$V(x, s) = (v_s(x), 0). \tag{2.9}$$

LEMMA 2.1. *Let the vector field V be the same as above. Suppose that there exists a vector field U such that*

$$U(x,s) = (h(x,s),1), \tag{2.10}$$
$$[U,V] \equiv 0. \tag{2.11}$$

Suppose that $\Omega_0 \subset \Omega$ is such that the phase flow transformations of the vector field U are well defined in $\Omega_0 \times \{0\}$ for all $t \in [0,1]$. Denote the image of the above domain under the time one transformation of this flow by $\Omega_1 \times \{1\}$. Then the fields v and w are smoothly equivalent to each other in the domains Ω_0, Ω_1 respectively.

PROOF. The vector field V determines a family-like equation

$$\dot{x} = v_s(x), \qquad \dot{s} = 0.$$

The planes $s = \text{const}$ are the invariant surfaces of this equation. The restrictions of the vector field V to the planes $s = 0$, $s = 1$ are equal to v and w respectively.

The vector field U shifts the planes $s = \text{const}$. The time one transformation of the phase flow of U maps the plane $s = 0$ to the plane $s = 1$.

The assumptions of the lemma, together with the previous theorem, imply that the phase flows of the fields U and V commute. Therefore the map g_U^1 conjugates the phase flow of V restricted to $\Omega_0 \times \{0\}$ with the phase flow of V restricted to $\Omega_1 \times \{1\}$. Hence the phase flows of v and w are smoothly conjugated in the corresponding domains. Therefore, the fields themselves are smoothly equivalent in the same domains.

Equation (2.11) for the fields (2.9) and (2.10) is the kernel of the homotopy method. It is known as the *homological equation.* To apply the homotopy method is to prove that the vector field U from Lemma 2.1 exists. This means that for V given by (2.9), equation (2.11) has a solution of the form (2.10). The investigation of the homological equation in each concrete problem is based on the properties of the given vector field V. This method will be applied to the proof of the Belitskii–Samovol theorem in §§3, 4.

2.4. Calculations: commutator and homological equation. We begin by finding the explicit expression for the commutator of two vector fields. It is sufficient to calculate the vector field w_t, see (2.5), up to terms of order higher than t. The transformation formula of a vector field by a diffeomorphism implies that

$$w_t = \left(\frac{\partial g_v^t}{\partial x} w\right) \circ g_v^{-t}.$$

If we introduce coordinates, then vector fields become vector-valued functions. Thus

$$g_v^t x = x + tv(x) + O(t^2), \qquad \frac{\partial g_v^t}{\partial x} = E + t\frac{\partial v}{\partial x} + O(t^2).$$

Here and below the dots stand for $O(t^2)$. Hence

$$w_t = \left[\left(E + t\frac{\partial v}{\partial x} + \cdots\right) w\right] \circ (x - tv(x)) + \cdots = w - t\frac{\partial w}{\partial x} v(x) + t\frac{\partial v}{\partial x} w(x) + \cdots.$$

Finally

$$[v,w] = \frac{\partial v}{\partial x} w - \frac{\partial w}{\partial x} v. \tag{2.12}$$

This is an important formula with numerous applications.

COROLLARY 2.1. *The commutator of two vector fields is linear with respect to both entries.*

COROLLARY 2.2. *The commutator of vector fields is skew symmetric.*

Both statements are obvious.

COROLLARY 2.3. *For the commutator of vector fields, the following Jacobi identity holds:*

$$[[u, v], w] + [[v, w], u] + [[w, u], v] \equiv 0.$$

This may be verified by a straightforward calculation using (2.12).

REMARK. Corollaries 2.1–2.3 imply that vector fields form a Lie algebra with the commutation operation given by formula (2.6) or, in coordinates, by (2.12). This Lie algebra structure will be used in Chapter 10 to obtain the so-called "embedding theorem in jets".

Formula (2.12) allows us to simplify the homological equation (2.11). Denote, as in (2.9) and (2.10), $V = (v_s, 0)$, $U = (h, 1)$; here v_s and h are the n-dimensional x-components, and 0 and 1 are one-dimensional s-components. Then

$$[V, U] = \left(\frac{\partial v_s}{\partial x} h - \frac{\partial h}{\partial x} v_s + \frac{\partial v_s}{\partial s} \right) \frac{\partial}{\partial x} .$$

Denote $H = (h, 0)$, $\bar{R} = (R, 0)$. Then the equation $[U, V] \equiv 0$ takes the form

$$[H, V] = \bar{R}. \tag{2.13}$$

This is the final form of the homological equation to be used below. The rest of the section will be devoted to the integrability problem and will not be applied in what follows.

2.5. Integrability criterion. Theorem 2.1 together with Proposition 2.1 implies that if the vector fields v and w are everywhere linearly independent and commute, then the field of planes spanned by these vector fields is integrable. The converse is not true, because two commuting vector fields after multiplication by functional coefficients cease to commute but span the same field of planes. The commutator of the new fields still belongs to their linear hull. We will prove that this example describes all integrable fields of planes.

DEFINITION 2.3. Two everywhere linearly independent vector fields are said to be *in involution* if their commutator is spanned by these fields.

THEOREM 2.2. *The family of planes generated by two everywhere linear independent vector fields is integrable if and only if these vector fields are in involution.*

PROOF. *Necessity.* Suppose that the field of planes is integrable. Let v and w be two vector fields spanning these planes. An integral surface passing through a point p has the form (2.4). By the theorem on the smooth dependence of a solution to a differential equation on the initial condition, the surfaces Γ_p are smooth and smoothly dependent on p. Therefore near every point p there exists a diffeomorphism rectifying the family of integral surfaces. This means that there exists a coordinate system

$$x = (x_1, \dots, x_n) = (x_1, x_2, x'), \qquad x' \in \mathbb{R}^{n-2},$$

such that the integral surfaces have the form $x' = c$, $c \in \mathbb{R}^{n-2}$. The vector fields v and w will be tangent to these surfaces. The same will be true for the field w_t and hence for $[v, w]$; see (2.5), (2.6). Therefore, $[v, w]$ is a linear combination of v and w with functional coefficients.

Sufficiency. Suppose that $[v, w]$ is spanned by v and w. We will prove that this implies the existence of commuting vector fields X and Y spanning the same field of planes. The situation is local: for every point of the phase space, the vector fields X and Y will be found in some neighborhood of this point. This will prove the integrability of the field of planes under consideration. Fix a point p and the splitting $\mathbb{R}^n = \mathbb{R}^2 \oplus \mathbb{R}^{n-2}$ with the projection $\pi \colon \mathbb{R}^n \to \mathbb{R}^2$. The splitting is chosen so that $d\pi(p)$ restricted to the plane of our field attached to p is an isomorphism. Choose a neighborhood G of p such that the same property holds for any $q \in G$. Choose two commuting vector fields e_1, e_2 in $\mathbb{R}^2$, for instance, $e_1 = \partial/\partial x_1$, $e_2 = \partial/\partial x_2$, where x_1, x_2 are the coordinates in $\mathbb{R}^2$. Lift these vector fields to the planes of our field. Denote the lifted fields by X and Y.

PROPOSITION 2.2. *The vector fields X and Y constructed above commute.*

PROOF. Extend e_1 and e_2 to $\mathbb{R}^n$ by the same formulas: $e_1 = \partial/\partial x_1$, $e_2 = \partial/\partial x_2$, and consider the two vector fields

$$X' = X - e_1, \qquad Y' = Y - e_2.$$

These fields are *vertical*, that is, tangent to the fibers of the projection π. Now

$$[X, Y] = [e_1 + X', e_2 + Y'] = [e_1, Y'] - [e_2, X'] + [X', Y'].$$

Suppose that this commutator is nonzero. Note that all the three terms on the right-hand side are vertical. Indeed, the transformation of a vertical vector field by the phase flow shift of either e_1 or e_2 is vertical itself, because both these flows commute with the projection π. Moreover, the commutator of two vertical vector fields is vertical. Hence if the right-hand side is nonzero, then it is transversal to the planes of the field. On the other hand, the vector fields X and Y, as well as v and w, are in involution. Hence their commutator belongs to the planes of our field. This contradiction proves the proposition, and yields the theorem.

2.6. Integrability criterion for fields of planes of arbitrary dimension. Suppose that a field of l-dimensional planes is defined by l everywhere linearly independent vector fields.

DEFINITION 2.4. A tuple of l everywhere linearly independent vector fields is said to be *in involution* if the commutator of any two of them is spanned by these vector fields.

THEOREM 2.3. *The field of planes defined by l everywhere linearly independent vector fields is integrable if and only if these vector fields are in involution.*

The proof is just the same as in the case $l = 2$.

§3. Belitskii–Samovol theorem for vector fields

In this section the theorem stated in 1.2 will be proved for vector fields.

3.1. Statement and outline of the proof. Recall the statement of the theorem.

THEOREM 3.1 (Belitskii–Samovol theorem for vector fields). *For any natural k and any tuple $\lambda \in \mathbb{C}^n$ there exists an integer $N = N(k, \lambda)$ such that the following holds. Suppose that two germs of C^∞ vector fields at a fixed point with the spectrum of the linearization equal to λ have a common center manifold, and their jets of order N coincide at all the points of the manifold. Then the germs are C^k equivalent. If, moreover, the equation is family-like, then a conjugating coordinate change preserves the leaves $\varepsilon = \text{const}$.*

The proof of the first statement will be split into four steps. First, we rectify the center manifold of the two fields, that is, transform it to a coordinate plane.

The second step is the splitting of the discrepancy, that is, the difference of the two given fields. This difference is small near the center manifold. We split it into the sum of two terms, the first small near the center unstable manifold, the second near the center stable one. Denote the given vector fields by v, w, the discrepancy by $R = v - w$, and the splitting by $R = R^+ + R^-$. We prove first that v and $v + R^+$ are C^k-equivalent, then that so are $v + R^+$ and w. This will imply the desired theorem. It is sufficient to study only one pair of fields; the equivalence for the other two is proved *verbatim*.

The third step is to globalize the vector fields. Each of them is replaced by a vector field coinciding with the initial one in some neighborhood of zero and with its linear part at zero outside some other compact neighborhood of zero. The higher terms of the globalized fields must be small. This allows us to obtain global phase flows for all the vector fields under consideration, and to approximate these flows by linear ones.

The fourth step is to prove that the pairs of globalized vector fields are C^k equivalent. The equivalence is proved by the homotopy method, and completes the proof of the first statement of Belitskii–Samovol theorem.

We shall not prove the second statement in detail. We just mention that every step of the proof below preserves the foliation $\varepsilon = \text{const}$ for a family-like equation.

3.2. Rectification. Let us apply the center manifold theorem to one of the vector fields, say v, mentioned in the Belitskii–Samovol theorem. Let

$$A = \frac{\partial v}{\partial x}(0)$$

and

$$\mathbb{R}^n = T^c \oplus T^s \oplus T^u \tag{3.1}$$

be the splitting corresponding to neutral, stable, and unstable parts of the spectrum of A (see 1.2.2). By Theorem 1.2.1, and the subsequent remark, the field v has five finitely smooth invariant manifolds: W_v^c, W_v^s, W_v^u, W_v^{cs}, W_v^{cu}. The rate of smoothness is arbitrarily high: the smaller the neighborhood of 0 on the invariant manifold, the smoother the corresponding piece of this manifold. Therefore, without loss of generality, one may assume that the common center manifold for v and w is T^c, and the vector fields v, w are C^M smooth for any prescribed M. Moreover, other invariant manifolds for v coincide with the coordinate planes

$$W_v^s = T^s, \quad W_v^u = T^u, \quad W_v^{cs} = T^c \oplus T^s, \quad W_v^{cu} = T^c \oplus T^u.$$

Denote by x, y, z the coordinates in T^c, T^s, T^u respectively, and let $\bar{x} = (x, y, z)$ be the coordinates in $\mathbb{R}^n$.

3.3. Splitting the discrepancy.

LEMMA 3.1. *Suppose that $R\colon \mathbb{R}^n \to \mathbb{R}^m$ is a smooth vector function. Let the space $\mathbb{R}^n$ be split as in* (3.1), *and*

$$j^{2N}_{\bar{x}} R = 0 \quad \text{for all } \bar{x} \in T^c. \tag{3.2}$$

Then there exists a splitting $R = R^+ + R^-$ such that

$$j^N_{\bar{x}} R^+ = 0 \quad \text{for all } \bar{x} \text{ such that } y(\bar{x}) = 0, \tag{3.3}$$

$$j^N_{\bar{x}} R^- = 0 \quad \text{for all } \bar{x} \text{ such that } z(\bar{x}) = 0. \tag{3.4}$$

PROOF. Decompose R in a Taylor series in the powers of z with coefficients depending on x and y and take a partial sum of this series:

$$\begin{gathered} R(\bar{x}) = \sum_{|\alpha| \leqslant N} a_\alpha(x, y)\, z^\alpha + R_N, \\ j^N_{\bar{x}} R_N = 0 \text{ for all } \bar{x} \text{ such that } z(\bar{x}) = 0. \end{gathered} \tag{3.5}$$

Denote $R^- = R_N$. Then (3.4) is fulfilled.

Now let

$$R^+ = \sum_{|\alpha| \leqslant N} a_\alpha(x, y)\, z^\alpha.$$

The assumption (3.2) implies that for $|\alpha| \leqslant N$ we have

$$j^N_{(x,0)} a_\alpha = 0 \quad \text{for all } x \in T^c.$$

Hence, (3.3) is fulfilled for R^+. Therefore, formula (3.5) provides the desired decomposition.

3.4. Globalization. Choose two balls centered at 0 of radius r and $2r$; the (small value) r will be specified later. Take a C^∞ function $\varphi\colon \mathbb{R}^+ \to \mathbb{R}$ such that $\varphi \equiv 1$ on $[0, 1]$ and $\varphi \equiv 0$ on $[2, \infty)$. Let $\varphi_r(\bar{x}) = \varphi(r^{-1}|\bar{x}|)$. Suppose that

$$v(\bar{x}) = A\bar{x} + f(\bar{x}), \qquad f = O(|\bar{x}|^2).$$

Define the globalized vector fields

$$v_r(\bar{x}) = A\bar{x} + \varphi_r f(\bar{x}), \quad R_r = \varphi_r R, \quad R_r^+ = \varphi_r R^+, \quad R_r^- = \varphi_r R^-, \quad w_r = v_r + R_r.$$

In what follows we omit the subscript r for simplicity.

PROPOSITION 3.1. *For any ε the radius r may be chosen so that*

$$\|d(\varphi_r f)\| < \varepsilon, \qquad \|d(\varphi_r R)\| < \varepsilon.$$

PROOF. Let us prove the proposition for f; for R the proof is similar.

Outside the ball of radius $2r$ centered at 0, the matrix on the left-hand side is zero. Inside this ball $|f|$ is of order r^2, $\|df\|$ of order r, and $d\varphi_r$ of order r^{-1}. Hence

$$\|d(\varphi_r f)\| = \|\varphi_r\, df + f\, d\varphi_r\|$$

is of order r.

THEOREM 3.2 (Generalized Version of the Belitskii–Samovol Theorem). *For any natural k and any tuple $\lambda \in \mathbb{C}^n$ there exist $N = N(k, \lambda)$ and $\varepsilon > 0$ such that the following holds. Suppose that two C^N smooth vector fields or diffeomorphisms in $\mathbb{R}^N$ have the fixed point 0 with the same linearization and with spectrum λ. Let* (3.1) *be the splitting of $\mathbb{R}^N$ related to this linearization, and $\mathbb{R}^+ = T^u \oplus T^c$. Suppose that the N-jets of these vector fields coincide at any point of $\mathbb{R}^+$. Moreover, let their discrepancy* (*the difference from the linear part*) *be smaller than ε in the C^1 norm and have compact support. Then these vector fields or diffeomorphisms are C^k equivalent in $\mathbb{R}^N$, and the conjugation map preserves* 0. *If, moreover, the vector fields or diffeomorphisms are family-like, then the conjugating coordinate change preserves the level surface of the parameter.*

The reduction of the local version of the Belitskii–Samovol theorem to the global one was carried out in the previous three subsections. The proof of the global version is given below by using the homotopy method. The case of vector fields is discussed in this section, and the case of maps in the next one.

3.5. Solution to the homological equation. Let v and w be the vector fields from the previous theorem. As in 2.3, let

$$R = w - v, \qquad v_s = v + sR, \quad s \in \mathbb{R},$$
$$V = (v_s, 0), \quad U = (h, 1), \quad H = (h, 0), \quad \bar{R} = (R, 0).$$

The vector fields U and H are to be found.

MAIN LEMMA. *Under the assumptions of Theorem* 3.2 *for the vector fields v and w, the homological equation*

$$[H, V] = \bar{R} \tag{3.6}$$

has a C^k solution H with the following properties: H is bounded in $\mathbb{R}^n \times [0, 1]$ and $H(0, s) \equiv 0$.

Theorem 3.2 for vector fields immediately follows from the main lemma. Indeed, consider the vector field $U = H + \partial/\partial s$ and the map $G = g_U^1$. This is the diffeomorphism $\mathbb{R}^n \times 0 \to \mathbb{R}^n \times 1$. It is well defined because H, hence U, is bounded. By Lemma 2.1, G conjugates the phase flows of v and w. On the other hand, G preserves zero, because $H(0, s) \equiv 0$.

PROOF OF THE MAIN LEMMA. The solution to the homological equation is given by a surprisingly simple formula.

PROPOSITION 3.2. *Let the integral*

$$H(\bar{x}) = -\int_0^\infty ((g_V^t)_* \bar{R})(\bar{x})\, dt \tag{3.7}$$

converge uniformly with respect to $\bar{x}$. Then H is a solution to the equation (3.6).

PROOF. Both the equation and formula (3.7) for its solution are in invariant form. To prove Proposition 3.2, it suffices to substitute (3.7) into (3.6) and obtain the identity. This will be done in the most convenient coordinates. For any point p take a flow box of the field V containing p and corresponding to infinite time $t \in [0, \infty)$. In this flow box, the vector field V may be rectified, i.e., transformed to a constant vector field e_1. Denote by G the rectifying diffeomorphism and let G_*V, G_*H, $G_*\bar{R}$ be the images of V, H, $\bar{R}$ under the map G. Then for any q in the flow box mentioned above,

$$G_*V = e_1, \qquad G_*H(q) = -\int_0^\infty G_*\bar{R}(q + te_1)\, dt.$$

Formula (2.12) for the commutator implies

$$[G_*V, G_*H] = L_{e_1} G_*H = -\int_0^\infty \frac{d}{dt} G_*\bar{R}(q + te_1)\, dt = G_*\bar{R}.$$

This proves the proposition.

3.6. Smoothness of the solution. In order to prove the C^k-smoothness of the solution H, we must study the properties of the phase flow of the vector field V.

PROPOSITION 3.3. *Let $F = g_V^1$, where V is defined via v and w as above, v and w satisfying the assumptions of Theorem 3.2. Then there exist μ and L depending on λ such that*

$$|y \circ g_V^t(\bar{x}, s)| < |y| \exp(-\mu t) \tag{3.8}$$

for any $\bar{x} \in \mathbb{R}^n$, $t \in \mathbb{R}^+$, $s \in [0,1]$,

$$\|dF\| \leqslant L, \qquad \|d(F^{-1})\| \leqslant L. \tag{3.9}$$

PROOF. Let $A = dv(0)$, and let $\mathbf{A}$ be the vector field $A\bar{x}$. Let 2μ be smaller than the moduli of real parts of all the eigenvalues of A restricted to T^s. Recall that all these real parts are negative. Then the derivative of $|y|$ along the vector field $\mathbf{A}$ may be estimated as $L_{\mathbf{A}}|y| \leqslant -2\mu|y|$. If ε in Theorem 3.2 is small enough, then $L_V|y| \leqslant -\mu|y|$. This implies (3.8).

To prove (3.9), choose a basis in $\mathbb{R}^n$ such that the matrix of A in this basis is close to diag λ. Chose the Euclidean structure in $\mathbb{R}^n$ so that this base is orthogonal. There exists a number L depending on λ only such that

$$\|e^A\| < L/2, \qquad \|e^{-A}\| < L/2.$$

If ε in Theorem 3.2 is small enough, then the vector fields $(\mathbf{A}, 0)$ and V are close to each other. Hence the previous inequalities imply (3.9).

Now we can prove the smoothness of the integral (3.7).

PROPOSITION 3.4. *Let V satisfy assumptions* (3.8), (3.9), *and $\bar{R}$ has the property*

$$j_{\bar{x}}^N \bar{R} = 0 \quad \textit{for all } \bar{x} \textit{ such that } y(\bar{x}) = 0, \tag{3.10}$$

where N is the same as in the main lemma. Then the integral (3.7) *is C^k smooth.*

PROOF. The proof is based on the following key arguments. The y coordinate decreases along the orbits of the vector field V with well estimated exponential rate (see (3.8)). Hence, the composition $\bar{R} \circ g_V^t$ decreases like the rapid exponent $\exp(-N\mu t)$, where N may be chosen arbitrarily large. On the other hand, the norm $\|g_V^{-t}\|$ grows like a well estimated exponent, too: for integer $t = l$, we have the relation $\|g_V^{-l}\| = \|F^{-l}\| \leqslant L^l$. Hence, the integrand in (3.7) decreases exponentially, and the integral converges.

The same is true for derivatives of moderate order. After every derivation of the integrand in (3.7), it still decreases, but not as fast as before, and finally stops decreasing. Hence, the solution (3.7) to the homological equation is only finitely smooth.

The formal proof of Proposition 3.4 uses reduction to discrete time. Let

$$\widetilde{R} = \int_0^1 (g_V^{-t})_* \bar{R}\, dt.$$

Then, by the group property of phase flows,

$$H = -\sum_{k=0}^{\infty} (F^{-k})_* \widetilde{R} = -\sum_{1}^{\infty} \frac{\partial}{\partial x}(F^{-k}) \widetilde{R} \circ F^k. \tag{3.11}$$

By the definition of $\widetilde{R}$, formula (3.10) holds with $\widetilde{R}$ replaced by R. Formula (3.8) implies

$$|y \circ F| \leqslant \exp(-\mu)|y|. \tag{3.12}$$

Thus the integral (3.7) is replaced by the discrete sum (3.11). It is convenient to claim the smoothness of this sum in a separate proposition.

PROPOSITION 3.5. *Let F satisfy assumptions* (3.12), (3.9), *and*

$$j_{\bar{x}}^N \widetilde{R} = 0 \quad \textit{for all } \bar{x} \textit{ such that } y(\bar{x}) = 0.$$

Then, for sufficiently large N depending on μ, L, k, the series (3.11) *converges to a C^k smooth vector field in the whole space.*

PROOF. We use the same key arguments as before. In more detail, for any multi-index α with $|\alpha| < N$ there exists an a such that

$$|D^\alpha F^l| \leqslant \exp al, \qquad l \geqslant 1. \tag{3.13}$$

Inequality (3.13) will be proved by induction on α by making use of the following partial ordering: $\alpha > \beta$ if $\alpha - \beta$ has nonnegative components only.

Base of induction. For $|\alpha| = 1$, (3.13) is a direct consequence of the chain rule. One may take $|\alpha| = \log L$, L from (3.9).

Induction step. By the chain rule,

$$|D^\alpha F^l| \leqslant |(D^\alpha F) \circ F^{l-1}| \|dF^{l-1}\|^\alpha + P(|D^\beta F^m| \mid \beta < \alpha,\, m < l), \tag{3.14}$$

where P is a polynomial in its variables.

The factor $|(D^\alpha F) \circ F^{l-1}|$ is bounded because F is linear and independent of s outside some compact set. The factor $\|dF^{l-1}\|^\alpha$ grows no faster than an exponent in l (see the base of induction). By the induction assumption, the same is true for the last term of (3.14). This proves (3.13).

Therefore, the derivatives of F^{-l} of order no greater than k grow exponentially with a well estimated power in the exponent. On the other hand, $\widetilde{R} \circ F^l$ decreases faster than $\exp(-N\mu l)$ with arbitrarily large N. More precisely, the number N depending on μ, L, and k may be chosen so that this decrease will be stronger than the increase of the derivatives of F^{-l} of order no greater than k. Hence, the series (3.12) converges in C^k. This proves the main lemma, and the Belitskii–Samovol theorem for vector fields along with it.

§4. Belitskii–Samovol theorem for maps

In this section Theorem 1.6 is proved for maps.

4.1. Statement and reductions. The theorem is formulated in 1.4 and repeated for vector fields in 3.1. For maps, the proof follows the same four steps as for vector fields: rectification of invariant manifolds; splitting the discrepancy; globalization; application of the homotopy method. The first three steps are the same for vector fields and maps. They reduce the theorem to the study of the following global situation. Suppose that two maps f and g, globally defined in $\mathbb{R}^n$, are C^M smoothly equivalent near 0 to the maps in the Belitskii–Samovol theorem. The value of M may be taken arbitrarily large. The maps f and g coincide with the same linear isomorphism A outside some ball. After a natural isomorphism between $\mathbb{R}^n$ and its tangent space at 0, the transformation A coincides with the common linear part of f and g at 0. The spectrum of A splits into three parts: neutral, stable and unstable, lying on, inside, and outside the unit circle respectively. The corresponding splitting of $\mathbb{R}^n$ is (3.1) with coordinates $x \in T^c$, $y \in T^s$, $z \in T^u$. Suppose that the N-jets of f and g at all the points of the center-unstable plane

$$T^c \oplus T^u = \{y = 0\}$$

coincide, i.e.,

$$f - g \equiv R = o(|y|^N), \qquad R \in C^N. \tag{4.1}$$

Suppose that all the orbits of all points of the maps f and g approach the center-unstable plane with well-estimated exponential rate. Namely, let

$$f_s = f + sR, \quad F(\bar{x}, s) = (f_s(\bar{x}), s), \quad p = (\bar{x}, s).$$

Then

$$y \circ F(p) \leqslant \nu|y(p)|, \qquad 0 < \nu < 1, \tag{4.2}$$

for some ν and all $p \in \mathbb{R}^n \times [0,1]$. Let

$$\|dF\| < L, \quad \|(dF)^{-1}\| < L, \quad L \leqslant 2(\|A\| + \|A^{-1}\|). \tag{4.3}$$

LEMMA 4.1. *Under the above assumptions, for any k, μ, L there exists a positive integer N such that the maps f and g are C^k conjugated in the whole space $\mathbb{R}^n$.*

This lemma implies the theorem. The lemma itself is proved by a kind of homotopy method modified for maps.

4.2. Homotopy method for maps. Let f and g be two diffeomorphisms of $\mathbb{R}^n$ onto itself. Take a homotopy joining f to g:

$$f_s = f + sR, \quad R = g - f, \quad s \in [0,1], \quad f_0 = f, \quad f_1 = g,$$

and suppose that f_s is a diffeomorphism of $\mathbb{R}^n$ onto itself. Let

$$F\colon [0,1] \times \mathbb{R}^n \to [0,1] \times \mathbb{R}^n, \qquad (s, \bar{x}) \to (s, f_s(\bar{x})).$$

The homotopy method is based on the following

LEMMA 4.2. *Let f, g, F be the same as above. Suppose that there exists a bounded C^k smooth field*

$$U = h\frac{\partial}{\partial x} + \frac{\partial}{\partial s}$$

(the notation is the same as in 2.4) such that F preserves U:

$$F_*U = U. \tag{4.4}$$

Then the maps f and g are C^k equivalent in the whole space $\mathbb{R}^n$.

PROOF. The phase flow of the vector field U is well defined in $\mathbb{R}^n$ for all values of time, because U is bounded. The phase flow transformation corresponding to time 1 takes the plane $s = 0$ to the plane $s = 1$. It commutes with the map F because this map preserves the vector field. Hence it conjugates the maps $f_0 = f$ and $f_1 = g$.

Lemma 4.2 reduces the proof of Lemma 4.1 to the solvability of equation (4.4). Let us prove this solvability. Denote $H = h\,\partial/\partial x = (h, 0)$, $\bar{R} = (0, R)$; here H, $\bar{R}$ are vector fields in $\bar{\mathbb{R}}^n \times [0, 1]$ tangent to the fibers $s = \text{const}$. In this notation, the homological equation has the following form:

$$H - F_*H = \bar{R}.$$

This is the so-called *transfer equation* to be studied in 10.4.5 in another context. The solution H to this equation is given by the formula

$$H = -\sum_1^\infty (F^{-k})_*\bar{R}. \tag{4.5}$$

This may be easily verified, provided that the series on the right-hand side converges:

$$F_*H = -\sum_0^\infty (F^{-k})_*\bar{R} = -\bar{R} - H.$$

But (4.5) is just the series (3.11) studied above. By Proposition 3.5, it converges in C^k. This proves Lemma 4.1, and hence, the Belitskii–Samovol theorem for maps.

CHAPTER 10

Normal Forms for Unfoldings of Saddlenodes

In this chapter we prove the theorem on the normal form for an unfolding of a saddlenode vector field, Theorem 5.3 of Chapter 2. Moreover, we prove the partial embedding theorem for unfoldings of saddlenode fixed points of maps, Theorem 2.5.4. The kernel of both proofs is the Takens theorem on smooth saddle suspensions, which is the subject of the next section.

§1. Takens theorem on the smooth saddle suspension

The well-known reduction principle of Shoshitaishvili, §1.1, reduces the study of the topology of a germ of a vector field to that of the restriction to its center manifold. The principal effect is that the influence of the hyperbolic variables is standard. They appear in the topological normal form as a linear factor of the Cartesian product of two systems. The other factor is the restriction of the initial germ to its center manifold. The Takens theorem gives a *finitely smooth* normal form for a germ of a vector field. This normal form turns out to be not Cartesian, but triangular. The restriction to the center manifold is still separated. The hyperbolic variables satisfy a linear system with operator depending on the point of the center manifold.

1.1. Statement of the theorem and plan of the proof. Let us state the first main result of this chapter.

THEOREM 1.1. *Let v be a smooth germ of a vector field at its singular point 0. Let the linearization of this field correspond to an operator A with invariant subspaces T^s, T^u, T^c and pairwise distinct eigenvalues. Let the restriction of A to these subspaces have its spectrum in the left half-plane, right half-plane, and on the imaginary axis respectively. Let A^+, A^-, and A^0 be the restrictions of A to these planes and λ^+, λ^-, μ be their spectra, $\lambda = (\lambda^+, \lambda^-)$. Let the real part of the vector λ be nonresonant*:

$$\operatorname{Re}\lambda_j \neq \operatorname{Re}(\lambda, k) \quad \textit{for any } k \in \mathbb{Z}_+^n,\ |k| \geqslant 2. \tag{1.1}$$

Then the germ v is finitely smoothly equivalent to

$$\dot{x} = w(x), \quad \dot{y} = A^+(x)y, \quad \dot{z} = A^-(x)z, \tag{1.2}$$

where $x \in T^c$, $y \in T^s$, $z \in T^u$, $A^+(0) = A^+$, $A^-(0) = A^-$.

Moreover, for any preassigned germ of the center manifold W^c of the vector field at 0, a coordinate change may be chosen so that this preassigned germ will become the germ of the x-plane at 0.

The theorem is proved by using the Belitskii–Samovol theorem. Namely, one rectifies the invariant manifolds of the initial germ, that is, transforms W^c, W^{cs},

W^{cu} to the planes T^c, $T^c \oplus T^s$, $T^c \oplus T^u$. After this the new germ is *normalized along the center manifold* in order to obtain a germ that differs from its normal form prescribed by the theorem up to a high power of the distance to the center manifold. This allows the application of the Belitskii–Samovol theorem and proves the desired equivalence. The normalization is carried out by means of the method of successive approximations. Each step requires the solution of some first order partial differential equation, which turns out to be the equation for the center manifold of some linear suspension over a nonlinear system. The nonresonance condition of the theorem allows us to describe this latter manifold.

1.2. Formal point of view. The Takens saddle suspension theorem is the smooth counterpart of a simple statement from the formal theory. Let us denote $T^h = T^s \oplus T^u$, where $u = (y, z)$ is a tuple of hyperbolic variables.

PROPOSITION 1.1. *A smooth germ of a vector field that satisfies the assumptions of Theorem* 1.1 *is formally equivalent to*

$$\hat{v}(x, u) = (\widehat{w}(x), \widehat{A}(x)u), \tag{1.3}$$

where $\widehat{w}(x)$ *and* $\widehat{A}$ *are formal series in* x *with coefficients in* T^c *and* $\operatorname{Hom}(T^h, T^h)$ (*the set of linear operators from* T^h *to itself*), *respectively.*

PROOF. Proposition 1.1 is a direct consequence of the Poincaré–Dulac theorem. This theorem claims that the formal normal form of a germ of a vector field consists of resonant terms only. Let us find all the resonant terms, that is, all resonant relations for the tuple (λ, μ). By definition, they have the form

$$\mu_j = (\lambda, k) + (\mu, l), \tag{1.4}$$
$$\lambda_j = (\lambda, k) + (\mu, l), \tag{1.5}$$

$(k, l) \in \mathbb{Z}_+^n$, $|(k, l)| \geqslant 2$. Taking real parts in (1.4), we get: $(\operatorname{Re}\lambda, k) = 0$. If such a relation holds for some $k \neq 0$, then $\operatorname{Re}\lambda$ is a resonant tuple. This contradicts assumption (1.1) of Theorem 1.1. Hence, (1.4) has the form $\mu_j = (\mu, l)$. In fact, a tuple of purely imaginary roots of real polynomial satisfies numerous resonant relations. The corresponding resonant monomials are $x^l \partial/\partial x_j$. Their formal sum gives the x-component of the normal form (1.3).

Similarly, taking real parts in (1.5), we get $\operatorname{Re}\lambda_j = \operatorname{Re}(\lambda, k)$. By (1.1), this is possible for $(0, \dots 0, 1_i, 0, \dots, 0)$, provided that $\operatorname{Re}\lambda_j = \operatorname{Re}\lambda_i$.

The corresponding resonant term has the form $u_i x^l \partial/\partial u_j$. The sum of these monomials is $\widehat{A}(x)u$, the u-component of the formal normal form (1.3).

REMARK. In fact,

$$\widehat{A}(x) = \begin{pmatrix} \widehat{A}^-(x) & 0 \\ 0 & \widehat{A}^+(x) \end{pmatrix},$$

where the block structure corresponds to the splitting $T^h = T^s \oplus T^u$. Indeed, the equality $\operatorname{Re}\lambda_j = \operatorname{Re}\lambda_i$ never holds for $\operatorname{Re}\lambda_i \leqslant 0$, $\operatorname{Re}\lambda_j \geqslant 0$; hence, the terms $y_i x^l \partial/\partial z_j$, $z_i x^l \partial/\partial y_j$ are never resonant. For any N, the initial germ in Theorem 1.1 can be reduced by a polynomial coordinate change to its formal normal form (1.2) modulo $o(r^N)$, where $r = |(x,u)|$; the new vector field is still denoted by v:

$$v(x,u) = (w(x), A(x)\,u) + o(r^N), \tag{1.6}$$

$$A(x) = \begin{pmatrix} A^-(x) & 0 \\ 0 & A^+(x) \end{pmatrix}. \tag{1.7}$$

For the germ thus obtained, center, center-stable and center-unstable invariant manifolds tend to the corresponding planes T^c, $T^c \oplus T^s$, $T^c \oplus T^u$ at 0 up to order N, as the following proposition claims.

1.3. Rectification of center manifolds.

PROPOSITION 1.2. *Let the invariant manifolds W^c, W^{cs}, W^{cu} of the vector field* (1.6) *be the graphs of the maps h^c, h^{cs}, h^{cu}, namely,*

$$W^c = \{u = h^c(x)\}, \quad W^{cs} = \{z = h^{cs}(x,y)\}, \quad W^{cu} = \{y = h^{cu}(x,z)\}.$$

Then $j_0^N h^c = 0$, $j_0^N h^{cs} = 0$, $j_0^N h^{cu} = 0$.

PROOF. Let us prove the proposition for the function h^{cs}. The same proof goes for h^{cu}. Geometrically, we will prove that the manifolds W^{cs} and W^{cu} are tangent to the corresponding planes at 0 up to order N. This implies the same statement for the center manifold, which is a transversal intersection of the form

$$W^c = W^{cs} \cap W^{cu}.$$

So, consider the function $h = h^{cs}$. The invariance of the graph $z = h(x,y)$ for the vector field (1.6) is equivalent to

$$A^+(x)\,h = \frac{\partial h}{\partial x}\,w + \frac{\partial h}{\partial y}\,A^-(x)\,y + o(r^N).$$

Let h_s be the principal nonzero homogeneous part of the Taylor decomposition of h. If $s > N$, we are done. If $s \leqslant N$, then h_s satisfies the equation

$$A^+(0)\,h_s - \left[\frac{\partial h_s}{\partial x}\,A^c(0)\,x + \frac{\partial h_s}{\partial y}\,A^-(0)\,y\right] = 0.$$

The operators $A^+(0)$, $A^c(0)$, $A^-(0)$ are diagonal with spectra equal to λ^+, μ, λ, respectively. The left-hand side of the last equation is the linear operator L applied to h_s. The monomials $x^k y^l \partial/\partial z_j$ are the eigenvectors of this operator. The corresponding eigenvalues are $\lambda_{j,k,l} = \lambda_j^+ - (\mu, k) - (\lambda^-, l)$. We have $\operatorname{Re}\lambda_j^+ > 0$, $\operatorname{Re}\mu_j = 0$, $\operatorname{Re}\lambda_j^- < 0$. Hence, $\operatorname{Re}\lambda_{j,k,l} > 0$, $\lambda_{j,k,l} \neq 0$. Consequently, the operator L is nondegenerate, and the equation $Lh_s = 0$ implies $h_s = 0$, a contradiction.

PROOF OF THEOREM 1.1. Let W^c be the preassigned germ of the center manifold, and W^{cs}, W^{cu} the germs of the center-stable and center unstable manifolds that contain W^c. Then, by Proposition 1.2, we can rectify the invariant manifolds W^c, W^{cs}, W^{cu} of the germ (1.6) changing the remainder term $o(r^N)$ only. Indeed, the change

$$(x,u) \mapsto (x, u - h^c(x))$$

takes W^c to T^c and preserves the N-jet of v at 0, because it is tangent to the identity up to order N.

The same is true for the rectification of W^{cs} and W^{cu}.

Now we can consider the initial vector field $v = (v^c, v^s, v^u)$ in the form (1.6) with the additional properties: $v^s = 0$ for $y = 0$, $v^u = 0$ for $z = 0$. This implies

$$v(x,u) = (w(x) + r^c, A(x)u + r^h), \quad r^c = O(|u|), \quad r^h = O(|u|^2), \tag{1.8}$$

$$j_0^N r^c = 0, \quad j_0^N r^h = 0, \tag{1.9}$$

where $A(x)$ has the form (1.7).

1.4. Normalization along the center manifold: the x-component (beginning). The proof of the Takens theorem will be finished in two more steps. Consider five vector fields; the first is just (1.8), with assumptions (1.9):

$$\begin{aligned}
v_1 &= (w(x) + O(|u|), A(x)u + O(|u|^2)),\\
v_2 &= (w(x) + O(|u|^N), A(x)u + O(|u|^2)),\\
v_3 &= (w(x), A(x)u + O(|u|^2)),\\
v_4 &= (w(x), A(x)u + O(|u|^N)),\\
v_5 &= (w(x), A(x)u).
\end{aligned}$$

The last vector field, modulo trivial modifications, is the desired normal form.

We shall prove that all these vector fields are finitely smoothly equivalent. That is, for any given k, they are C^k equivalent if $N = N(k)$ is sufficiently large. The equivalence of v_2 and v_3, as well as v_4 and v_5, follows from the Belitskii–Samovol theorem. The equivalence of v_1 and v_2 will be proved in this subsection, and that of v_3 and v_4, in the next one.

The proof uses the method of successive approximations. That is, we assume that the difference between the two fields, the discrepancy, is of order $|u|^s$ and transform the field to the new one having a discrepancy of higher order. The transformation is finitely smooth. Moreover, the change at any step does not perturb the terms normalized at the previous step. The finite composition of these transformations provides the desired equivalence.

We now turn to the detailed proof of the equivalence of the vector fields $v = v_1$ and v_2. For simplicity we slightly change the notation. Let

$$v(x,u) = (w(x) + f_s(x,u) + \cdots, A(x)u + O(|u|)^2). \tag{1.10}$$

Here f_s is a homogeneous vector polynomial of degree s in u with finitely smooth coefficients depending on x. Let us say, for brevity, that f_s is *s-homogeneous.* Perform the smooth coordinate change

$$(x,u) \mapsto (x + h_s(x;u), u) = (x_1, u). \tag{1.11}$$

Here h_s is s-homogeneous as well as f_s, both polynomials taking values in T^c. Our goal is to find the coordinate change in such a way that the field v in the new coordinates will have a discrepancy of higher order. Therefore, we shall be interested in the terms of order s in u only. Higher order terms will be replaced by dots.

PROPOSITION 1.3. *Let the vector field* (1.10) *satisfy the assumptions of the Takens theorem. Then there exists a finitely smooth coordinate change* (1.11) *that takes the germ* (1.10) *to the form given by the same formula with $f_s \equiv 0$.*

PROOF. Note that the transformation inverse to (1.11), up to negligible higher order terms, has the following simple form:

$$(x_1, u) \mapsto (x_1 - h_s(x_1, u) + \cdots, u) = (x, u). \tag{1.12}$$

Here and in the next subsection dots replace terms of order $o(|u|^s)$. The x_1-component of the initial vector field written in the new coordinates gives the equation

$$\dot{x}_1 = \dot{x} + \frac{\partial h_s}{\partial x}\dot{x} + \frac{\partial h_s}{\partial u}\dot{u} = \left(E + \frac{\partial h_s}{\partial x}\right)w(x) + \frac{\partial h_s}{\partial u}A(x)u + f_s(x, u) + \cdots.$$

Replacing x by $x_1 - h_s(x_1, u)$ (see (1.12)), we get

$$\dot{x}_1 = \left(E + \frac{\partial h_s}{\partial x_1}\right)\left(w(x_1) - \frac{\partial w}{\partial x}h_s(x_1, u)\right) + \frac{\partial h_s}{\partial u}A(x_1)u + f_s(x_1, u) + \cdots.$$

The u_1-component of the same vector field satisfies the equation

$$\dot{u} = A(x_1)u + O(|u|^2).$$

We should like the x-component of the new vector field to be equal to $w(x_1)$ up to higher order terms $o(|u|^s)$. This gives the following equation to the x dependent s-homogeneous vector polynomial h_s in u:

$$-\frac{\partial h_s}{\partial x}(w(x)) + \frac{\partial w}{\partial x}h_s - \frac{\partial h_s}{\partial u}Au - f_s = 0. \tag{1.13}$$

The proof of the solvability of this equation requires a certain idea, which is central in the proof. It is presented in the next subsection.

1.5. Appendix: center manifolds for linear extensions. Consider a nonhomogeneous linear extension of the nonlinear system

$$\dot{x} = w(x), \qquad \dot{U} = L(x)U + b(x), \tag{1.14}$$

where L is a linear operator depending on x. Let $w(0) = 0$, $b(0) = 0$. Then, 0 is a singular point of (1.14).

Suppose that the operator $L(0)$ is hyperbolic. On the contrary, let $A^c = \partial w/\partial x(0)$ have all its spectrum on the imaginary axes. Let $x \in T^c$, $U \in T$. Then the system (1.14) has the center manifold at the point zero transversal to T and having the same dimension as T^c.

Indeed, the linearization of the system at 0 is given by

$$\begin{pmatrix} A^c & 0 \\ * & L(0) \end{pmatrix}.$$

This matrix has its center plane complementary to T. The application of the center manifold theorem completes the proof.

The center manifold above is the graph of a smooth map

$$h\colon T^c \to T. \tag{1.15}$$

Now let us consider the inverse problem. The condition for the graph of the map (1.15) to be the invariant manifold of equation (1.14) is

$$(U - h(x))'|_{U=h(x)} = 0.$$

Here the prime means the derivative along the vector field of the equation (1.14). This implies

$$-\frac{\partial h}{\partial x}\, w(x) + L(x)\, h + b(x) = 0. \tag{1.16}$$

PROPOSITION 1.4. *Let L and w be the same as at the beginning of the subsection, in particular, assume that the eigenvalues of $A^c = \partial\omega/\partial x(0)$ lie on the imaginary axis and $L(0)$ is hyperbolic and $b(0) = 0$. Then equation* (1.16) *has the finitely smooth solution* (1.15).

PROOF. Consider equation (1.14). It has a center manifold that may be represented as the graph of the map (1.15). This map is a solution to (1.16).

1.6. Normalization along the center manifold: the x-component (conclusion). To prove the solvability of equation (1.13), we must recognize it as a particular case of (1.16), and then check the assumption of Proposition 1.4 for this equation. Let T be the space of all s-homogeneous vector polynomials in u taking values $h(u)$ in T^c. Then h_s in (1.13) is a map $T^c \to T$. Define the operator $L(0)\colon T \to T$, $h \to h'$ by setting

$$h'(x) = \frac{\partial w}{\partial x}\, h(x) - \frac{\partial h}{\partial u}\, A(x)\, u.$$

The polynomial h' is again s-homogeneous in u, hence $L(x)$ really maps T to T. Equation (1.13) corresponds to the system

$$\dot{x} = w(x), \qquad \dot{U} = L(x)U + f_s(x)$$

in the same way as (1.16) corresponds to (1.14).

Let us now verify the assumptions of Proposition 1.4. The matrix $A^c = \partial w/\partial x(0)$ has purely imaginary eigenvalues by the assumptions of Theorem 1.1.

Now we must check that the operator $L(0)$ is hyperbolic. Complexify the tangent space (x, u) at zero, and suppose that the basis is formed by the eigenvectors of the operator A. This is possible, because the eigenvalues of $A = \partial v_1/\partial(x, u)(0)$ are pairwise distinct by the assumption of Theorem 1.1. In these coordinates the operator $L(0)$ has the eigenvectors

$$u^k \frac{\partial}{\partial x_j}, \qquad |k| = s,\ j = 1, \dots, n,$$

with the eigenvalues $\lambda_{k,j} = (\lambda, k) - \mu_j$. The numbers μ_j lie on the imaginary axes, while the eigenvalues $\lambda_{k,j}$ do not. Indeed,

$$\operatorname{Re} \lambda_{k,j} = \operatorname{Re}(\lambda, k) \neq 0,$$

because the tuple $\operatorname{Re} \lambda$ is nonresonant by assumption.

Let us check that $f_s(0, u) \equiv 0$. Indeed, any term of f_s has the form $a(x)u^\alpha$, $|\alpha| = s$, where the coefficient a, taking values in T^c, is a derivative of r^c in u of order α. Since $s \leqslant N$, we have, by (1.9), the relation $a(0) = 0$. Hence,

$$f_s(0, u) \equiv 0.$$

This allows us to apply Proposition 1.4, which proves Proposition 1.3. The method of successive approximations, with Proposition 1.3 applied at every step, proves the finite smoothly equivalence of the vector fields v_1 and v_2.

1.7. Continuation: the u-component. In this subsection we prove the finitely smoothly equivalence of the vector fields v_3 and v_4. This is also done via the normalization procedure along the center manifold by the successive approximation method. Hence it is sufficient to prove the following analog of the previous proposition.

PROPOSITION 1.5. *Let the vector field*

$$v(x, u) = (w(x), A(x)u + f_s(x, u) + \cdots) \tag{1.17}$$

satisfy the assumptions of the Takens theorem. Then there exists a finitely smooth coordinate change

$$(x, u) \mapsto (x, u + h_s(x, u)) = (x, u_1) \tag{1.18}$$

that brings the germ v to the form (1.17) *with $f_s \equiv 0$. Here f_s and h_s are s-homogeneous in u.*

PROOF. The proof follows the same lines as the previous one. The coordinate change inverse to (1.18) has the form

$$(x, u_1) \mapsto (x, u) = (x, u_1 - h_s(x, u_1) + \cdots).$$

Hence the vector field (1.17) in the new coordinates yields the following equation, which may be obtained in the same way as in 1.4:

$$\begin{aligned} \dot{x} = w(x), \qquad \dot{u}_1 &= \left(E + \frac{\partial h_s}{\partial u}\right)\dot{u} + \frac{\partial h_s}{\partial x}\,\dot{x} \\ &= \left(E + \frac{\partial h_s}{\partial u_1}\right)(A(x)u_1 - A(x)h_s + f_s) + \frac{\partial h_s}{\partial x}\,w(x) + \cdots. \end{aligned}$$

We want the u_1-component of the new vector field to be equal to $A(x)u_1 + \cdots$. This gives the following equation for the unknown vector polynomial h_s (we omit the subscript 1 for simplicity):

$$\frac{\partial h_s}{\partial x}\,w(x) + \frac{\partial h_s}{\partial u}\,A(x)u - A(x)h_s + f_s = 0. \tag{1.19}$$

This equation also has type (1.16). This can be verified in the same way as for equation (1.13). Namely, let T^c be the same as before, and let T be the space of all s-homogeneous vector polynomials in u taking values in $T^u \oplus T^s$. Define the operator $L(x)\colon T \to T$, $h \to h'$, by setting

$$h'(u) = \frac{\partial h}{\partial u}\,A(x)u - A(x)h.$$

Suppose, as before, that the operator $A = A(0)$ is diagonalized. Then the operator $L(0)$ has the eigenvectors $u^k\partial/\partial u_j$ with the eigenvalues $(\lambda, k) - \lambda_j$. These eigenvalues lie outside the imaginary axis because the vector $\operatorname{Re}\lambda$ is nonresonant by the assumption of the theorem. The application of Proposition 1.4 now proves Proposition 1.5.

Propositions 1.3 and 1.5 imply the finitely smooth equivalence of the vector fields v_1 and v_2, v_3 and v_4, respectively. The C^k smooth equivalence of the fields v_2 and v_3, as well as v_4 and v_5, follows from the Belitskii–Samovol theorem for sufficiently large N, as mentioned before. Hence, the fields v_1 and v_5 are finitely smoothly equivalent. Rectifying the center-stable and center-unstable manifolds, we can split the linear system $\dot{u} = A(x)\,u$ into two: the contracting and the expanding one:

$$\dot{y} = A^{-}(x)\,y, \qquad \dot{z} = A^{+}(x)\,z.$$

This completes the proof of Theorem 1.1.

Now we shall apply this theorem to obtain some concrete normal forms.

§2. Unfoldings of saddlenodes

In this section Theorem 2.5.3 is proved together with its multiparameter generalizations. These generalizations are not used in the book, but they are obtained by the same tools and have numerous applications.

2.1. Normal forms for unfoldings of saddlenodes. Recall that a saddlenode of a vector field is a singular point with one zero eigenvalue and others lying outside the imaginary axis. Normal forms for unfoldings of such germs are given by the following

THEOREM 2.1. *Suppose that a germ of a saddlenode vector field satisfies the Łojasiewicz condition, that is, the modulus of the vector field decreases no faster than some power of the distance to the singular point. Suppose that the nonzero eigenvalues are pairwise distinct and their real parts form a nonresonant tuple. Then any smooth finite parameter deformation of this germ is finitely smoothly equivalent to the local family given by the system of equations*

$$\dot{x} = P(x,\varepsilon), \quad x \in (\mathbb{R}^1, 0), \qquad \dot{y} = A(x,\varepsilon)y, \quad y \in (\mathbb{R}^{n-1}, 0),\ \varepsilon \in (\mathbb{R}^p, 0). \tag{2.1}$$

Here P is a polynomial in x of degree no greater than $2\mu - 1$, namely $P(x,\varepsilon) = \sum_1^{2\mu-1} a_i(\varepsilon)x^i$, the integer $\mu \geqslant 2$ being the multiplicity of the singular point of the initial saddlenode germ. The coefficients of P are finitely smooth in ε and the matrix A is finitely smooth with respect to x and ε. Moreover, $a_i(0) = 0$ for $0 \leqslant i \leqslant \mu - 1$.

REMARKS. 1. The Łojasiewicz condition implies that the restriction of the unperturbed germ to its center manifold is $ax^\mu + \cdots$, $a \neq 0$ for some μ.

2. In fact, a better result is true: one may claim that the polynomial in the theorem above has a gap in the sense that $a_j(\varepsilon) \equiv 0$ for intermediate values of j: $\mu < j < 2\mu - 1$. In the case $\mu = 1$, which plays the principal role for the applications in this book, there is no difference between the initial and improved statements of Theorem 2.1, because in this case there is no intermediate term at all.

For $\mu = 1$, Theorem 2.1 implies

COROLLARY 2.1 (Theorem 2.5.3). *In generic one-parameter families of vector fields only those germs of saddlenode vector fields occur whose unfoldings in these families are finitely smoothly equivalent to the following normal form:*

$$\dot{x} = (x^2 + \varepsilon)(1 + a(\varepsilon)x)^{-1}, \quad \dot{y} = A(x,\varepsilon)y, \quad \dot{z} = B(x,\varepsilon)\,z. \tag{2.2}$$

The matrices $A(0,0)$ and $B(0,0)$ have their eigenvalues in the open left and right half-planes respectively.

This corollary will be proved in 2.3 below.

For planar vector fields the normal form in Theorem 2.1 may be simplified.

COROLLARY 2.2. *If in Theorem 2.1 the dimension of the phase space equals 2, then an orbital finitely smooth coordinate change may bring the initial family to the polynomial normal form*

$$\dot{x} = P(x, \varepsilon), \quad \dot{y} = y, \qquad x, y \in (\mathbb{R}^1, 0),$$

$P(x, \varepsilon)$ *is the same as in* (2.1).

PROOF. In the planar case, the matrix A in (2.1) is just a function. By the definition of saddlenodes, $A(0,0) \neq 0$. Hence, one may divide the vector field by A in some neighborhood of zero. The equation for x will be a new family, which has lost its canonical form. But it may be reduced to this form by Lemma 2.1 below. This proves Corollary 1.2.

This corollary is important for investigations of bifurcations of planar polycycles in multiparameter families of vector fields.

2.2. Normal forms for saddlenode families of vector fields on the line.

LEMMA 2.1. *Let the smooth family of vector fields on the line*

$$\dot{x} = f(x, \varepsilon), \qquad x \in (\mathbb{R}, 0), \ \varepsilon \in (\mathbb{R}^p, 0), \tag{2.3}$$

be an unfolding of a germ of a vector field $(x^\mu + \text{higher terms})\partial/\partial x$. *Then for any* k *there exists a positive integer* $N = N(k)$ *such that the family* (2.3) *with* $f \in C^N$ *is* C^k *smoothly equivalent to a polynomial one in* x *with coefficients depending on* ε. *The degree of this polynomial is no greater than* $2\mu - 1$, *and* μ *lower terms in* x *for* $\varepsilon = 0$ *are zeros.*

REMARK. Together with the Takens theorem (Theorem 1.1), this lemma proves Theorem 2.1 as is shown below.

PROOF OF LEMMA 2.1. Recall that two classical results of local complex analysis, the Weierstrass preparation and division theorems, have finitely smooth counterparts. They are as follows.

DIVISION THEOREM. *Let a finitely smooth function* $f(x, \varepsilon)$ *restricted to the line* $\varepsilon = 0$ *have a zero value of multiplicity* μ *at zero. Then any function* g *of the same rate of smoothness may be represented in some neighborhood of zero as*

$$g(x, \varepsilon) = f(x, \varepsilon)q(x, \varepsilon) + R(x, \varepsilon). \tag{2.4}$$

Here the residue R *is a polynomial in* x *of degree no greater than* $\mu - 1$ *with coefficients depending on* ε. *The rate of smoothness of these coefficients, as well as of the quotient* q, *tends to infinity together with that of the function* f.

WEIERSTRASS PREPARATION THEOREM. *Let* f *be the same as in the division theorem. Then* f *can be represented as the product of a polynomial in* x *of degree* μ *with coefficients depending on* ε *to a nonzero function* Ω:

$$f(x, \varepsilon) = W(x, \varepsilon)\Omega(x, \varepsilon), \qquad W(x, \varepsilon) = \sum_{i=0}^{\mu} a_i(\varepsilon)x^i. \tag{2.5}$$

Here $a_\mu(\varepsilon) \equiv 1$, $\Omega(0,0) \neq 0$, $a_i(0) = 0$ for $0 \leqslant i \leqslant \mu - 1$ and the rate of smoothness of the coefficients a_i and the invertible factor Ω tends to infinity together with that of the function f.

The polynomial W in the Weierstrass preparation theorem is called the *Weierstrass polynomial.*

REMARK. The Weierstrass preparation theorem immediately follows from the division theorem. One must only apply the last theorem to the function $g = x^\mu$.

These two theorems imply Lemma 2.1 in the following way. Decompose f into the product (2.5), using the Weierstrass preparation theorem. Take $\Omega = g$ and represent this function in the form (2.4), using the division theorem. This gives the following expression for f (we omit the arguments for simplicity):

$$f = W(R + fq). \tag{2.6}$$

We shall prove that the initial family

$$\dot{x} = f \tag{2.7}$$

is finitely smoothly equivalent to the truncated one (compare with (2.6)):

$$\dot{x} = WR. \tag{2.8}$$

The product on the right-hand side is a polynomial in x of degree no greater than $2\mu - 1$. Moreover, μ lower terms in W for $\varepsilon = 0$ are zeros. Hence the last family is the desired normal form.

The equivalence of the family (2.7) and its truncation (2.8) is proved by the homotopy method. The conjugating map will have the rate of smoothness k equal to the minimum rate of the families (2.7) and (2.8). Let

$$f_s = WR + sWfq.$$

Then $f_0 = WR$, $f_1 = f$. We shall prove that for all $s \in [0,1]$ the families $\dot{x} = f_s$ are C^k-equivalent. The homotopy method reduces this statement to the solution of the homological equation (9.2.13) with respect to the unknown function h. In the one-dimensional case this equation has the form

$$f_s'h - f_sh' = Wfq. \tag{2.9}$$

This is a linear nonhomogeneous equation. The corresponding homogeneous equation has the solution $h = f_s$. The nonhomogeneous equation (2.9) may be solved by the method of variation of constants. If $h = f_sc$, then $c' = Wfqf_s^{-2}$. But the function on the right-hand side is in fact smooth, although the denominator takes zero values. Indeed

$$R(0) = \Omega(0) \neq 0, \qquad f_s = W(R + sfq).$$

Hence $c' = \Omega q/(R + sfq)^2$. The last denominator is nonzero and all the functions above are at least C^k smooth.

This proves the lemma.

In fact we have proved more than claimed.

COROLLARY 2.2. *The polynomial normal form in Lemma 2.1 is equal to the product of a Weierstrass polynomial of degree μ and a polynomial of degree less than μ with nonzero free term.*

2.3. Proof of the normal form theorem and the one-parameter case.

PROOF OF THEOREM 2.1. The proof of Theorem 2.1 can now be carried out in a few lines. Replace the family, as usual, by the family-like equation, adding the equation $\dot{\varepsilon} = 0$. The new equation will have a center manifold with coordinates $x \in R^1$, and ε. All the coordinate changes in the Takens Theorem preserve the family-like type of equation.

Hence, by Theorem 1.1 (the Takens smooth saddle suspension theorem) the unfolding of a saddlenode germ satisfying the condition of Theorem 2.2 is finitely smoothly equivalent to the germ

$$\dot{x} = w(x, \varepsilon), \quad \dot{y} = A^+(x, \varepsilon)y, \quad \dot{z} = A^-(x, \varepsilon)\, z, \quad \dot{\varepsilon} = 0.$$

The first equation gives a finite parameter family of vector fields on the line. By Lemma 2.1, changing the coordinate x only, one can reduce this family to the normal form from Theorem 2.2. After this we forget the trivial equation for the parameter and obtain the desired normal form.

Now we prove Corollary 1.1. The proof is based on the constructions used in the previous subsection. First of all, we list the genericity assumptions:

1. All the eigenvalues of a singular point of the unperturbed germ are pairwise distinct.
2. The tuple of real parts of the hyperbolic eigenvalues is nonresonant.
3. The singular point has multiplicity two.
4. The saddlenode in the family splits in a transversal manner (this last assumption will be clarified later).

REMARK. The appearance of a saddlenode is itself a degeneration. It never occurs for generic vector fields but is unavoidable in one-parameter families. The violation of any of the assumptions above is an extra degeneracy. Degeneracies of codimension two never occur in typical one-parameter families.

This remark allows us to assume that generic one-parameter families satisfy the assumptions 1–4.

Assumptions 1 and 2 allow us to apply Theorem 1.1. Hence, our local family is finitely smoothly equivalent to

$$\dot{x} = P(x, \varepsilon), \quad \dot{y} = A^+(x, \varepsilon)y, \quad \dot{z} = A^-(x, \varepsilon)\, z. \tag{2.10}$$

Let us now normalize P. Assumption 3 implies that P is of degree 3. Moreover, by Corollary 2.2, $P = WR$, $\deg W = 2$, $\deg R = 1$, $R(0,0) \neq 0$. Then the family on the center plane is

$$\begin{aligned} \dot{x} &= P(x, \varepsilon), & P &= W(x)(b_0(\varepsilon) + b_1(\varepsilon)x), \\ b_0(0) &\neq 0, & W(x) &= x^2 + a_1(\varepsilon)x + a_0(\varepsilon). \end{aligned}$$

By the shift $x \mapsto x_1 = x + a_1/2$ one can get rid of a_1; in what follows we assume that $a_1 \equiv 0$. By rescaling $x \mapsto x_1 = b_0(\varepsilon)x$, one can obtain $b_0 \equiv 1$. We get the family

$$\dot{x} = (x^2 + \tilde{a}(\varepsilon))(1 + \tilde{b}(\varepsilon)x).$$

By assumption 4, the free term of the Weierstrass polynomial has a noncritical point 0 as a function of the parameter. No matter what term, a_0 or $\tilde{a}_0$, is meant: the assumptions $a_0'(0) \neq 0$, $\tilde{a}_0'(0) \neq 0$ are equivalent, because $a_1(0) = 0$, $b_0(0) \neq 0$. Thus we get the family

$$\dot{x} = (x^2 + \varepsilon)(1 + a(\varepsilon)x).$$

We must prove now that the last product may be replaced by the fraction (2.2). In the same way we can obtain the normal form (2.2) for the generic perturbation of the saddlenode germ of a vector field of multiplicity 2 at zero. This is done in the following way.

Let f, W, and Ω be the same as before and let $\Omega_1 = \Omega^{-1}$. Then Ω_1 is a smooth germ, because Ω is invertible. By the division theorem, $\Omega_1 = R + fq$, with $\deg R < \deg W$ (in our particular case, $\deg R \leqslant 1$). Note that R is invertible because so is Ω_1, and $f(0) = 0$. We claim that the families $\dot{x} = f$ and $\dot{x} = W/R$ are finitely smoothly equivalent. Indeed,

$$f - \frac{W}{R} = Wfq_1, \qquad q_1 = -\frac{q}{R\Omega_1}.$$

Both Ω and R are invertible. Hence q_1 is smooth.

Smooth equivalence of the families $\dot{x} = f$ and $\dot{x} = f - Wfq_1$ with smooth q_1 is already proved above. For $\deg W = 2$, $\deg R \leqslant 1$, we have

$$\frac{W}{R} = \frac{x^2 + a_1(\varepsilon)x + a_0(\varepsilon)}{b_0(\varepsilon) + b_1(\varepsilon)x}.$$

By affine transformations in x and an appropriate parameter change we get, as before, $a_1 \equiv 0$, $a_0 \equiv \varepsilon$, $b_0 \equiv 1$, $b_1 = a(\varepsilon)$. This gives the normal form (2.2).

Thus we have completed the proof of the second main result of this chapter, Theorem 2.1, together with Corollary 2.1.

§3. Takens smooth saddle suspension theorem for maps

In this section the counterpart of Theorem 1.1 for discrete time is proved. The proof follows the very same lines as for continuous time.

3.1. Statement and sketch of the proof.

THEOREM 3.1. *Let F be a smooth germ of a diffeomorphism at its fixed point 0. Let the linearization of this germ be the operator A with invariant subspaces T^c, T^s, T^u and pairwise distinct eigenvalues. Let the restriction of A to these subspaces have its spectrum inside, outside, and on the unit circle, respectively. Let A^-, A^+, and A^c be restrictions of A to these planes, and let λ^-, λ^+, μ be their spectra, $\lambda = (\lambda^-, \lambda^+)$. Let the moduli of the components of the vector λ form a multiplicatively nonresonant vector $|\lambda|$:*

$$|\lambda_j| \neq |\lambda|^k \quad \textit{for any } k \in \mathbb{Z}_+^n,\ |k| \geqslant 2,$$

where $|\lambda| = (|\lambda_1|, \ldots, |\lambda_n|)$. Then the germ F is finitely smoothly equivalent to

$$(x, y, z) \mapsto (w(x), A^-(x)\,y, A^+(x)\,z), \tag{3.1}$$

where $x \in T^c$, $y \in T^s$, $z \in T^u$, $A^-(0) = A^-$, $A^+(0) = A^+$. If the germ F is family-like, then its normal form (3.1) is also family-like.

Moreover, for any preassigned germ of the center manifold W^c of the vector field at 0, the coordinate change may be chosen so that this preassigned germ will become the germ of the x-plane at 0.

PROOF. In the very same way as this was done in 1.2, we prove that the initial germ of the map is formally equivalent to

$$(x, y, z) \mapsto (\widehat{w}(x), \widehat{A}^-(x)\, y, \widehat{A}^+(x)\, z).$$

This implies that for any N there exists a polynomial coordinate change that brings the initial germ to the form

$$F(x, y, z) = (w(x), A^-(x)\, y, A^+(x)\, z) + o(r^N), \tag{3.2}$$

where $r = |(x, y, z)|$.

A counterpart of Proposition 1.2 holds: any center, center-stable, and center-unstable manifolds of the germ F are tangent at 0 to the planes T^c, T^{cs}, T^{cu}, respectively, at least up to order $N + 1$. Hence, the rectification of these manifolds does not change the form (3.2) of the germ F. Let $u = (x, y)$ and $T^h = T^s \oplus T^u$.

Let W^c be the preassigned germ of the center manifold, and W^{cs}, W^{cu} the germs of the center stable and center unstable manifolds that contain W^c. Then the germ F with the rectified manifolds $W^c = T^c$, $W^{cs} = T^{cs}$, $W^{cu} = T^{cu}$, has the form

$$\begin{aligned} F(x, y, z) &= (w(x) + r^c, A^-(x)y + r^s, A^+(x)z + r^u), \qquad (3.3)\\ r^c &= O(|u|), \qquad (r^s, r^u) = O(|u|^2),\\ j_0^N r &= 0, \quad \text{where } r = (r^s, r^c, r^u). \qquad (3.4) \end{aligned}$$

The initial germ F transformed to (3.3), (3.4) coincides with the first of the following five germs:

$$\begin{aligned} F_1 &= (w(x) + O(|u|), A(x)\, u + O(|u|^2)),\\ F_2 &= (w(x) + O(|u|^N), A(x)\, u + O(|u|^2)),\\ F_3 &= (w(x), A(x)\, u + O(|u|^2)),\\ F_4 &= (w(x), A(x)\, u + O(|u|^N)),\\ F_0 &= (w(x), A(x)\, u). \end{aligned}$$

The last map is the desired normal form. We will prove that all these germs are finitely smoothly equivalent. For two pairs, F_2 and F_3, F_4 and F_0, this follows from the Belitskii–Samovol theorem for maps (see §9.4). We must prove the equivalence of F_1 and F_2, then of F_3 and F_4. Both statements are proved by the method of successive approximations. At each step of this method, one must solve a functional equation that turns out to be the equation for the center manifold of some linear nonhomogeneous suspension over a nonlinear map. The existence of the solution to the last equation follows from the hyperbolicity of a certain linear operator. This hyperbolicity, in turn, is a consequence of the nonresonance assumption in Theorem 3.1.

3.2. Center manifolds for linear extensions. This subsection is parallel to 1.5. It contains a key tool used for the normalization of a map along the center manifold.

PROPOSITION 3.1. *Consider a linear extension over a nonlinear map:*

$$\begin{aligned} (x, U) &\mapsto (x', U'), \qquad x, x' \in T^c,\ U, U' \in T, \\ x' &= w(x), \qquad U' = L(x)U + b(x). \end{aligned} \tag{3.5}$$

Let $A^c = \partial w/\partial x(0)$ be totally nonhyperbolic: all its eigenvalues lie on the unit circle. Let $L(0)$ be hyperbolic, and $b(0) = 0$. Then the equation

$$h \circ w - L(x)h(x) + b(x) = 0 \tag{3.6}$$

with respect to an unknown map $h\colon T^c \to T$ has a finitely smooth solution defined near 0. The surface $U = h(x)$ is the center manifold for the map (3.5).

PROOF. As in 1.5, the linearization of the map (3.5) at 0 has its center plane complementary to T and of the same dimension as T^c. Hence, (3.5) has a center manifold tangent to T^c at 0 that may be represented as the graph of a map $h\colon T^c \to T$. The condition of invariance of this graph under the map (3.5) is just equation (3.6). Hence, this equation has the desired solution.

3.3. Normalization along the center manifold. First we prove that the germs F_1 and F_2 are equivalent. The method of successive approximations reduces this equivalence to the following

PROPOSITION 3.2. *Let a germ*

$$F(x, u) = (w(x) + f_s(x; u) + \cdots, A(x)u + O(|u|^2)) \tag{3.7}$$

satisfy the assumptions of Theorem 3.1, *together with* (3.3), (3.4). *Let f_s and h_s be homogeneous vector polynomials in u of degree s with coefficients depending on x. Then there exists a finite smooth coordinate change*

$$(x, u) \mapsto (x + h_s(x; u), u) = (x_1, u) \tag{3.8}$$

that brings the germ (3.7) *to the form given by the same formula with $f_s \equiv 0$.*

PROOF. Formula (1.12) for the change inverse to (3.8) is valid. Recall that F_0 is the map $(x, u) \mapsto (w(x), A(x)u)$. The initial map in the new coordinates takes the form

$$\begin{aligned} (x_1, u) \mapsto (x_1', u'), \qquad x_1' &= w(x_1) - \frac{\partial w}{\partial x} h_s + f_s + h_s \circ F_0 + \cdots, \\ u' &= A(x_1)u + O(|u|^2). \end{aligned}$$

Let us omit the subscript 1 for simplicity. The annihilation of the terms of degree s in u is equivalent to

$$h_s \circ F_0 + f_s - \frac{\partial w}{\partial x} h_s = 0. \tag{3.9}$$

Now we must recognize that this equation is a particular case of (3.6). Let $c = \dim T^c$. Let T, as in §1, be the space of s-homogeneous vector polynomials $h(u)$, $u \in T^h$, with c components. Then the unknown h_s in (3.9) can be regarded as a map $h\colon T^c \to T$. Note that $f_s(0, u) \equiv 0$ by (3.4). The explanation is the same as at the end of 1.6.

For any linear operator $B\colon T^h \to T^h$ the map $\mathbf{B}\colon h \to h \circ B$ is linear. Indeed, a monomial of degree s, after a linear substitution, becomes a homogeneous polynomial of the same degree. If, moreover, the operator B is diagonal: $B = \operatorname{diag} \lambda$, then the monomial $u^k\partial/\partial x_j$ becomes an eigenvector of $\mathbf{B}$ corresponding to the eigenvalue ν^k. If $|\mu|$ is multiplicatively nonresonant, then the operator $\mathbf{B}$ is hyperbolic, i.e., $\bar{\lambda}^k \neq 1$.

We apply this argument to $B = A(x)$. It gives

$$h \circ F_0(x,u) = h(w(x), A(x)u) = \mathbf{A(x)}h(w(x),u).$$

The operator $\mathbf{A(0)}$ is hyperbolic by the above reasoning. Hence, $\mathbf{A(0)}^{-1}(0)$ is also hyperbolic. Now (3.9) takes the form

$$\mathbf{A(x)}h_s \circ w + f_s - \frac{\partial w}{\partial x} h_s = 0$$

or

$$h_s \circ w = \mathbf{A(x)}^{-1}(x)\left(\frac{\partial w}{\partial x} h_s\right) - \mathbf{A(x)}^{-1} \circ f_s.$$

This is equation (3.6) with elements

$$h = h_s, \quad L(x) = \mathbf{A(x)}^{-1}\frac{\partial w}{\partial x}(x), \quad f = \mathbf{A(x)}^{-1} f_s$$

belonging to T.

The operator $L(0)$ is hyperbolic. Indeed, it has the eigenvectors $u^k\partial/\partial x_j$ provided that (x,u) are the coordinates in the complexification of $\mathbb{R}^N$ for which the linear part A of the initial germ is diagonal. The corresponding eigenvalues are $\lambda_{jk} = \lambda^{-k}\mu_j$. We have: $|\lambda_{jk}| = |\lambda^{-k}| = \bar{\lambda}^{-k} \neq 1$ by the assumption of Theorem 3.1. This proves the hyperbolicity of $L(0)$.

By Proposition 3.1, equation (3.6) has a finitely smooth solution. This proves Proposition 3.2.

The method of successive approximations now proves the desired equivalence of the germs F_1 and F_2. The equivalence of F_3 and F_4 is proved in the same way below.

PROPOSITION 3.3. *Let the germ*

$$F(x,u) = (w(x), A(x)u + f_s(x,u) + \cdots) \tag{3.10}$$

satisfy the assumptions of Theorem 3.1. *Let* f_s *and* h_s *be homogeneous vector polynomials in* u *of degree* s *with the coefficients depending on* x.

Then there exists a finitely smooth coordinate change

$$(x,u) \to (x, u + h_s(x,u)) = (x, u_1)$$

that brings the germ (3.9) *to the form given by the same formula with* $f_s \equiv 0$.

PROOF. A simple calculation shows that the absence of a term of degree s in the expression for the map (3.10) implies that in the new coordinates it has the form

$$h_s \circ F_0 - A(x)h_s + f_s = 0. \tag{3.11}$$

Define the operator $\bar{A}(x)\colon T \to T$ by $h \mapsto A(x)h$. The operator $A(0)$ is diagonalizable by the assumption of Theorem 3.1. It has the spectrum λ. The eigenvectors of $\bar{A}(0)$ are $u^k\partial/\partial x_j$ with eigenvalues λ_j.

The same argument as in the proof of Proposition 3.2 shows that equation (3.11) has the form

$$\mathbf{A}(x)\,h_s \circ w - \bar{A}(x)\,h_s + f_s = 0.$$

It is equivalent to (3.6) with $h = h_s$, $L(x) = \mathbf{A}(\mathbf{x})^{-1}\bar{A}(x)$, $b = \mathbf{A}(\mathbf{x})^{-1} f_s$.

Moreover, the operator $L(0)$ is hyperbolic. This is verified as in the proof of Proposition 3.2 with the only difference that the eigenvalues of $L(0)$ are $\lambda_{jk} = \lambda^{-k}\lambda_j$. They are hyperbolic, i.e., $|\lambda_{jk}| \neq 1$, because the tuple $|\lambda|$ is multiplicatively nonresonant by the assumption of Theorem 3.1. This proves the hyperbolicity of $L(0)$.

Proposition 3.1 now implies Proposition 3.3.

Thus the finite smooth equivalence of the germs F_1 and F_2, as well as F_3 and F_4, is proved. This completes the proof of the Takens theorem for maps.

§4. Partial embedding theorem for saddlenode families of maps

In this section finitely smooth normal forms for unfoldings of germs of saddlenode maps will be obtained. As a corollary, Theorem 4 from §5 in Chapter 2 will be proved. This theorem was an important tool in the studies of Chapters 5 and 6.

4.1. Statement of the partial embedding theorem. This is the second main result of this chapter:

THEOREM 4.1. *In generic one-parameter families of maps only those germs of saddlenode maps occur whose perturbation in the family corresponding to small nonnegative values of an appropriately chosen parameter ε is finitely smoothly equivalent to*

$$F(x,y,z,\varepsilon) = (x',y',z'), \quad x' = f(x,\varepsilon), \quad y' = A(x,\varepsilon)\,y, \quad z' = B(x,\varepsilon)\,z,$$
$$\operatorname{Sp} A \subset \{\lambda \in \mathbb{C} : |\lambda| < 1\}, \quad \operatorname{Sp} B \subset \{\lambda \in \mathbb{C} : |\lambda| > 1\}.$$

Here f is a time one shift along the orbits of a vector field (2.2):

$$f(x,\varepsilon) = g^1_{w_\varepsilon} x, \qquad w_\varepsilon(x) = (x^2+\varepsilon)(1+a(\varepsilon)x)^{-1}.$$

REMARKS. 1. A smooth map in a half-neighborhood of a point with the boundary included is, by definition, a map that may be smoothly extended to the entire neighborhood. The main example is the C^k function defined on one side of a hyperplane and having a zero k-jet on the hyperplane itself. It may be extended to the other half-space as a C^k function by setting it identically equal to zero.

2. The fact that the equivalence takes place in the half-neighborhood only gives rise to the word "partial" in the name of the theorem. In fact, a stronger statement holds true, the *sectorial embedding theorem* (see 4.10 below).

3. The finitely smooth classification in the whole neighborhood of zero has functional moduli described at the end of the section.

4.2. Genericity assumptions and the one-dimensional partial embedding theorem. Suppose that the family in Theorem 4.1 satisfies the following genericity assumptions.

1. The hyperbolic multipliers of the unperturbed germ satisfy the conditions of the Takens smooth saddle suspension theorem for maps (Theorem 3.1). Namely,

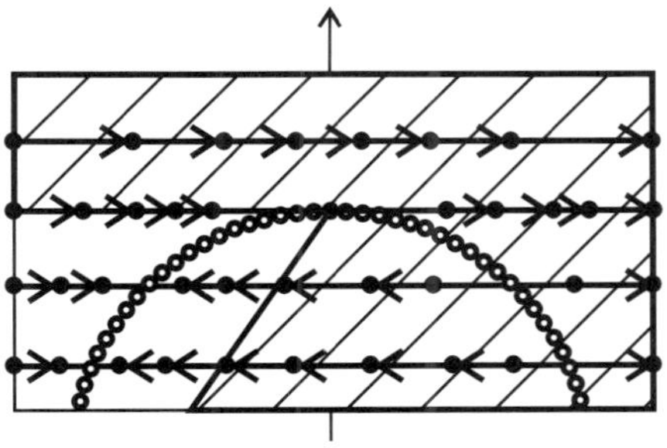

FIGURE 10.1. Local saddlenode family. The sectorial embedding theorem concerns the shadowed domain.

all these multipliers are pairwise distinct and their moduli form a multiplicatively nonresonant tuple.

2. The restriction of the family to its center manifold is the local family of one-dimensional maps of the form

$$x \mapsto f(x,\varepsilon) = x + \alpha(\varepsilon)x^2 + a(\varepsilon) + \cdots \tag{4.1}$$

with the following properties:

2a. The unperturbed germ has multiplicity two: $\alpha(0) \neq 0$.

2b. The family transversally intersects the hypersurface of saddlenode jets of maps at the point corresponding to the zero value of ε: $a(0) = 0$, $a'(0) \neq 0$.

The partial embedding theorem (Theorem 4.1) is an immediate consequence of the Takens theorem (Theorem 3.1) and the following

THEOREM 4.2 (One-dimensional Partial Embedding Theorem). *In a generic one-parameter family of maps of the line onto itself only those germs of saddlenode maps occur whose perturbation in the family corresponding to small nonnegative values of an appropriately chosen parameter ε is finitely smoothly equivalent to the time one shift along the orbits of the vector field* (2.2).

The genericity assumptions in Theorem 4.2 are just assumptions 2a, 2b above.

It remains to prove Theorem 4.2. This proof will form the main body of the rest of this section.

4.3. Sketch of the proof of the one-dimensional partial embedding theorem. Perturbations of the saddlenode germ of a map of the line are called *saddlenode families*. As usual, we replace the saddlenode family of maps by a single family-like map

$$F\colon (x,\varepsilon) \mapsto (f(x,\varepsilon),\varepsilon), \tag{4.2}$$

where $f(x,\varepsilon)$ is the same as in (4.1) (see Figure 10.1).

DEFINITION 4.1. A germ of a map at a fixed point is *embeddable* if there exists a germ of a vector field with the same equilibrium point such that the time one phase flow transformation of the latter germ is equal to the given germ. The germ of the vector field in this case is called the *generator* of the germ of the map.

DEFINITION 4.2. If a map has zero k-jet at any point of some subset of its domain, then this map is called *k-flat* on this subset.

The proof of Theorem 4.2 consists of four steps. Step by step, we find the "approximate generator" whose time one phase flow transformation coincides with the initial germ F with more and more accuracy. At the last step this coincidence becomes exact. Let r be the polar radius in the (x, ε)-plane. The accuracy at the subsequent steps, in some neighborhood of zero in a half-plane $\varepsilon \geqslant 0$, is r^N, εr^N, ε^N, zero. The first step uses the following

LEMMA 4.1 (Embedding in Jets). *Suppose that a germ of a C^k map at a fixed point has unipotent linear part. Then there exists a germ of a C^k vector field with nilpotent linear part such that its time one phase flow transformation coincides with the initial germ of the map up to a k-flat correction. The k-jet of this field is uniquely determined. Here k is either a natural number, or infinity.*

For any N not higher than the rate of smoothness of F, this lemma provides a vector field whose time one shift coincides with F up to r^N.

The next step provides a generator on the line $\varepsilon = 0$. Denote this line by L.

LEMMA 4.2 (Takens Embedding Theorem). *A smooth germ of a map*

$$f\colon (\mathbb{R}, 0) \to (\mathbb{R}, 0)$$

with multiplier one and fixed point of finite multiplicity is embeddable. For any k there exists an $N = N(k)$ such that if the germ f is of class C^N, then its generator is of class C^k.

For any N described in its statement, this lemma provides a vector field whose time one shift coincides with F up to εr^{N-1}.

Hence, we already have constructed the exact generator on L near 0. Next, at any point of L near zero we will construct a jet of a generator of high order.

LEMMA 4.3 (embedding in jets along the line $\varepsilon = 0$). *Let F be the map* (4.2), (4.1) *satisfying the genericity assumptions* 2a, 2b *from* 4.2. *Then for any k there exists an $N = N(k)$ with the following property. Let $F \in C^N$. Then there exists a germ of a C^k smooth vector field whose time one phase flow transformation coincides with the initial germ f up to ε^k in some neighborhood of 0 in the half-plane $\varepsilon \geqslant 0$.*

The last step is to prove

LEMMA 4.4. *Let F be the same as in Lemma 4.3. Then for any k there exists an $N = N(k)$ such that if the germ F coincides with the embeddable germ of class C^N with accuracy $o(\varepsilon^N)$, then these two germs are C^k equivalent.*

4.4. Embedding at a fixed point. Here we prove Lemma 4.1 for germs of maps defined in a space of arbitrary dimension. Our goal is the following. For any germ $f \in C^k$

$$f\colon (\mathbb{R}^n, 0) \to (\mathbb{R}^n, 0), \quad \text{where } df(0) \text{ is unipotent}, \tag{4.3}$$

to find a germ of a vector field $v \in C^k$ on $(\mathbb{R}^n, 0)$ such that

$$j_0^k g_v^1 = j_0^k f, \quad \text{where } dv(0) \text{ is nilpotent}. \tag{4.4}$$

This will be done by means of Lie group theory. Namely, the k-jets of germs (4.3) form a group G^k under "truncated composition" defined as follows:

$$j^k f \circ j^k g = j^k(f \circ g).$$

All the jets in this subsection are taken at zero, and we omit the subscript 0 for brevity. The group G^k has a faithful linear representation. That is, for some linear space V an isomorphic embedding

$$j^k f \mapsto T(j^k f) \in GL(V) \tag{4.5}$$

exists. The space V consists, by definition, of k-jets of functions $\varphi\colon (\mathbb{R}^n, 0) \to (\mathbb{R}, 0)$. The map (4.5) is defined as the following shift of the argument:

$$T(j^k f)\colon j^k\varphi \mapsto j^k(\varphi \circ f^{-1}). \tag{4.6}$$

Obviously, for any germs f, g of type (4.3) we have

$$T(j^k f \circ j^k g) = T(j^k f)T(j^k g).$$

Hence, (4.6) is really a representation. This representation is faithful. Indeed, the jet $j^k f$ may be reconstructed if the action of T is known: the component $(f^{-1})_i$ is just the image of $\varphi \equiv x_i$.

The operators of the representation (4.5) are unipotent. Indeed, let the operator $df(0)$ have lower triangular Jordan normal form J, and let the corresponding coordinates be $(x_1, \dots, x_n) = x$. Introduce an order in the set of monomials x^l: $l = (l_1, \dots, l_n) \in \mathbb{Z}^n$, $|l| = l_1 + \dots + l_n$. Say that x^l is *older* than x^m, $x^l > x^m$, if $|l| > |m|$ or $|l| = |m|$, and the vector l is lexicographically older than m. This order agrees with the Jordan structure of J: for any monomial x^l the difference $J(x^l) = x^l - (Jx)^l$ is a sum of monomials that are all older than x^l. Hence, the operator J is nilpotent in the space of polynomials of degree no greater than k. Moreover, the difference $T(j^k f)(x^l) - x^l = j^k((f^{-1})^l - x^l)$ has a polynomial representative. All the monomials of this polynomial are older than x^l. Hence, the operator $T - \mathrm{id}$ is nilpotent, and T is unipotent.

The group $\mathbf{T}$ of unipotent operators (4.5) is a Lie group. Its Lie algebra τ consists of nilpotent operators, and the exponential map $\exp\colon \tau \to \mathbf{T}$ is bijective. This is a general fact of Lie group theory. We will not repeat the proof here. Let us mention only that any unipotent linear operator has a unique nilpotent logarithm given by the formula

$$L = \sum_1^\infty (-1)^l (T - E)^l / l.$$

The series on the right-hand side is convergent. In fact, it is polynomial, because the operator $T - E$ is nilpotent. Moreover, the operator L is nilpotent, because the polynomial mentioned above has no free term.

Now take an arbitrary jet $j^k f \in \mathbf{T}$. We will find a vector field v satisfying (4.4). This will prove Lemma 4.1.

Take the corresponding operator $T = T(j^k f)$ and its logarithm L written above. Consider the one-parameter subgroup of $\mathbf{T}$, $T^t = \exp tL \in \mathbf{T}$. Any T^t corresponds to some jet $j^k f^t$ because the representation (4.5) is faithful. Take a jet of a vector field defined by the formula

$$j^k v = \frac{d}{dt} j^k f^t |_{t=0}.$$

The operator

$$\tau_v : j^k\varphi \mapsto j^k L_v\varphi = \frac{d}{dt}\varphi \circ f^t|_{t=0}$$

belongs to the Lie algebra τ because it is tangent to the subgroup T^t at the point E.

Consider the one-parameter subgroup in G^k generated by $j^k v$:

$$t \mapsto j^k g_v^t.$$

The corresponding family $\widetilde{T}^t = T(g_v^t) \in \mathbf{T}$ is the one-parameter group of linear operators with the same tangent vector as $\{T^t\}$.

But vectors of the Lie algebra and one-parameter subgroups of the Lie group are in one-to-one correspondence. Hence, the two subgroups constructed above coincide:

$$\widetilde{T}^t = T^t \quad \text{and} \quad j^k f^t = j^k g_v^t.$$

Substituting $t = 1$, we obtain (4.4).

The uniqueness of $j_0^k(v)$ follows from the fact that the representation (4.6) is faithful, and the exponential map exp: $\tau \to T$ is bijective.

This proves Lemma 4.1.

4.5. Generator on the line. Here Lemma 4.2 is proved for maps with nonzero quadratic term, that is, having a fixed point of multiplicity two:

$$f(x) = x + ax^2 + \cdots, \qquad a \neq 0.$$

Here we give only a sketch of the proof. A more sophisticated statement is proved in detail in 4.7 below.

Suppose that f is N-smooth. Then, by Lemma 4.1, there exists a vector field v such that

$$j_0^N f = j_0^N g_v^1, \quad f = g_v^1 + R, \quad j_0^N R = 0. \tag{4.7}$$

We want to kill the discrepancy R by a finitely smooth coordinate change. First normalize the germ v. By Corollary 2.1, for any M, there exists an M-smooth coordinate change that takes the germ of v at zero to the normal form (4.11) below. Now rectify the vector field (4.11) by the coordinate change $x \mapsto t = x^{-1} + a\log|x|$. The map f will take the form

$$\tilde{f} : t \mapsto t + 1 + \widetilde{R}$$

with $\widetilde{R}$ rapidly decreasing with k derivatives at $+\infty$ provided that $N = N(k)$ in (4.7) is large enough. We want to conjugate the map $\tilde{f}$ with the shift $t \mapsto t + 1$. Denote by $\mathrm{id} + \tilde{h}$ the required coordinate change. We obtain $\tilde{h} - \tilde{h} \circ \tilde{f} = \widetilde{R}$. Hence, $\tilde{h} = \sum \widetilde{R} \circ \tilde{f}^j$. The proof of the convergence of this series in C^M for M depending on N and tending to infinity together with N is proved in a straightforward way; see 4.7 below. Here we only note that the sequence $\tilde{f}^j(t)$ grows faster than some arithmetic progression, say $t + j/2$. Indeed, we have $\tilde{f}(t) > t + 1/2$ near infinity. Hence, $\tilde{f}^j(t) > t + j/2$. This completes the proof of Lemma 4.2.

4.6. Normalization of jets along the line L. In this and the next three subsections we prove Lemma 4.4.

Let F be the map (4.2), (4.1) satisfying assumptions 2a, 2b of 4.2. A linear change of the coordinate x allows us to assume that $\alpha(\varepsilon) \equiv 1$ in (4.1). The linear part of F at zero is a unipotent Jordan block:

$$dF(0) = \begin{pmatrix} 1 & 1 \\ 0 & 1 \end{pmatrix}.$$

Lemma 4.1 allows us to represent this map in the form $F = g_V^1 + R$. Here V is a family-like vector field. Moreover,

$$dV(0) = \begin{pmatrix} 0 & 1 \\ 0 & 0 \end{pmatrix}, \quad j_0^N R = 0, \quad V(x, \varepsilon) = (x^2 + \varepsilon + \cdots, 0).$$

As before, denote by L the line $\varepsilon = 0$. By Lemma 4.2 (the Takens embedding theorem), the restriction of F to L is embeddable. The generator of this restriction has the same N-jet at zero as the restriction of V to L, by the uniqueness property asserted in Lemma 4.1. Hence, one can find a vector field V' with the same N-jet at zero, which coincides with the generator of $F|_L$.

Moreover, the vector field V' is finitely smoothly equivalent to $W = (w, 0)$, where w is the right-hand side in (2.2). Hence, by Lemmas 4.1, 4.2, we can reduce F to the form

$$\begin{gathered} F = G + R, \quad G = g_W^1, \quad R = O(\varepsilon), \quad j_0^N R = 0, \quad W = (w, 0), \\ w = (x^2 + \varepsilon)(1 + a(\varepsilon)x)^{-1}. \end{gathered} \tag{4.8}$$

REMARK. Formula (4.8) comparing F and G claims that F is embeddable with accuracy $r^{N-1}\varepsilon$.

Our goal is to prove that this accuracy may be improved up to a high power of ε.

This will be done in three steps. Note first that the generator of the map is invariant under this map. The inverse statement is wrong: an invariant vector field is not, in general, the generator of a map. Yet for diffeomorphisms on the line, invariant vector fields and generators are closely related. The same is true for families of maps.

Step 1. Find an approximate generator of the map F from (4.8) with accuracy $O(\varepsilon^N)$ provided that the vector field invariant under F with the same accuracy is known.

Step 2. Solve the so-called *transfer equation* on the line:

$$f'h - h \circ f = R \circ f.$$

This equation has the following invariant meaning. Given a vector field R and a map f, find a vector field h such that the difference between the fields f_*h and h (the image of h under f) is equal to R.

Step 3. Using the solution found in Step 2, find a vector field invariant under the map F from (4.8) with accuracy $O(\varepsilon^N)$.

This will prove Lemma 4.3, because the reduction from the approximately invariant vector field to the approximate generator has already been performed in Step 1.

The desired reduction is given by

PROPOSITION 4.1. *Let F be the same as in (4.8), and let $V = (v, 0)$ be a family-like vector field with the following properties*:

$$dFV - V \circ F = o(\varepsilon^N), \tag{4.9}$$

$$v|_L = w|_L, \qquad |V| > \varepsilon/2, \tag{4.10}$$

where w is the same as above. Then there exists a function $\tau(\varepsilon)$ such that

$$F = g^1_{\tau V} + \widetilde{R}, \qquad j_p^{N-2}\widetilde{R} = 0$$

for any $p \in (L, 0)$. If $F, V \in C^k$, then $\tau \in C^k$.

REMARK. The proof of Proposition 4.1 exploits the following idea:

For a map of the line without fixed points, the solution to the transfer equation is proportional to the generator of the map with a constant coefficient.

Indeed, let f be the map, and v be the solution of the corresponding transfer equation. Then the time $T(x)$ of motion from x to $f(x)$ along the phase curves of the vector field v does not depend on x. Indeed, if t is a time function of v, $t' = 1/v$, then

$$t(f(x)) - t(x) = t(f(y)) - t(y)$$

for all x, y. Indeed,

$$t(y) - t(x) = t(f(y)) - t(f(x)),$$

because v is f-invariant. Hence, $T(x) \equiv T$, $f(x) = g_v^T x$. The vector field Tv proportional to v is the generator of f.

PROOF OF PROPOSITION 4.1. Consider the "time distance" between the source and target of the map f:

$$T(x, \varepsilon) = \int_x^{f(x,\varepsilon)} \frac{d\xi}{v(\xi, \varepsilon)}.$$

This function is independent of x with accuracy $O(\varepsilon^{N-2})$, i.e., there exists a function $\tau(\varepsilon)$ such that

$$T(x, \varepsilon) = \tau(\varepsilon) + O(\varepsilon^{N-2}).$$

Indeed, fix a small $a \neq 0$ and let $\tau(\varepsilon) = T(a, \varepsilon)$. The function τ is of class C^k provided that f and v are C^k smooth. This follows from the explicit formula above. Then

$$T(x, \varepsilon) - \tau(\varepsilon) = \int_a^x \frac{d\xi}{v(\xi, \varepsilon)} - \int_{f(a,\varepsilon)}^{f(x,\varepsilon)} \frac{d\xi}{v(\xi, \varepsilon)}.$$

But the field V is F-invariant with absolute accuracy ε^N (see (4.9)). Hence,

$$f_x v = v \circ f + \varepsilon^N, \qquad f_x = \frac{\partial f}{\partial x}.$$

By (4.10),

$$\frac{1}{v} = \frac{f_x}{v \circ f} + O(\varepsilon^{N-2}).$$

Hence,

$$\int_{f(a,\varepsilon)}^{f(x,\varepsilon)} \frac{d\xi}{v(\xi, \varepsilon)} = \int_a^x \frac{f_x(\xi, \varepsilon)}{v(f(\xi, \varepsilon), \varepsilon)}\, d\xi = \int_a^x \frac{d\xi}{v(\xi, \varepsilon)} + O(\varepsilon^{N-2}).$$

Then

$$T(x, \varepsilon) - \tau(\varepsilon) = O(\varepsilon^{N-2}).$$

This proves Proposition 4.1.

4.7. Transfer equation on the line. Step 2 is formalized by the following

LEMMA 4.5 (Solution to the Transfer Equation). *Let f be the time one shift along the vector field $x^2 + \cdots$. Then for any k there exists an $N = N(k)$ such that if the germ R on the right-hand side of the transfer equation is of class C^N and is N-flat at zero, then the transfer equation has a C^k smooth solution h with $j_0^k h = 0$.*

PROOF. The vector field in the statement of Lemma 4.3 is finitely smoothly equivalent to the normal form (2.2) for $\varepsilon = 0$, namely, to

$$v_0 = x^2(1 + ax)^{-1} \frac{\partial}{\partial x}. \tag{4.11}$$

The map f in the statement of this lemma may be now regarded as the time one shift generated by this standard vector field. The invariant character of the transfer equation allows us to choose the most convenient coordinate system and replace f by a simple shift, preserving the form of the equation. Rectify the vector field v_0. The rectifying map has the form

$$x \mapsto t(x) = -\frac{1}{x} + a \log |x|. \tag{4.12}$$

This map takes the negative half-neighborhood of zero to the positive half-neighborhood of infinity, and the vector field v_0 to $\partial/\partial t$. Denote the vector fields h, R written in the new coordinates by $\tilde{h}$, $\widetilde{R}$, and the time one shift, that is, the map f in the new coordinates, by $\tilde{f}\colon t \mapsto t + 1$. The transfer equation takes the form

$$\tilde{h}(t) - \tilde{h}(t+1) = \widetilde{R}(t+1). \tag{4.13}$$

Formally the solution to the last equation may be obtained very easily. Shift the argument in the previous equation by n for an arbitrary n:

$$\tilde{h}(t+n) - \tilde{h}(t+n+1) = \widetilde{R}(t+n+1),$$

and then sum the results, obtaining $\tilde{h} = \sum \widetilde{R}(t+n)$. In order to prove the convergence of this series and the smoothness of the result, it is sufficient to prove that a jet of some high order of the function $\widetilde{R}$ at infinity is zero. To do this, note that for any M there exists a number N with the following property. As before, let the vector field $\widetilde{R}$ be obtained from the vector field R by the coordinate change (4.12). Let $j_0^N R = 0$. Then

$$\widetilde{R}^{(k)}(t) = o(t^{-M}) \tag{4.14}$$

at infinity for all $k < M$. This is evident because all the derivatives of the rectifying map (4.12) have only polynomial growth at 0.

In what follows, we assume that $M \geqslant 2$. The series for $\tilde{h}$ converges in C^M, because of the last equality.

IMPORTANT REMARK. The above argument provides a solution to the transfer equation *in a negative half-neighborhood of zero only.* For sure, the same argument allows us to solve the transfer equation in a positive half-neighborhood. The gluing problem then arises: do these two solutions constitute the same C^M smooth function? The positive answer follows from the fact that M-jets of the solution at x tend to zero as x tends to zero. The glued function has a zero M-jet at zero.

This remark completes the proof of Lemma 4.5. Different versions of this argument will be used below.

4.8. Approximate solution to the transfer equation. In this subsection the final step of the proof of Lemma 4.3 is carried out.

PROPOSITION 4.2. *For any map F of the form* (4.8) *there exists a solution to system* (4.9), (4.10). *For any k, the solution V is C^k smooth provided that N is sufficiently large.*

We shall study the discrepancy $u = v - w$. It will be a polynomial in ε with coefficients depending on x that will be found recurrently. Any coefficient satisfies the one-dimensional transfer equation and may be found using Lemma 4.5. The loss of smoothness at each step is controlled. Hence, if the smoothness of the data was high enough, the solutions will have the required smoothness.

Let us give a more detailed exposition. By assumption, (4.8) holds. The field W is G-invariant. Substituting $V = W + U$, $U = (u, 0)$ into the transfer equation and taking $R = (r, 0)$, $r_x = \partial r/\partial x$, we get

$$f_x u - u \circ F = h + o(\varepsilon^N), \quad \text{where } h = -r_x w + [w \circ (G + R) - w \circ G].$$

Let

$$f(x, \varepsilon) = \sum_0^N f_j(x)\varepsilon^j + o(\varepsilon^N), \qquad f_0 = G|_L,$$

$$h(x, \varepsilon) = \sum_0^N h_j(x)\varepsilon^j, \qquad u(x, \varepsilon) = \sum_0^N u_j(x)\varepsilon^j.$$

The function h has a zero $N-1$ jet at 0, because R has a zero N-jet at 0 (see (4.8)). Hence, $j_0^M h_j = 0$ for $M < (N-1)/2$. The coefficients u_j of u will be found recurrently as the solutions to the transfer equation on the line with the right-hand side having a zero jet of high order at 0. By Lemma 4.5, this equation has a relatively smooth solution.

In more detail, let $g = G|_L$. Then the coefficient u_1 satisfies the equation

$$g_x u_1 - u_1 \circ g = h_1.$$

By Lemma 4.5, this equation has a k-smooth solution k-flat at 0 with k tending to infinity together with N. Similarly, the coefficient u_j satisfies the transfer equation for the same g, i.e., $g_x u_j - u_j \circ g = r_j$, where r_j is a polynomial in u_l with $l < j$, in f_l with $l \leqslant j$, and in the derivatives of f_l of order no greater than j. All these functions have a zero jet at 0 of order depending on j. For any k there exists an N such that if $j_0^N R = 0$, then $j_0^{N(k)} r_j = 0$, where $N(k)$ is the same as in Lemma 4.5. Hence, the latter transfer equation has k-smooth solution. This proves Proposition 4.2. Together with Proposition 4.1, it implies Lemma 4.4.

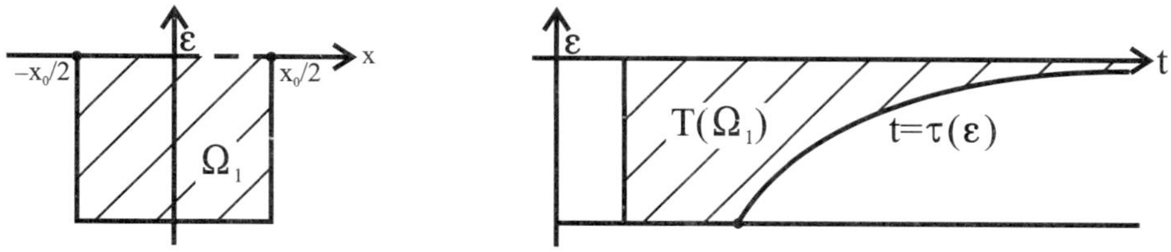

FIGURE 10.2. Rectification of the approximate generator.

Now we pass to the proof of Lemma 4.4, which completes the proof of the partial embedding theorem.

4.9. Explicit embedding. Lemma 4.4 claims that the map F under consideration is embeddable up to an N-flat correction on L. We shall conjugate it with an embeddable map by an M-smooth coordinate change, where $M = M(N)$ tends to infinity together with N. In what follows, different functions $M(N)$ will be considered, all satisfying this condition.

We must conjugate the C^N maps

$$F = g_W^1 + R \quad \text{and} \quad G = g_W^1, \tag{4.15}$$

where

$$W = (w, 0), \quad w = (x^2 + \varepsilon)(1 + a(\varepsilon)x)^{-1}, \quad R = o(\varepsilon^N).$$

As usual, we begin with globalization. It is specific: we need to change the vector field W outside some neighborhood of zero in order to have a globally defined phase flow. After that, we globalize the discrepancy R in such a way that it vanishes outside the neighborhood where the field W remains unchanged. Formally, fix a small $x_0 > 0$ and let φ and ψ be nonnegative C^∞ functions of x such that:

$$\begin{array}{lll} \varphi \equiv 0 \text{ for } |x| \geqslant x_0, & \varphi > 0 \text{ for } |x| < x_0, & \varphi \equiv 1 \text{ for } |x| \leqslant x_0/2, \\ \psi \equiv 0 \text{ for } |x| \geqslant x_0/2, & \psi > 0 \text{ for } |x| < x_0/2, & \psi \equiv 1 \text{ for } |x| \leqslant x_0/4. \end{array}$$

Let

$$w' = \varphi w, \quad W' = (w', 0), \quad G' = g_{W'}^1, \quad R' = \psi R.$$

We shall find a map conjugating F' and G' in a half-neighborhood of zero (with boundary included) using the special chart T. This chart rectifies the vector field W', that is, brings it to the form $\partial/\partial t$ and transforms the time one shift G' along the orbits of W' to the standard shift $\widetilde{G}\colon (t, \varepsilon) \mapsto (t + 1, \varepsilon)$. The chart is given by the formula

$$T\colon (x, \varepsilon) \mapsto (t(x, \varepsilon), \varepsilon), \qquad t(x, \varepsilon) = \int_{-x_0/2}^{x} \frac{ds}{w'(s, \varepsilon)} + 1,$$

where x_0 is a small positive constant. The map T is well defined in

$$\Omega = \{(x, 0) \mid x \in [-x_0/2, 0)\} \cup \{(x, \varepsilon) \mid x \in [-x_0/2, x_0),\ \varepsilon \in (0, \varepsilon_0)\} \cap \{\varphi > 0\},$$

where the small ε_0 will be chosen below (see Figure 10.2). The image is the half-strip

$$\Pi^+ = \{(t, \varepsilon) \mid t \geqslant 1,\ \varepsilon \in [0, \varepsilon_0]\}.$$

As we mentioned before, the vector field W' in the new chart takes the form $\partial/\partial t$ and G' becomes the standard shift $\widetilde{G}\colon (t,\varepsilon)\mapsto(t+1,\varepsilon)$. Denote the map F written in the new chart by

$$\widetilde{F}=(\tilde{f},\varepsilon),\qquad \tilde{f}=t+1+\widetilde{R}.$$

Derivatives of the new coordinate function t tend to infinity as $\varepsilon\to 0$ no faster than some negative power of ε depending on the order of the derivative. Consider two maps of Ω: F' and G', whose difference is N-flat on the line L and zero outside $\Omega_1=\{\psi>0\}\cap\Omega$. Then their difference $\widetilde{R}$ in the new chart is small in the following sense:

$$|D^\alpha\widetilde{R}|<C\varepsilon^M,\qquad |\alpha|<M,\ M=M(N).$$

On the other hand, let $\{T(x_0/2,\varepsilon)\mid \varepsilon\in(0,\varepsilon_0]\}$ be the graph of the function $t=\tau(\varepsilon)$ (see Figure 10.2). The explicit calculation of t shows that $\tau(\varepsilon)\leqslant C'\varepsilon^{-1/2}$. Hence, on $T(\Omega_1)$ we have

$$\varepsilon\leqslant C''\varepsilon^{1/2}t^{-1}.$$

Therefore, there exists a constant C such that in $T(\Omega_1)$, hence in Π^+, we have

$$|D^\alpha\widetilde{R}|<C\varepsilon^{M/2}t^{-M},\qquad |\alpha|<M,\ M=M(N).\tag{4.16}$$

Conversely, if the discrepancy $\widetilde{R}$ of some map H (the difference between H and the identity) satisfies (4.16), then $T\circ H\circ T^{-1}-\mathrm{id}$ is K-flat on L for some $K=K(M)$, where $K\to\infty$ as $M\to\infty$.

Our goal is to find a coordinate change with discrepancy satisfying (4.16) when substituted for $\widetilde{R}$. This will prove the existence of a map conjugating F and G near zero in the half-plane $\varepsilon\geqslant 0$ with discrepancy K-flat on L. Hence, this map is K smooth: it may be K-smoothly extended by the identity to the whole neighborhood of zero. This will prove Lemma 4.4. It remains to prove

LEMMA 4.6. *Let $\widetilde{F}=(\tilde{f},\varepsilon)$, $\tilde{f}=t+1+\widetilde{R}$ with $\widetilde{R}$ satisfying* (4.16). *Then there exists a solution $H=(t+h,\varepsilon)$ to the equation $H\circ\widetilde{F}=H+1$ with the property*

$$|D^\alpha h|\leqslant C'\varepsilon^M t^{1-M},\qquad |\alpha|<M.\tag{4.17}$$

PROOF. The unknown function h satisfies the equation $h-h\circ\widetilde{F}=\widetilde{R}$. The formal series

$$h=\sum_{j=0}^{\infty}\widetilde{R}\circ\widetilde{F}^j\tag{4.18}$$

provides the solution to this equation. Thus we have reduced the proof of Lemma 4.6 to the following

PROPOSITION 4.3. *The series* (4.18) *converges in $C^N(\Pi^+)$ provided that* (4.16) *holds and $\tilde{R}\equiv 0$ in $\Pi^+\setminus T(\Omega_1)$. Its sum h satisfies inequalities* (4.17).

PROOF. First we prove the uniform convergence of the series. Denote

$$\widetilde{F^j}=(F_j,\varepsilon),\qquad F_j=t+j+R_j,\ j=0,1,\dots.$$

Then $R_j=R_{j-1}+\widetilde{R}\circ\widetilde{F^j}$ and $R_1=\widetilde{R}$. Hence, $R_j=\sum_{l=0}^{j-1}\widetilde{R}\circ\widetilde{F}^l$. Note that this is a partial sum for the series (4.18). The latter formula implies that for ε_0 small enough we have the following "*a priori* estimates":

$$R_j>-j/2,\qquad F_j>t+j/2.$$

Hence,

$$h(t, \varepsilon) \leqslant C\varepsilon^M \sum (t + j/2)^{-M} \tag{4.19}$$

(see (4.16)). Hence, the series (4.18) uniformly converges in Π^+ and, moreover, satisfies (4.17) with $\alpha = 0$.

Now let us prove the uniform convergence of the derivatives of (4.18). The proof follows the same lines as before: first we give an *a priori* estimate for all the derivatives of R_j up to order M; then majorize the series for the same derivatives of h.

PROPOSITION 4.4. *If the strip Π^+ is sufficiently narrow and* (4.16) *holds, then the absolute values of all the derivatives of R_j up to order M are less than or equal to* 1.

Proposition 4.3 immediately follows from the one just stated. Indeed,

$$D^\alpha h = \sum_j \sum_{0<\beta\leqslant\alpha} (D^\beta \widetilde{R}) \circ F^j P_\beta(D^\gamma R_j \mid \gamma \leqslant \alpha), \tag{4.20}$$

where the factors P_β are polynomials. Here $\gamma \leqslant \alpha$ means $\alpha - \gamma \in \mathbb{Z}_+^2$. Note that not all the derivatives $D^\gamma R_j$ for $\gamma \leqslant \alpha$ are really contained in the expression P_β. By Proposition 4.4, all the arguments in these polynomials range over a bounded domain. Hence, the polynomials in (4.20) take bounded values. Therefore, the series (4.20) is majorized by the right-hand side of (4.19). This proves Proposition 4.3.

PROOF OF PROPOSITION 4.4. The proof goes by induction on j. The base of induction is given by the above *a priori* estimate. Together with the formula for R_j, it implies that R_j is majorized by the jth partial sum of the series on the right-hand side of (4.19), which is already estimated above. We have $|R_j| < 1$ in Π^+ provided that the width ε_0 of Π^+ is sufficiently small.

Now let us proceed to the induction step. Let

$$|D^\gamma R_j| < 1 \quad \text{for } \gamma \leqslant \alpha,\ i \leqslant j \leqslant l-1.$$

We must prove that the same is true for $j = l$. As mentioned before, R_j is the partial sum of the series (4.18). Hence, the derivative $D^\alpha R_l$ is given by (4.20) with summation over j from 0 to $l-1$. By the induction assumption, for $j \leqslant l-1$ we have

$$\sum_{\beta\leqslant\alpha} P_\beta(D^\gamma R_j \mid \gamma \leqslant \alpha) < C(\alpha),$$

where $C(\alpha)$ is a universal constant depending on α only. Hence, by (4.16),

$$|D^\alpha R_l| < \varepsilon_0^M C(\alpha) \sum (j/2)^{-M} < 1$$

for sufficiently small ε. This proves Proposition 4.4, hence, Lemmas 4.6 and 4.4, and Theorem 4.2 as well.

4.10. Sectorial embedding theorem and functional moduli of saddlenode families. The C^1-classification of perturbations of saddlenode families of the line have functional moduli. They are constructed using the following

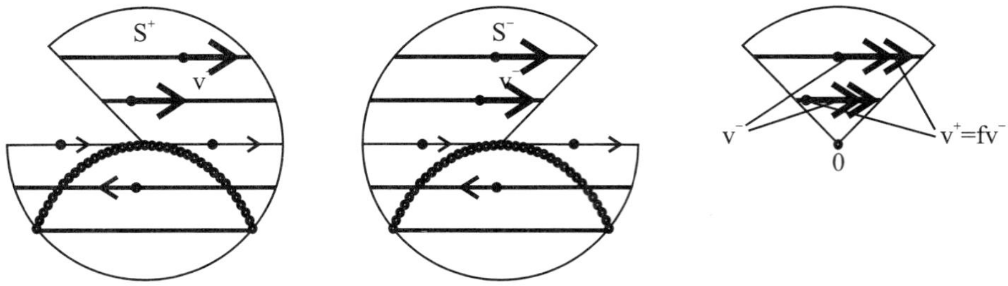

FIGURE 10.3. Sectors S^+, S^-, generators and functional moduli of the local saddlenode family.

SECTORIAL EMBEDDING THEOREM [**IY2**]. *Consider a generic one-parameter perturbation of a saddlenode map of the line which is in fact a single map*

$$F(x, \varepsilon) = (x + x^2 + \varepsilon + \cdots, \varepsilon).$$

For any $k \geqslant 1$ there exists an N such that if $F \in C^N$, then there is a C^k smooth vector field v in a neighborhood of 0 on the (x, ε)-plane for which $F = g^1_{v^+}$ in a sector S^+:

$$S^+ = \{(x, \varepsilon) \mid x^2 + \varepsilon^2 \leqslant r^2,\ \arg(x + i\varepsilon) \in [-\pi, \pi/2 + \alpha],\ 0 < \alpha < \pi/2\}.$$

Here α can take an arbitrary value in $(0, \pi/2)$; the larger is α, the smaller is r, hence, the smaller is the sector S^+ in which v is the generator for F.

The same statement: $F = g^1_{v^-}$ holds for the sector

$$S^- = \{(x, \varepsilon) \mid x^2 + \varepsilon^2 \leqslant r^2,\ \arg(x + i\varepsilon) \in [-3\pi/2 - \alpha, 0],\ 0 < \alpha < \pi/2\}$$

(*see Figure* 10.3).

The two generators v^+ and v^- are uniquely determined in the intersections $S^+ \cap \{\varepsilon \leqslant 0\}$ and $S^- \cap \{\varepsilon \leqslant 0\}$, respectively. They are tangent to the lines $\varepsilon = \text{const}$. Therefore, there exists a function f such that $v^+ = fv^-$. This function is uniquely determined and constitutes a functional modulus for the C^k classification of local saddlenode families of maps of the line for $k \geqslant 1$.

This explains why the partial embedding theorem cannot be improved to become an embedding theorem in the entire neighborhood of zero on the (x, ε)-plane for the map F.

Bibliography

[AAIS] V. S. Afraimovich, V. I. Arnold, Yu. S. Il′yashenko, and L. P. Shil′nikov, *Bifurcation theory*, Dynamical systems, Sovrem. Probl. Mat. Fund. Naprav., vol. 5, VINITI, Moscow, 1986, pp. 5–218; English transl., Encyclopaedia of Math. Sci., vol. 5, Springer-Verlag, Berlin–New York, 1989.

[ABS] V. S. Afraimovich, V. V. Bykov, and L. P. Shil′nikov, *On structurally unstable attracting limit sets of Lorenz attractor type*, Trans. Moscow Math. Soc. **2** (1983), 153–216.

[ACL] V. Afraimovich, S. N. Chow, and W. Liu, *Lorenz-type attractors from codimension one bifurcation*, J. Dynam. Differential Equations **7** (1995), no. 2, 375–407.

[ALY] V. Afraimovich, W. S. Liu, and T. Young, *Convectional multipliers for homoclinic orbits*, Nonlinearity **9** (1996), no. 1, 115–136.

[AS] V. S. Afraimovich and M. A. Shereshevsky, *The Hausdorff dimension of attractors appearing by saddle-node bifurcations*, Internat. J. Bifur. Chaos Appl. Sci. Engrg. **1** (1991), no. 2, 309–325.

[ALGM] A. A. Andronov, E. A. Leontovich, I. I. Gordon, and A. G. Maier, *Theory of bifurcations of dynamic systems on a plane*, Israel Program for Sci. Transl., Wiley, New York, 1973.

[AP] A. A. Andronov and L. Pontryagin, *Systèmes grossiers*, Dokl. Akad. Nauk. SSSR **14** (1937), 247–251.

[An] D. V. Anosov, *Geodesic flows on closed Riemannian manifolds of negative curvature*, Trudy Mat. Inst. Steklov. **90** (1967).

[A] V. I. Arnold, *Geometrical methods in the theory of ordinary differential equations*, Springer-Verlag, New York–Heidelberg–Berlin, 1982.

[AI] V. I. Arnold and Yu. S. Ilyashenko, *Ordinary differential equations*, Dynamical systems, Sovrem. Probl. Mat. Fund. Naprav., vol. 1, VINITI, Moscow, 1985, pp. 7–150; English transl., Encyclopaedia of Math. Sci., vol. 1, Springer-Verlag, Berlin–New York, 1988.

[BLMP] R. Bamón, R. Labarca, R. Mañé, and M. J. Pacifico, *The explosion of singular cycles*, Inst. Hautes Études Sci. Publ. Math. **78** (1993), 207–232.

[Be] G. R. Belitskii, *Normal forms, invariance and local maps*, Naukova Dumka, Kiev, 1979.

[BP] J. Beloqui and M. J. Pacifico, *Quasi-transversal saddle-node bifurcation on surfaces*, Ergodic Theory Dynam. Systems **10** (1990), no. 1, 63–88.

[BS] V. S. Biragov and L. P. Shil′nikov, *On the bifurcation of the saddle-focus loop in a three-dimensional conservative dynamical system*, Methods in Qualitative Theory and Bifurcation Theory, Gor′kov. Gos. Univ., Gor′kiĭ, 1989, pp. 25–34. (Russian)

[Bi] G. D. Birkhoff, *Dynamical systems*, Amer. Math. Soc., Providence, RI, 1966.

[BF] L. Block and J. Franke, *Existence of periodic points for maps of S^1*, Invent. Math. **22** (1973), 69–73.

[Br] A. D. Brjuno, *Local methods in non-linear differential equations*, Springer Ser. Soviet Math., Springer-Verlag, Berlin–Heidelberg–New York, 1989.

[By] V. V. Bykov, *Bifurcations of separatrix contours and chaos*, Methods in qualitative theory and bifurcation theory, Nizhegorod. Gos. Univ., Nizhniĭ Novgorod, 1991. (Russian)

[Ch] V. E. Chernyshev, *Bifurcations of the contours of singular trajectories*, Vestnik St. Petersburg Univ. Math. **25** (1992), no. 2, 50–54.

[CDT] S. N. Chow, B. Deng, and D. Terman, *The bifurcation of homoclinic and periodic orbits from two heteroclinic orbits*, SIAM J. Math. Anal. **21** (1990), no. 1, 179–204.

[CH] S. N. Chow and J. Hale, *Methods of bifurcation theory*, Springer-Verlag, New York–Heidelberg–Tokyo, 1982.

[CLW] S. N. Chow, C. Li, and D. Wang, *Normal forms and bifurcations of planar vector fields*, Cambridge University Press, 1994.

[CL] S. N. Chow and X. B. Lin, *Bifurcation of a homoclinic orbit with a saddle-node equilibrium*, Differential Integral Equations **3** (1990), no. 3, 435–466.

[D] G. J. Davis, *Infinitely many coexisting sinks from degenerate homoclinic tangencies*, Trans. Amer. Math. Soc. **323** (1991), 727–748.

[De1] Deng Bo, *Exponential expansion with Shil′nikov's saddle-focus*, J. Differential Equations **82** (1989), no. 1, 156–173.

[De2] ———, *Homoclinic twisting bifurcations and cusp horseshoe maps*, J. Differential Equations **5** (1993), no. 3, 417–467.

[Di] L. Diaz, *Robust nonhyperbolic dynamics and heterodimensional cycles*, Ergodic Theory Dynam. Systems **15** (1995), no. 2, 291–315.

[DR] L. Diaz and J. Rocha, *Nonconnected heterodimensional cycles: bifurcations and stability*, Nonlinearity **5** (1995), no. 6, 1315–1341.

[DRV] L. Diaz, J. Rocha, and M. Viana, *Strange attractors in saddlenode cycles: prevalence and globality*, Invent. Math. **125** (1996), 37–74.

[DU] L. Diaz and R. Ures, *Persistence of cycles and nonhyperbolic dynamics at heteroclinic bifurcations*, Nonlinearity **8** (1995), no. 5, 693–713.

[DKO] F. Dumortier, H. Kokubu, and H. Oka, *A degenerate singularity generating geometric Lorenz attractors*, Ergodic Theory Dynam. Systems **15** (1995), no. 5, 833–856.

[F] N. Fenichel, *Persistence and smoothness of invariant manifolds for flows*, Indiana Univ. Math. J. **21** (1971), 193–225.

[FHTKM] R. Fujimoto, A. Hotta, R. Tokunaga, M. Komuro, and T. Matsumoto, *Bifurcation phenomena in nonlinear systems and theory of dynamical systems (Kyoto, 1989)*, World Sci. Adv. Ser. Dynam. Systems **8** (1990), World Sci. Publishing, Teaneck, NJ., 125–142.

[Ga] P. Gaspard, *Local birth of homoclinic chaos*, Physica D **62** (1993), no. 1–4, 94–122.

[GS1] N. K. Gavrilov and L. P. Shil′nikov, *On three-dimensional dynamical systems close to systems with a structurally unstable homoclinic curve, I*, Math. USSR-Sb. **17** (1972), 467–485.

[GS2] ———, *On three-dimensional dynamical systems close to systems with a structurally unstable homoclinic curve,* II, Math. USSR-Sb. **19** (1973), 139–156.

[Go] E. P. Gomozov, *Finitely smooth equivalence of families of diffeomorphisms*, Vest. Khar′kov. Univ. Ser. Mech.-Mat. **41** (1976), 95–104. (Russian)

[GTS1] S. V. Gonchenko, D. V. Turaev, and L. P. Shil′nikov, *On models with a structurally unstable homoclinic Poincaré curve*, Methods in Qualitative Theory and Bifurcation Theory, Nizhegorod. Gos. Univ., Nizhniĭ Novgorod, 1991, pp. 36–61. (Russian)

[GTS2] ———, *Dynamical phenomena in multidimensional systems with a structurally unstable homoclinic Poincaré curve*, Russian Acad. Sci. Dokl. Math. **47** (1993), no. 3, 410–415.

[GTS3] ———, *Dynamical phenomena in systems with structurally unstable Poincaré homoclinic orbit*, Chaos **6** (1996), no. 1, 15–31.

[GI] A. Gorodetski and Yu. S. Ilyashenko, *Minimal and strange attractors*, Internat. J. Bifur. Chaos Appl. Sci. Engrg. **6** (1996), 1177–1183.

[GH] J. Guckenheimer and P. J. Holmes, *Nonlinear oscillations, dynamical systems and bifurcations of vector fields*, Appl. Math. Sci., vol. 42, Springer-Verlag, Berlin–New York, 1983.

[H] P. Hartman, *Ordinary differential equations*, Wiley, New York, 1964.

[Ha] F. Hausdorff, *Set theory*, Chelsea, New York, 1962.

[HPS] M. W. Hirsch, C. C. Pugh, and M. Shub, *Invariant manifolds*, Lecture Notes in Math., vol. 583, Springer-Verlag, New York–Heidelberg–Berlin, 1977.

[HS] M. W. Hirsch and S. Smale, *Differential equations, dynamical systems, and linear algebra*, Academic Press, New York, 1974.

[HKK] A. J. Homburg, H. Kokubu, and M. Krupa, *The cusp horseshoe and its bifurcations in the unfolding of an inclination flip homoclinic orbit*, Ergodic Theory Dynam. Systems **14** (1994), no. 4, 667–693.

[HSY] B. Hunt, T. Sauer, and J. A. Yorke, *Prevalence: a translation invariant "almost every" on infinite-dimensional spaces*, Bull. Amer. Math. Soc. **27** (1992), 217–238.

[I1] Yu. S. Ilyashenko, *Normal forms for local families and nonlocal bifurcations*, Complex Analytic Methods in Dynamical Systems (Rio de Janeiro, 1992), Astérisque **222** (1994), 233–258.

[I2] Yu. S. Ilyashenko (ed.), *Differential equations with real and complex time*, Proc. Steklov Inst. Math., vol. 213, 1996. (Russian)

[I3] ______, *Embedding theorems for local maps, slow-fast systems and bifurcation from Morse–Smale to Smale–Williams*, Amer. Math. Soc. Transl. (2) **180** (1997), 127–139.

[IY1] Yu. S. Ilyashenko and S. Yu. Yakovenko, *Finitely smooth normal forms of local families of diffeomorphisms and vector fields*, Russian Math. Surveys **46** (1991), 1–43.

[IY2] ______, *Nonlinear Stokes phenomena in smooth classification problems*, Adv. Soviet Math., vol. 14, 1993, pp. 235–287.

[IY3] Yu. S. Ilyashenko and S. Yu. Yakovenko (eds.), *Concerning Hilbert 16th problem*, Transl. Amer. Math. Soc. (2), vol. 165, Amer. Math. Soc., Providence, RI, 1995.

[Ir] M. C. Irwin, *Smooth dynamical systems*, Academic Press, London–New York, 1980.

[Jo] P. Joyal, *The generalized homoclinic bifurcation*, J. Differential Equations **108** (1994), no. 2, 217–261.

[K] V. Kaloshin, *Some prevalent properties of smooth dynamical systems*, Differential Equations with Real and Complex Time, Proc. Steklov Inst. Math., vol. 213, 1996, pp. 115–140. (Russian)

[KH] A. Katok and B. Hasselblatt, *Introduction to the modern theory of dynamical systems*, Cambridge Univ. Press, Cambridge, 1995.

[Kok] H. Kokubu, *Heteroclinic bifurcations associated with different saddle indices*, Dynamical Systems and Related Topics (Nagoya, 1990), Adv. Ser. Dynam. Systems, vol. 9, World Sci. Publishing, River Edge, NJ, 1991, pp. 236–260.

[KKO] H. Kokubu, M. Kisaka, and H. Oka, *Bifurcations of N-homoclinic orbits and N-periodic orbits in vector fields*, J. Dynam. Differential Equations **5** (1993), no. 2, 305–357.

[Ko] A. N. Kolmogorov, *General theory of dynamical systems and classical mechanics*, Proc. Int. Congress of Math. (1954), North-Holland, pp. 315–333.

[KS] A. Kotova and V. Stanzo, *On few-parameter generic families of vector fields on the two-dimensional sphere*, Concerning the Hilbert 16th Problem, Amer. Math. Soc. Transl. (2), vol. 165, Amer. Math. Soc., Providence, RI, 1965, pp. 155–201.

[L] R. Labarca, *Bifurcation of contracting singular cycles*, Ann. Sci. École Norm. Sup. (4) **28** (1995), no. 6, 705–745.

[LP] R. Labarca and M. J. Pacifico, *Stability of Morse–Smale vector fields on manifolds with boundary*, Topology **29** (1990), 57–81.

[LPl] R. Labarca and S. Plaza, *Global stability of families of vector fields*, ETDS **13** (1993), no. 4, 737–766.

[Le] G. A. Leonov, *Estimates for the Hausdorff dimension of attractors*, Vestnik Leningrad. Univ. Mat. Mekh. Astronom. **24** (1991), no. 3, 38–41; English transl. in Vestnik Leningrad Univ. Math.

[LS] L. M. Lerman and L. P. Shil′nikov, *Homoclinical structures in nonautonomous systems: nonautonomous chaos*, Chaos **2** (1992), no. 3, 447–454.

[Li] W. Li, *On a class of semilocal bifurcations of Lorenz type*, Acta Math. Sinica (N.S.) **8** (1992), no. 2, 158–176.

[LLZ] W. Li, C. Li, and Z. F. Zhang, *Unfolding of the critical homoclinic orbit of a class of degenerate equilibrium points*, Symposium of the Special Year on O.D.E. and Dynamical Systems in Nankai Univ. 1990 (1992), World Scientific, River Edge, NJ, pp. 99–110.

[LZ] W. Li and Z. F. Zhang, *The "blue sky" catastrophe on closed surfaces*, Adv. Ser. Dynam. Systems **9**, World Scientific, River Edge, NJ, 1991, pp. 316–332.

[Lin] X. B. Lin, *Using Mel′nikov's method to solve Šilnikov's problems*, Proc. Roy. Soc. Edinburgh Sect. A **116** (1990), no. 3–4, 295–325.

[Lo] E. N. Lorenz, *The local structure of a chaotic attractor in four dimensions*, Phys. D **13** (1984), 90–104.

[Ma] J. C. Matin-Rivas, *Homoclinic bifurcations and cascades of period doubling bifurcations in higher dimensions*, IMPA thesis, 1991.

[Me] V. S. Medvedev, *On a new type of bifurcation on manifolds*, Mat. Sb. **113** (1980), no. 3, 487–492; English transl. in Math. USSR-Sb. **41** (1982).

[Mi] J. Milnor, *On the concept of attractor*, Comm. Math. Phys. **99** (1985), 177–195.

[Mo] J. Moser, *Stable and random motions in dynamical systems*, Princeton Univ. Press, Princeton, NJ, 1973.

[Mou] A. Mourtada, *Degenerate and nontrivial hyperbolic polycycles with two vertices*, J. Differential Equations **113** (1994), no. 1, 68–83.

[MP] I. P. Malta and J. Palis, *Families of vector fields with finite modulus of stability*, Lecture Notes in Math., vol. 898, Springer-Verlag, New York–Berlin, 1981, pp. 212–229.

[MV] L. Mora and M. Viana, *Abundance of strange attractors*, Acta Math. **171** (1993), no. 1, 1–71.

[N] S. Newhouse, *Diffeomorphisms with infinitely many sinks*, Topology **13** (1974), 9–18.

[NP] S. Newhouse and J. Palis, *Bifurcations of Morse–Smale dynamical systems*, Dynamical Systems (M. M. Peixoto, ed.), Academic Press, New York–London, 1973.

[NPT] S. Newhouse, J. Palis, and F. Takens, *Bifurcations and stability of families of diffeomorphisms*, Inst. Hautes Études Sci. Publ. Math. **57** (1983), 5–71.

[Ni] Z. Nitecki, *Differentiable dynamics*, MIT Press, Cambridge, 1971.

[PR] M. J. Pacifico and A. Rovella, *Unfolding contracting singular cycles*, Ann. Sci. École Norm. Sup. (4) **26** (1993), no. 6, 691–700.

[P] J. Palis, *On Morse–Smale dynamical systems*, Topology **8** (1969), 385–405.

[PM] J. Palis and W. de Melo, *Geometric theory of dynamical systems: An introduction*, Springer-Verlag, New York–Heidelberg–Berlin, 1982.

[PP] J. Palis and C. C. Pugh, *Fifty problems in dynamical systems*, Lecture Notes in Math., vol. 468, Springer-Verlag, New York–Heidelberg–Berlin, 1975, pp. 345–353.

[PT] J. Palis and F. Takens, *Hyperbolicity and sensitive chaotic dynamics at homoclinic bifurcations*, Cambridge University Press, Cambridge, 1992.

[PV1] J. Palis and M. Viana, *Continuity of Hausdorff dimension and limit capacity for horseshoes*, Dynamical Systems, Lecture Notes in Math., vol. 1331, Springer-Verlag, New York–Heidelberg–Berlin, 1988, pp. 150–160.

[PV2] ———, *High dimension diffeomorphisms displaying infinitely many sinks*, Ann. of Math. (2) **140** (1994), no. 1, 207–250.

[PY] J. Palis and J. C. Yoccoz, *Homoclinic tangencies for hyperbolic sets of large Hausdorff dimension*, Acta Math. **172** (1994), no. 1, 91–136.

[Pa] K. J. Palmer, *Bifurcations, chaos and fractals*, Nonlinear Dynamics and Chaos (Canberra, 1991), World Sci., River Edge, NJ, 1992, pp. 91–133.

[Pe] M. M. Peixoto, *Structural stability on two-dimensional manifolds*, Topology **1** (1962), 101–120.

[Per] L. M. Perko, *Homoclinic loop and multiple limit cycle bifurcation surface*, Trans. Amer. Math. Soc. **344** (1994), no. 1, 101–130.

[Pl] V. A. Pliss and G. R. Sell, *Perturbations of attractors of differential equations*, J. Differential Equations **92** (1991), no. 1, 100–124.

[PoT] A. Posthumus Rense and F. Takens, *Homoclinic tangencies, moduli and topology of separatrices*, Ergodic Theory Dynam. Systems **13** (1993), no. 2, 369–385.

[R1] C. Robinson, *Bifurcation to infinitely many sinks*, Comm. Math. Phys. **90** (1983), 433–459.

[R2] ———, *Homoclinic bifurcation to a transitive attractor of Lorenz type*, Nonlinearity **2** (1989), no. 4, 495–518.

[Rom] N. Romero, *Persistence of homoclinic tangencies in higher dimensions*, Ergodic Theory Dynam. Systems **15** (1995), no. 4, 735–757.

[Rou] R. Roussarie, *On the number of limit cycles which appear by perturbation of separatrix of planar vector fields*, Bol. Soc. Brasil. Mat. **17** (1986), 67–101.

[RR] R. Roussarie and C. Rousseau, *Almost planar homoclinic loops in* $\mathbb{R}^3$, J. Differential Equations **126** (1996), no. 1, 1–47.

[Rov] A. Rovella, *The dynamics of the perturbations of the contracting Lorenz attractor*, Bol. Soc. Brasil. Mat. (N.S.) **24** (1993), no. 2, 233–259.

[RT] D. Ruelle and F. Takens, *On the nature of turbulence*, Comm. Math. Phys. **20** (1971), 167–192.

[Ry] M. Rychlik, *Lorenz attractors through Šilnikov-type bifurcation*, Ergodic Theory Dynam. Systems **10** (1990), 793–821.

[Sa] V. S. Samovol, *Equivalence of system of differential equations in the neighborhood of a singular point*, Trans. Moscow Math. Soc. **44** (1982), 213–234.

[Sha] M. V. Shashkov, *On bifurcations of separatrix contours with two saddles*, Internat. J. Bifur. Chaos Appl. Sci. Engrg. **2** (1992), no. 4, 911–915.

[Shi1] L. P. Shil′nikov, *A case of the existence of a denumerable set of periodic motions*, Soviet Math. Dokl. **6** (1965), 163–166.

[Shi2] ______, *On a Poincaré–Birkhoff problem*, Math. USSR-Sb. **3** (1967), 353–371.

[Shi3] ______, *The existence of a denumerable set of periodic motions in four-dimensional space in an extended neighborhood of a saddle-focus*, Soviet Math. Dokl. **8** (1967), 54–58.

[Shi4] ______, *On the generation of a periodic motion from a trajectory doubly asymptotic to an equilibrium state of saddle type*, Math. USSR-Sb. **6** (1968), 428–438.

[Shi5] ______, *Structure of the neighborhood of a homoclinic tube of an invariant torus*, Soviet Math. Dokl. **9** (1968), 624–628.

[Shi6] ______, *On a new type bifurcation of multi-dimensionally dynamical systems*, Soviet Math. Dokl. **10** (1969), 59–62.

[Shi7] ______, *A contribution to the problem of the structure of an extended neighborhood of a rough equilibrium state of saddle-focus type*, Math. USSR-Sb. **10** (1970), 91–102.

[Shi8] ______, *Strange attractors and dynamical models*, J. Circuits Systems Comput. **3** (1993), no. 1, 1–10.

[SS] A. L. Shil′nikov and L. P. Shil′nikov, *On the nonsymmetrical Lorenz model*, Internat. J. Bifur. Chaos Appl. Sci. Engrg. **1** (1991), no. 4, 773–776.

[SST] A. L. Shil′nikov, L. P. Shil′nikov, and D. V. Turaev, *Normal forms and Lorenz attractors*, Internat. J. Bifur. Chaos Appl. Sci. Engrg. **3** (1993), no. 5, 1123–1139.

[ST] L. P. Shil′nikov and D. V. Turaev, *Blue sky catastrophes*, Dokl. Akad. Nauk **342** (1995), no. 5, 596–599; English transl. in Russian Acad. Sci. Dokl. Math. **51** (1995).

[Sho] A. N. Shoshitaishvili, *Bifurcation of topological type of singular points of parameterized vector fields*, Functional Anal. Appl. **2** (1972), 169–170.

[Sm1] S. Smale, *Diffeomorphisms with many periodic points*, Differential and Combinatorial Topology (S. S. Cairns, ed.), Princeton University Press, Princeton, 1965, pp. 63–80.

[Sm2] ______, *Differentiable dynamical systems*, Bull. Amer. Math. Soc. **73** (1967), 747–817.

[So1] J. Sotomayor, *Generic one-parameter families of vector fields on two-dimensional manifolds*, Publ. Math. I.H.E.S. **43** (1974), 5–46.

[So2] ______, *Licoes de equacoes diferenciais ordinarias*, Projeto Euclides, CNPQ, 1979.

[Sp] C. Sparrow, *The Lorenz equations*, Springer-Verlag, New York–Heidelberg–Berlin, 1982.

[Ste] S. Sternberg, *On the structure of local homeomorphisms of Euclidean n-space,* II, Amer. J. Math. **80** (1958), 623–631.

[Str] S. J. van Strien, *Centre manifolds are not C^∞*, Math. Z. **166** (1979), 143–145.

[Ta1] F. Takens, *Normal forms for certain singularities of vector fields*, Ann. Inst. Fourier (Grenoble) **23** (1973), no. 2, 163–195.

[Ta2] ______, *Partially hyperbolic fixed point*, Topology **10** (1971), 133–147.

[Ta3] ______, *Homoclinic bifurcations*, Proc. Int. Congress of Math. (Berkeley, 1986), Amer. Math. Soc., Providence, RI, 1988, pp. 1229–1236.

[TS] J. C. Tatjer and C. Simó, *Basins of attraction near homoclinic tangencies*, Ergodic Theory Dynam. Systems **14** (1994), no. 2, 351–390.

[To] Y. Togawa, *Bifurcations from Šilnikov systems*, Topology and Computer Science (Atami, 1986), Kinokuniya, Tokyo, 1987, pp. 377–384.

[Tr] S. I. Trifonov, *Cyclicity of elementary polycycles in generic smooth vector fields*, Proc. Steklov. Inst. Math. **213** (1996), 141–199.

[U] R. Ures, *Approximating a Hénon-like strange attractor by a homoclinic tangency and an attracting cycle*, IMPA thesis, 1992.

[V] Viana, *Strange attractors in higher dimensions*, Bol. Soc. Brasil Mat. **24** (1993), no. 1, 13–62.

[W] S. Wiggins, *Global bifurcations and chaos: Analytical methods*, Springer-Verlag, New York–Berlin–Heidelberg, 1988.

[Wi] R. F. Williams, *The structure of Lorenz attractors*, Inst. Hautes Études Sci. Publ. Math. **50** (1979), 101–152.

[Y] J. C. Yoccoz, *Recent developments in dynamics*, Proc. Intern. Congress Math. (Zürich, 1994), Birkhäuser, Basel, 1995, pp. 246–265.

[Yo1] Todd Young, *Partially hyperbolic sets from a co-dimension one bifurcation*, Discrete Contin. Dynam. Systems **I** (1995), no. 2, 253–275.

[Yo2] ———, C^k*-smoothness of invariant curves in a global saddle-node*, J. Differential Equations **126** (1996), no. 1, 62–86.

[ZDHD] Z. F. Zhang, T. R. Ding, W. Z. Huang, and Z. X. Dong, *Qualitative theory of differential equations*, Transl. Math. Monographs, vol. 101, Amer. Math. Soc., Providence, RI, 1992.

Selected Titles in This Series

(Continued from the front of this publication)